P9-DOB-225

ELEMENTS OF DYNAMIC OPTIMIZATION

ELEMENTS OF DYNAMIC OPTIMIZATION

Alpha C. Chiang

Professor of Economics
The University of Connecticut

McGraw-Hill, Inc.

New York St. Louis San Francisco Auckland Bogotá
Caracas Lisbon London Madrid Mexico Milan Montreal
New Delhi Paris San Juan Singapore Sydney Tokyo Toronto

This book was set in Century Schoolbook by Science Typographers, Inc.
The editor was Scott D. Stratford;
the production supervisor was Denise L. Puryear.
The cover was designed by John Hite.
Project supervision was done by Science Typographers, Inc.
Art was prepared electronically by Science Typographers, Inc.
R. R. Donnelley & Sons Company was printer and binder.

This book is printed on recycled paper containing a minimum of 50% total recycled fiber with 10% postconsumer de-inked fiber.

ELEMENTS OF DYNAMIC OPTIMIZATION

2 3 4 5 6 7 8 9 0 DOC DOC 9 0 9 8 7 6 5 4 3 2

ISBN 0-07-010911-7

Library of Congress Cataloging-in-Publication Data

Chiang, Alpha C., (date).
Elements of dynamic optimization / Alpha C. Chiang.
p. cm.
Includes bibliographical references and index.
ISBN 0-07-010911-7
1. Mathematical optimization. 2. Economics, Mathematical.
I. Title.
HB143.7.C45 1992
330'.01'51—dc20 91-37803

To my parents
Rev. & Mrs. Peh-shi Chiang
with gratitude

Alpha C. Chiang received his Ph.D. from Columbia University in 1954 after earning the B.A. in 1946 at St. John's University (Shanghai, China), and the M.A. in 1948 at the University of Colorado. He has been a professor of economics at the University of Connecticut since 1964, but had also taught at Denison University, New Asia College (Hong Kong), and Cornell University. A holder of Ford Foundation and National Science Foundation fellowships, he served as Chairman of the Department of Economics at Denison from 1961 to 1964. His publications include *Fundamental Methods of Mathematical Economics*, 3d ed., McGraw-Hill, New York, 1984.

CONTENTS

PREFACE

In recent years I have received many requests to expand my *Fundamental Methods of Mathematical Economics* to include the subject of dynamic optimization. Since the existing size of that book would impose a severe space constraint, I decided to present the topic of dynamic optimization in a separate volume. Separateness notwithstanding, the present volume can be considered as a continuation of *Fundamental Methods of Mathematical Economics*.

As the title *Elements of Dynamic Optimization* implies, this book is intended as an introductory text rather than an encyclopedic tome. While the basics of the classical calculus of variations and its modern cousin, optimal control theory, are explained thoroughly, differential games and stochastic control are not included. Dynamic programming is explained in the discrete-time form; I exclude the continuous-time version because it needs partial differential equations as a prerequisite, which would have taken us far afield.

Although the advent of optimal control theory has caused the calculus of variations to be overshadowed, I deem it inadvisable to dismiss the topic of variational calculus. For one thing, a knowledge of the latter is indispensable for understanding many classic economic papers written in the calculus-of-variations mold. Besides, the method is used even in recent writings. Finally, a background in the calculus of variations facilitates a better and fuller understanding of optimal control theory. The reader who is only interested in optimal control theory may, if desired, skip Part 2 of the present volume. But I would strongly recommend reading at least the following: Chap. 2 (the Euler equation), Sec. 4.2 (checking concavity/convexity), and Sec. 5.1 (methodological issues of infinite horizon, relevant also to optimal control theory).

Certain features of this book are worth pointing out. In developing the Euler equation, I supply more details than most other books in order that the reader can better appreciate the beauty of the logic involved (Sec. 2.1).

In connection with infinite-horizon problems, I attempt to clarify some common misconceptions about the conditions for convergence of improper integrals (Sec. 5.1). I also try to argue that the alleged counterexamples in optimal control theory against infinite-horizon transversality conditions may be specious, since they involve a failure to recognize the presence of implicit fixed terminal states in those examples (Sec. 9.2).

To maintain a sense of continuity with *Fundamental Methods of Mathematical Economics*, I have written this volume with a comparable level of expository patience, and, I hope, clarity and readability. The discussion of mathematical techniques is always reinforced with numerical illustrations, economic examples, and exercise problems. In the numerical illustrations, I purposely present the shortest-distance problem—a simple problem with a well-known solution—in several different alternative formulations, and use it as a running thread through the book.

In the choice of economic examples, my major criterion is the suitability of the economic models as illustrations of the particular mathematical techniques under study. Although recent economic applications are natural candidates for inclusion, I have not shied away from classic articles. Some classic articles are not only worth studying in their own right, but also turn out to be excellent for illustrative purposes because their model structures are uncluttered with secondary complicating assumptions. As a by-product, the juxtaposition of old and new economic models also provides an interesting glimpse of the development of economic thought. For instance, from the classic Ramsey growth model (Sec. 5.3) through the neoclassical growth model of Cass (Sec. 9.3) to the Romer growth model with endogenous technological progress (Sec. 9.4), one sees a progressive refinement in the analytical framework. Similarly, from the classic Hotelling model of exhaustible resources (Sec. 6.3) to the Forster models of energy use and pollution (Sec. 7.7 and Sec. 8.5), one sees the shift in the focus of societal concerns from resource-exhaustion to environmental quality. A comparison of the classic Evans model of dynamic monopolist (Sec. 2.4) with the more recent model of Leland on the dynamics of a revenue-maximizing firm (Sec. 10.2) also illustrates one of the many developments in microeconomic reorientation.

In line with my pedagogical philosophy, I attempt to explain each economic model in a step-by-step manner from its initial construction through the intricacies of mathematical analysis to its final solution. Even though the resulting lengthier treatment requires limiting the number of economic models presented, I believe that the detailed guidance is desirable because it serves to minimize the trepidation and frustration often associated with the learning of mathematics.

In the writing of this book, I have benefited immensely from the numerous comments and suggestions of Professor Bruce A. Forster of the University of Wyoming, whose keen eyes caught many sins of commission and omission in the original manuscript. Since I did not accept all his suggestions, however, I alone should be held responsible for the remaining imperfections. Many of my students over the years, on whom I tried the earlier drafts of this book, also helped me with their questions and reactions. Scott D. Stratford, my editor, exerted the right amount of encouragement and pressure at critical moments to keep me going. And the cooperative efforts of Joseph Murphy at McGraw-Hill, and Sarah Roesser, Cheryl Kranz, and Ellie Simon at Science Typographers, Inc., made the production process smooth as well as pleasant. Thanks are also due to Edward T. Dowling for ferreting out some typographical errors that lurked in the initial printing of the book. Finally, my wife Emily again offered me unstinting assistance on manuscript preparation. To all of them, I am deeply grateful.

Alpha C. Chiang

ELEMENTS OF DYNAMIC OPTIMIZATION

PART 1

INTRODUCTION

CHAPTER 1

THE NATURE OF DYNAMIC OPTIMIZATION

Optimization is a predominant theme in economic analysis. For this reason, the classical calculus methods of finding free and constrained extrema and the more recent techniques of mathematical programming occupy an important place in the economist's everyday tool kit. Useful as they are, such tools are applicable only to static optimization problems. The solution sought in such problems usually consists of a *single* optimal magnitude for every choice variable, such as the optimal level of output per week and the optimal price to charge for a product. It does not call for a schedule of optimal sequential action.

In contrast, a *dynamic* optimization problem poses the question of what is the optimal magnitude of a choice variable in each period of time within the planning period (discrete-time case) or at each point of time in a given time interval, say $[0, T]$ (continuous-time case). It is even possible to consider an infinite planning horizon, so that the relevant time interval is $[0, \infty)$—literally "from here to eternity." The solution of a dynamic optimization problem would thus take the form of an *optimal time path* for every choice variable, detailing the best value of the variable today, tomorrow, and so forth, till the end of the planning period. Throughout this book, we shall use the asterisk to denote optimality. In particular, the optimal time path of a (continuous-time) variable y will be denoted by $y^*(t)$.

1.1 SALIENT FEATURES OF DYNAMIC OPTIMIZATION PROBLEMS

Although dynamic optimization is mostly couched in terms of a sequence of *time*, it is also possible to envisage the planning horizon as a sequence of *stages* in an economic process. In that case, dynamic optimization can be viewed as a problem of multistage decision making. The distinguishing feature, however, remains the fact that the optimal solution would involve more than one single value for the choice variable.

Multistage Decision Making

The multistage character of dynamic optimization can be illustrated with a simple discrete example. Suppose that a firm engages in transforming a certain substance from an *initial state* A (raw material state) into a *terminal state* Z (finished product state) through a five-stage production process. In every stage, the firm faces the problem of choosing among several possible alternative subprocesses, each entailing a specific cost. The question is: How should the firm select the sequence of subprocesses through the five stages in order to minimize the total cost?

In Fig. 1.1, we illustrate such a problem by plotting the *stages* horizontally and the *states* vertically. The initial state A is shown by the leftmost point (at the beginning of state 1); the terminal state Z is shown by the rightmost point (at the end of stage 5). The other points $B, C, \ldots, K$ show the various intermediate states into which the substance may be transformed during the process. These points (A, $B, \ldots, Z$) are referred to as *vertices*. To indicate the possibility of transformation from state A to state B, we draw an *arc* from point A to point B. The other arc AC shows

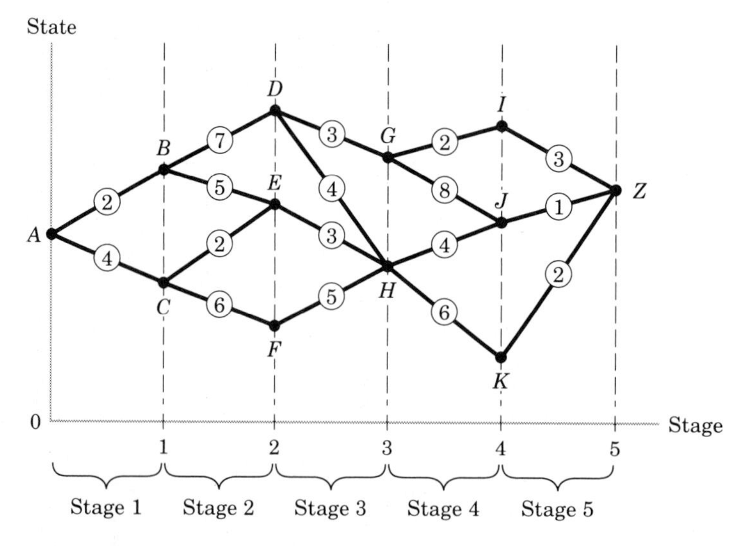

FIGURE 1.1

that the substance can also be transformed into state C instead of state B. Each arc is assigned a specific *value*—in the present example, a cost—shown in a circle in Fig. 1.1. The first-stage decision is whether to transform the raw material into state B (at a cost \$2) or into state C (at a cost of \$4), that is, whether to choose arc AB or arc AC. Once the decision is made, there will arise another problem of choice in stage 2, and so forth, till state Z is reached. Our problem is to choose a connected sequence of arcs going from left to right, starting at A and terminating at Z, such that the sum of the values of the component arcs is minimized. Such a sequence of arcs will constitute an *optimal path*.

The example in Fig. 1.1 is simple enough so that a solution may be found by enumerating all the admissible paths from A to Z and picking the one with the least total arc values. For more complicated problems, however, a systematic method of attack is needed. This we shall discuss later when we introduce dynamic programming in Sect. 1.4. For the time being, let us just note that the optimal solution for the present example is the path $ACEHJZ$, with \$14 as the minimum cost of production. This solution serves to point out a very important fact: A myopic, one-stage-at-a-time optimization procedure will *not* in general yield the optimal path! For example, a myopic decision maker would have chosen arc AB over arc AC in the first stage, because the former involves only half the cost of the latter; yet, over the span of five stages, the more costly first-stage arc AC should be selected instead. It is precisely for this reason, of course, that a method that can take into account the entire planning period needs to be developed.

The Continuous-Variable Version

The example in Fig. 1.1 is characterized by a discrete stage variable, which takes only integer values. Also, the state variable is assumed to take values belonging to a small finite set, $\{A, B, \ldots, Z\}$. If these variables are continu-

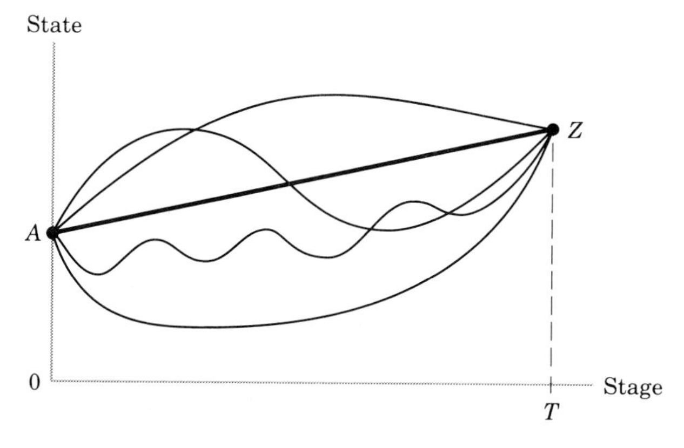

FIGURE 1.2

ous, we may instead have a situation as depicted in Fig. 1.2, where, for illustration, we have drawn only five possible paths from A to Z. Each possible path is now seen to travel through an infinite number of stages in the interval $[0, T]$. There is also an infinite number of states on each path, each state being the result of a particular choice made in a specific stage.

For concreteness, let us visualize Fig. 1.2 to be a map of an open terrain, with the stage variable representing the longitude, and the state variable representing the latitude. Our assigned task is to transport a load of cargo from location A to location Z at minimum cost by selecting an appropriate travel path. The cost associated with each possible path depends, in general, not only on the distance traveled, but also on the topography on that path. However, in the special case where the terrain is completely homogeneous, so that the transport cost per mile is a constant, the least-cost problem will simply reduce to a shortest-distance problem. The solution in that case is a straight-line path, because such a path entails the lowest total cost (has the lowest path value). The straight-line solution is, of course, well known—so much so that one usually accepts it without demanding to see a proof of it. In the next chapter (Sec. 2.2, Example 4), we shall prove this result by using the *calculus of variations*, the classical approach to the continuous version of dynamic optimization.

For most of the problems discussed in the following, the stage variable will represent *time*; then the curves in Fig. 1.2 will depict *time paths*. As a concrete example, consider a firm with an initial capital stock equal to A at time 0, and a predetermined target capital stock equal to Z at time T. Many alternative investment plans over the time interval $[0, T]$ are capable of achieving the target capital at time T. And each investment plan implies a specific capital path and entails a specific potential profit for the firm. In this case, we can interpret the curves in Fig. 1.2 as possible capital paths and their path values as the corresponding profits. The problem of the firm is to identify the investment plan—hence the capital path—that yields the maximum potential profit. The solution of the problem will, of course, depend crucially on how the potential profit is related to and determined by the configuration of the capital path.

From the preceding discussion, it should be clear that, regardless of whether the variables are discrete or continuous, a simple type of dynamic optimization problem would contain the following basic ingredients:

1 a given *initial point* and a given *terminal point*;

2 a set of *admissible paths* from the initial point to the terminal point;

3 a set of *path values* serving as performance indices (cost, profit, etc.) associated with the various paths; and

4 a specified objective—either to maximize or to minimize the path value or performance index by choosing the *optimal path*.

The Concept of a Functional

The relationship between paths and path values deserves our close attention, for it represents a special sort of mapping—not a mapping from real numbers to real numbers as in the usual *function*, but a mapping from *paths* (curves) to real numbers (performance indices). Let us think of the paths in question as time paths, and denote them by $y_{\mathrm{I}}(t)$, $y_{\mathrm{II}}(t)$, and so on. Then the mapping is as shown in Fig. 1.3, where V_{I}, V_{II} represent the associated path values. The general notation for the mapping should therefore be $V[y(t)]$. But it must be emphasized that this symbol fundamentally differs from the composite-function symbol $g[f(x)]$. In the latter, g is a function of f, and f is in turn a function of x; thus, g is in the final analysis a function of x. In the symbol $V[y(t)]$, on the other hand, the $y(t)$ component comes as an integral unit—to indicate time paths—and therefore we should not take V to be a function of t. Instead, V should be understood to be a function of "$y(t)$" as such.

To make clear this difference, this type of mapping is given a distinct name: *functional*. To further avoid confusion, many writers omit the "(t)" part of the symbol, and write the functional as $V[y]$ or $V\{y\}$, thereby underscoring the fact that it is the change in the position of the entire y path—the *variation* in the y path—as against the change in t, that results in a change in path value V. The symbol we employ is $V[y]$. Note that when the symbol y is used to indicate a certain state, it is suffixed, and appears as, say, $y(0)$ for the initial state or $y(T)$ for the terminal state. In contrast, in the path connotation, the t in $y(t)$ is not assigned a specific value. In the following, when we want to stress the specific time interval involved in a

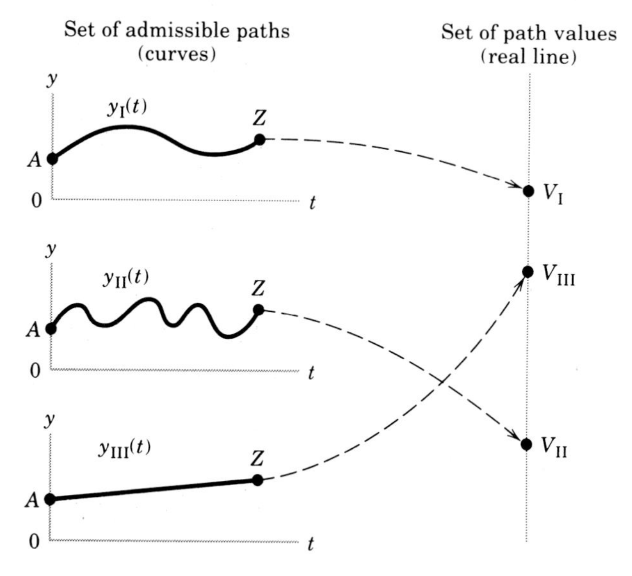

FIGURE 1.3

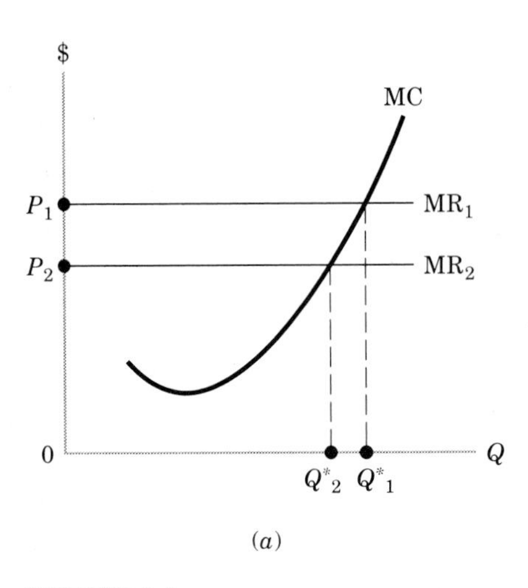

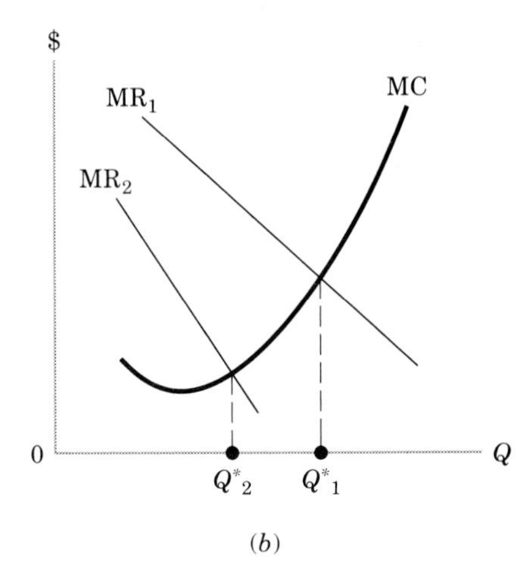

FIGURE 1.4

path or a segment thereof, we shall use the notation $y[0, T]$ or $y[0, \tau)$. More often, however, we shall simply use the $y(t)$ symbol, or use the term "y path". The optimal time path is then denoted by $y^*(t)$, or the y^* path.

As an aside, we may note that although the concept of a functional takes a prominent place primarily in dynamic optimization, we can find examples of it even in elementary economics. In the economics of the firm, the profit-maximizing output is found by the decision rule MC = MR (marginal cost = marginal revenue). Under purely competitive conditions, where the MR curves are horizontal as in Fig. 1.4a, each MR curve can be represented by a specific, exogenously determined price, P_0. Given the MC curve, we can therefore express the optimal output as $Q^* = Q^*(P_0)$, which is a function, mapping a real number (price) into a real number (optimal output). But when the MR curves are downward-sloping under imperfect competition, the optimal output of the firm with a given MC curve will depend on the specific position of the MR curve. In such a case, since the output decision involves a mapping from *curves* to real numbers, we in fact have something in the nature of a *functional*: $Q^* = Q^*[\text{MR}]$. It is, of course, precisely because of this inability to express the optimal output as a *function* of price that makes it impossible to draw a supply curve for a firm under imperfect competition, as we can for its competitive counterpart.

1.2 VARIABLE ENDPOINTS AND TRANSVERSALITY CONDITIONS

In our earlier statement of the problem of dynamic optimization, we simplified matters by assuming a *given* initial point [a given ordered pair $(0, A)$] and a *given* terminal point [a given ordered pair (T, Z)]. The assumption of a given initial point may not be unduly restrictive, because, in the usual problem, the optimizing plan must start from some specific initial position, say, the current position. For this reason, we shall retain this assumption throughout most of the book. The terminal position, on the other hand, may very well turn out to be a flexible matter, with no inherent need for it to be predetermined. We may, for instance, face only a fixed terminal time, but have complete freedom to choose the terminal state (say, the terminal capital stock). On the other hand, we may also be assigned a rigidly specified terminal state (say, a target inflation rate), but are free to select the terminal time (when to achieve the target). In such a case, the terminal point becomes a part of the optimal choice. In this section we shall briefly discuss some basic types of variable terminal points.

We shall take the stage variable to be continuous time. We shall also retain the symbols 0 and T for the initial *time* and terminal *time*, and the symbols A and Z for the initial and terminal *states*. When no confusion can arise, we may also use A and Z to designate the initial and terminal *points* (ordered pairs), especially in diagrams.

Types of Variable Terminal Points

As the first type of variable terminal point, we may be given a fixed terminal time T, but a free terminal state. In Fig. 1.5a, while the planning horizon is fixed at time T, any point on the vertical line $t = T$ is acceptable as a terminal point, such as Z_1, Z_2, and Z_3. In such a problem, the planner obviously enjoys much greater freedom in the choice of the optimal path and, as a consequence, will be able to achieve a better—or at least no worse—optimal path value, V^*, than if the terminal point is rigidly specified.

This type of problem is commonly referred to in the literature as a *fixed-time-horizon problem*, or *fixed-time problem*, meaning that the terminal time of the problem is fixed rather than free. Though explicit about the time horizon, this name fails to give a complete description of the problem, since nothing is said about the terminal state. Only by implication are we to understand that the terminal state is free. A more informative characterization of the problem is contained in the visual image in Fig. 1.5a. In line with this visual image, we shall alternatively refer to the fixed-time problem as the *vertical-terminal-line problem*.

To give an economic example of such a problem, suppose that a monopolistic firm is seeking to establish a (smooth) optimal price path over a given planning period, say, 12 months, for purpose of profit maximization. The current price enters into the problem as the initial state. If there is no legal price restriction in force, the terminal price will be completely up to the firm to decide. Since negative prices are inadmissible, however, we must eliminate from consideration all $P < 0$. The result is a *truncated vertical terminal line*, which is what Fig. 1.5a in fact shows. If, in addition, an

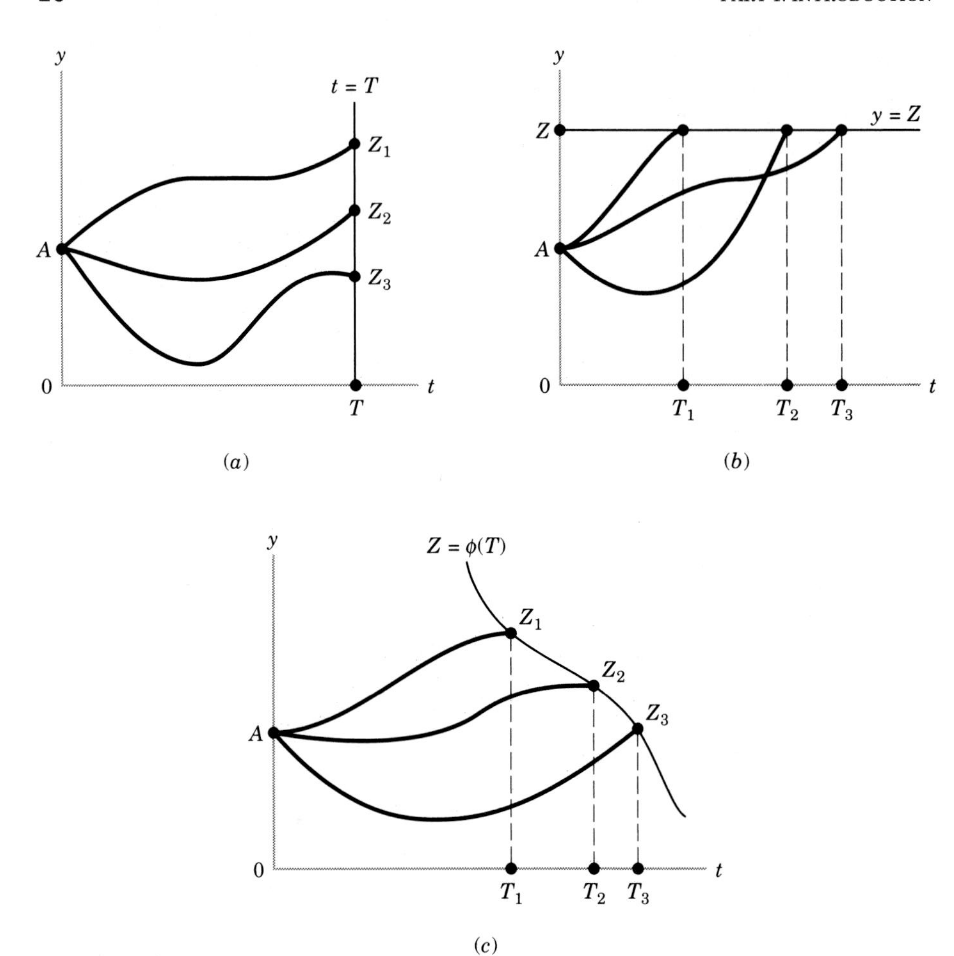

FIGURE 1.5

official price ceiling is expected to be in force at the terminal time $t = T$, then further truncation of the vertical terminal line is needed.

The second type of variable-terminal-point problem reverses the roles played by the terminal time and terminal state; now the terminal state Z is stipulated, but the terminal time is free. In Fig. 1.5*b*, the horizontal line $y = Z$ constitutes the set of admissible terminal points. Each of these, depending on the path chosen, may be associated with a different terminal time, as exemplified by T_1, T_2, and T_3. Again, there is greater freedom of choice as compared with the case of a fixed terminal point. The task of the planner might, for instance, be that of producing a good with a particular quality characteristic (steel with given tensile strength) at minimum cost, but there is complete discretion over the length of the production period. This permits the rise of a lengthier, but less expensive production method that might not be feasible under a rushed production schedule.

This type of problem is commonly referred to as a *fixed-endpoint problem*. A possible confusion can arise from this name because the word "endpoint" is used here to designate only the terminal state Z, not the entire endpoint in the sense of the ordered pair (T, Z). To take advantage of the visual image of the problem in Fig. 1.5*b*, we shall alternatively refer to the fixed-endpoint problem as the *horizontal-terminal-line problem*.

Turning the problem around, it is also possible in a problem of this type to have minimum production time (rather than minimum cost) as the objective. In that case, the path with T_1 as the terminal time becomes preferable to the one ending at T_2, regardless of the relative cost entailed. This latter type of problem is called a *time-optimal problem*.

In the third type of variable-terminal-point problem, neither the terminal time T nor the terminal state Z is individually preset, but the two are tied together via a constraint equation of the form $Z = \phi(T)$. As illustrated in Fig. 1.5*c*, such an equation plots as a *terminal curve* (or, in higher dimension, a *terminal surface*) that associates a particular terminal time (say, T_1) with a corresponding terminal state (say, Z_1). Even though the problem leaves *both* T and Z flexible, the planner actually has only one degree of freedom in the choice of the terminal point. Still, the field of choice is obviously again wider than if the terminal point is fully prescribed. We shall call this type of problem the *terminal-curve* (or *terminal-surface*) *problem*.

To give an economic example of a terminal curve, we may cite the case of a custom order for a product, for which the customer is interested in having both (1) an early date of completion, and (2) a particular quality characteristic, say, low breakability. Being aware that both cannot be attained simultaneously, the customer accepts a tradeoff between the two considerations. Such a tradeoff may appear as the curve $Z = \phi(T)$ in Fig. 1.5*c*, where y denotes breakability.

The preceding discussion pertains to problems with finite planning horizon, where the terminal time T is a finite number. Later we shall also encounter problems where the planning horizon is infinite ($T \to \infty$).

Transversality Condition

The common feature of variable-terminal-point problems is that the planner has one more degree of freedom than in the fixed-terminal-point case. But this fact automatically implies that, in deriving the optimal solution, an extra condition is needed to pinpoint the exact path chosen. To make this clear, let us compare the boundary conditions for the optimal path in the fixed- versus the variable-terminal-point cases. In the former, the optimal path must satisfy the boundary (initial and terminal) conditions

$$y(0) = A \qquad \text{and} \qquad y(T) = Z \qquad (T, A, \text{and } Z \text{ all given})$$

In the latter case, the initial condition $y(0) = A$ still applies by assumption.

But since T and/or Z are now variable, the terminal condition $y(T) = Z$ is no longer capable of pinpointing the optimal path for us. As Fig. 1.5 shows, all admissible paths, ending at Z_1, Z_2, or other possible terminal positions, equally satisfy the condition $y(T) = Z$. What is needed, therefore, is a terminal condition that can conclusively distinguish the optimal path from the other admissible paths. Such a condition is referred to as a *transversality condition*, because it normally appears as a description of how the optimal path crosses the terminal line or the terminal curve (to "transverse" means to "to go across").

Variable Initial Point

Although we have assumed that only the terminal point can vary, the discussion in this section can be adapted to a variable *initial* point as well. Thus, there may be an initial curve depicting admissible combinations of the initial time and the initial state. Or there may be a vertical initial line $t = 0$, indicating that initial time 0 is given, but the initial state is unrestricted. As an exercise, the reader is asked to sketch suitable diagrams similar to those in Fig. 1.5 for the case of a variable initial point.

If the initial point is variable, the characterization of the optimal path must also include another transversality condition in place of the equation $y(0) = A$, to describe how the optimal path crosses the initial line or initial curve.

EXERCISE 1.2

1. Sketch suitable diagrams similar to Fig. 1.5 for the case of a variable *initial* point.
2. In Fig. 1.5*a*, suppose that y represents the price variable. How would an official price ceiling that is expected to take effect at time $t = T$ affect the diagram?
3. In Fig. 1.5*c*, let y denote heat resistance in a product, a quality which takes longer production time to improve. How would you redraw the terminal curve to depict the tradeoff between early completion date and high heat resistance?

1.3 THE OBJECTIVE FUNCTIONAL

The Integral Form of Functional

An optimal path is, by definition, one that maximizes or minimizes the path value $V[y]$. Inasmuch as any y path must perforce travel through an interval of time, its total value would naturally be a sum. In the discrete-stage framework of Fig. 1.1, the path value is the sum of the values of its component arcs. The continuous-time counterpart of such a sum is a definite integral, $\int_0^T$ (arc value) dt. But how do we express the "arc value" for the continuous case?

To answer this, we must first be able to identify an "arc" on a continuous-time path. Figure 1.1 suggests that three pieces of information are needed for arc identification: (1) the starting stage (time), (2) the starting state, and (3) the direction in which the arc proceeds. With continuous time, since each arc is infinitesimal in length, these three items are represented by, respectively: (1) t, (2) $y(t)$, and (3) $y'(t) \equiv dy/dt$. For instance, on a given path y_I, the arc associated with a specific point of time t_0 is characterized by a unique value $y_I(t_0)$ and a unique slope $y_I'(t_0)$. If there exists some function, F, that assigns arc values to arcs, then the value of the said arc can be written as $F[t_0, y_I(t_0), y_I'(t_0)]$. Similarly, on another path, y_{II}, the height and the slope of the curve at $t = t_0$ are $y_{II}(t_0)$ and $y_{II}'(t_0)$, respectively, and the arc value is $F[t_0, y_{II}(t_0), y_{II}'(t_0)]$. It follows that the general expression for arc values is $F[t, y(t), y'(t)]$, and the path-value functional—the sum of arc values—can generally be written as the definite integral

$$V[y] = \int_0^T F[t, y(t), y'(t)]\, dt \tag{1.1}$$

It bears repeating that, as the symbol $V[y]$ emphasizes, it is the variation in the *y path* (y_I versus y_{II}) that alters the magnitude of V. Each different y path consists of a different set of arcs in the time interval $[0, T]$, which, through the arc-value-assigning function F, takes a different set of arc values. The definite integral sums those arc values on each y path into a path value.

If there are two state variables, y and z, in the problem, the arc values on both the y and z paths must be taken into account. The objective functional should then appear as

$$V[y, z] = \int_0^T F[t, y(t), z(t), y'(t), z'(t)]\, dt \tag{1.2}$$

A problem with an objective functional in the form of (1.1) or (1.2) constitutes the standard problem. For simplicity, we shall often suppress the time argument (t) for the state variables and write the integrand function more concisely as $F(t, y, y')$ or $F(t, y, z, y', z')$.

A Microeconomic Example

A functional of the standard form in (1.1) may arise, for instance, in the case of a profit-maximizing, long-range-planning monopolistic firm with a dynamic demand function $Q_d = D(P, P')$, where $P' \equiv dP/dt$. In order to set $Q_s = Q_d$ (to allow no inventory accumulation or decumulation), the firm's output should be $Q = D(P, P')$, so that its total-revenue function is

$$R \equiv PQ = R(P, P')$$

Assuming that the total-cost function depends only on the level of output, we can write the composite function

$$C = C(Q) = C[D(P, P')]$$

It follows that the total profit also depends on P and P':

$$\pi \equiv R - C = R(P, P') - C[D(P, P')] = \pi(P, P')$$

Summing π over, say, a five-year period, results in the objective functional

$$\int_0^5 \pi(P, P')\, dt$$

which conforms to the general form of (1.1), except that the argument t in the F function happens to be absent. However, if either the revenue function or the cost function can shift over time, then that function should contain t as a separate argument; in that case the π function would also have t as an argument. Then the corresponding objective functional

$$\int_0^5 \pi(t, P, P')\, dt$$

would be exactly in the form of (1.1). As another possibility, the variable t can enter into the integrand via a discount factor $e^{-\rho t}$.

To each price path in the time interval [0, 5], there must correspond a particular five-year profit figure, and the objective of the firm is to find the optimal price path $P^*[0, 5]$ that maximizes the five-year profit figure.

A Macroeconomic Example

Let the social welfare of an economy at any time be measured by the utility from consumption, $U = U(C)$. Consumption is by definition that portion of output not saved (and not invested). If we adopt the production function $Q = Q(K, L)$, and assume away depreciation, we can then write

$$C = Q(K, L) - I = Q(K, L) - K'$$

where $K' \equiv I$ denotes net investment. This implies that the utility function can be rewritten as

$$U(C) = U[Q(K, L) - K']$$

If the societal goal is to maximize the sum of utility over a period $[0, T]$, then its objective functional takes the form

$$\int_0^T U[Q(K, L) - K']\, dt$$

This exemplifies the functional in (1.2), where the two state variables y and z refer in the present example to K and L, respectively.

Note that while the integrand function of this example does contain both K and K' as arguments, the L variable appears only in its natural form unaccompanied by L'. Moreover, the t argument is absent from the F function, too. In terms of (1.2), the F function contains only three arguments in the present example: $F[y(t), z(t), y'(t)]$, or $F[K, L, K']$.

Other Forms of Functional

Occasionally, the optimization criterion in a problem may not depend on any intermediate positions that the path goes through, but may rely exclusively on the position of the terminal point attained. In that event, no definite integral arises, since there is no need to sum the arc values over an interval. Rather, the objective functional appears as

$$V[y] = G[T, y(T)] \tag{1.3}$$

where the G function is based on what happens at the terminal time T only.

It may also happen that both the definite integral in (1.1) and the terminal-point criterion in (1.3) enter simultaneously in the objective functional. Then we have

$$V[y] = \int_0^T F[t, y(t), y'(t)]\, dt + G[T, y(T)] \tag{1.4}$$

where the G function may represent, for instance, the scrap value of some capital equipment. Moreover, the functionals in (1.3) and (1.4) can again be expanded to include more than one state variable. For example, with two state variables y and z, (1.3) would become

$$V[y, z] = G[T, y(T), z(T)] \tag{1.5}$$

A problem with the type of objective functional in (1.3) is called a *problem of Mayer*. Since only the terminal position matters in V, it is also known as a *terminal-control problem*. If (1.4) is the form of the objective functional, then we have a *problem of Bolza*.

Although the problem of Bolza may seem to be the more general formulation, the truth is that the three types of problems—standard, Mayer, and Bolza—are all convertible into one another. For example, the functional (1.3) can be transformed into the form (1.1) by defining a new variable

$$z(t) \equiv G[t, y(t)] \qquad \text{with initial condition } z(0) = 0 \tag{1.6}$$

It should be noted that it is "t" rather than "T" that appears in the G function in (1.6). Since

$$\begin{aligned} \int_0^T z'(t)\,dt &= z(t)\Big|_0^T = z(T) - z(0) = z(T) && \text{[by the initial condition]} \\ &= G[T, y(T)] && \text{[by (1.6)]} \end{aligned} \tag{1.7}$$

the functional in (1.3), $G[T, y(T)]$, can be replaced by the integral in (1.7) in the new variable $z(t)$. The integrand, $z'(t) \equiv dG[t, y(t)]/dt$, is easily recognized as a special case of the function $F[t, z(t), z'(t)]$, with the arguments t and $z(t)$ absent; that is, the integral in (1.7) still falls into the general form of the objective functional (1.1). Thus we have transformed a problem of Mayer into a standard problem. Once we have found the optimal z path, the optimal y path can be deduced through the relationship in (1.6).

By a similar procedure, we can convert a problem of Bolza into a standard problem; this will be left to the reader. An economic example of this type of problem can be found in Sec. 3.4. In view of this convertibility, we shall deem it sufficient to couch our discussion primarily in terms of the standard problem, with the objective functional in the form of an integral.

EXERCISE 1.3

1 In a so-called "time-optimal problem," the objective is to move the state variable from a given initial value $y(0)$ to a given terminal value $y(T)$ in the least amount of time. In other words, we wish to minimize the functional $V[y] = T - 0$.

(*a*) Taking it as a standard problem, write the specific form of the F function in (1.1) that will produce the preceding functional.

(*b*) Taking it as a problem of Mayer, write the specific form of the G function in (1.3) that will produce the preceding functional.

2 Suppose we are given a function $D(t)$ which gives the *desired* level of the state variable at every point of time in the interval $[0, T]$. All deviations from $D(t)$, positive or negative, are undesirable, because they inflict a negative payoff (cost, pain, or disappointment). To formulate an appropriate dynamic minimization problem, which of the following objective functionals would be acceptable? Why?

(*a*) $\int_0^T [y(t) - D(t)]\,dt$

(*b*) $\int_0^T [y(t) - D(t)]^2\,dt$

(*c*) $\int_0^T [D(t) - y(t)]^3\,dt$

(*d*) $\int_0^T |D(t) - y(t)|\,dt$

3 Transform the objective functional (1.4) of the problem of Bolza into the format of the standard problem, as in (1.1) or (1.2). [*Hint:* Introduce a new variable $z(t) \equiv G[t, y(t)]$, with initial condition $z(0) = 0$.]

4 Transform the objective functional (1.4) of the problem of Bolza into the format of the problem of Mayer, as in (1.3) or (1.5). [*Hint:* Introduce a new variable $z(t)$ characterized by $z'(t) \equiv F[t, y(t), y'(t)]$, with initial condition $z(0) = 0$.]

1.4 ALTERNATIVE APPROACHES TO DYNAMIC OPTIMIZATION

To tackle the previously stated problem of dynamic optimization, there are three major approaches. We have earlier mentioned the calculus of variations and dynamic programming. The remaining one, the powerful modern generalization of variational calculus, goes under the name of optimal control theory. We shall give a brief account of each.

The Calculus of Variations

Dating back to the late 17th century, the calculus of variations is the classical approach to the problem. One of the earliest problems posed is that of determining the shape of a surface of revolution that would encounter the least resistance when moving through some resisting medium (a surface of revolution with the minimum area).[1] Issac Newton solved this problem and stated his results in his *Principia*, published in 1687. Other mathematicians of that era (e.g., John and James Bernoulli) also studied problems of a similar nature. These problems can be represented by the following general formulation:

$$\begin{aligned} &\text{Maximize or minimize} && V[y] = \int_0^T F[t, y(t), y'(t)]\,dt \\ (1.8)\qquad &\text{subject to} && y(0) = A \qquad (A \text{ given}) \\ &\text{and} && y(T) = Z \qquad (T, Z \text{ given}) \end{aligned}$$

---

[1]This problem will be discussed in Sec. 2.2, Example 3.

Such a problem, with an integral functional in a single state variable, with completely specified initial and terminal points, and with no constraints, is known as the *fundamental problem* (or *simplest problem*) of calculus of variations.

In order to make such problems meaningful, it is necessary that the functional be integrable (i.e., the integral must be convergent). We shall assume this condition is met whenever we write an integral of the general form, as in (1.8). Furthermore, we shall assume that all the functions that appear in the problem are continuous and continuously differentiable. This assumption is needed because the basic methodology underlying the calculus of variations closely parallels that of the classical differential calculus. The main difference is that, instead of dealing with the differential dx that changes the value of $y = f(x)$, we will now deal with the "variation" of an entire curve $y(t)$ that affects the value of the functional $V[y]$. The study of variational calculus will occupy us in Part 2.

Optimal Control Theory

The continued study of variational problems has led to the development of the more modern method of *optimal control theory*. In optimal control theory, the dynamic optimization problem is viewed as consisting of *three* (rather than two) types of variables. Aside from the time variable t and the state variable $y(t)$, consideration is given to a control variable $u(t)$. Indeed, it is the latter type of variable that gives optimal control theory its name and occupies the center of stage in this new approach to dynamic optimization.

To focus attention on the control variable implies that the state variable is relegated to a secondary status. This would be acceptable only if the decision on a control path $u(t)$ will, once given an initial condition on y, unambiguously determine a state-variable path $y(t)$ as a by-product. For this reason, an optimal control problem must contain an equation that relates y to u:

$$\frac{dy}{dt} = f[t, y(t), u(t)]$$

Such an equation, called an *equation of motion* (or *transition equation* or *state equation*), shows how, at any moment of time, given the value of the state variable, the planner's choice of u will drive the state variable y over time. Once we have found the optimal control-variable path $u^*(t)$, the equation of motion would make it possible to construct the related optimal state-variable path $y^*(t)$.

The optimal control problem corresponding to the calculus-of-variations problem (1.8) is as follows:

$$\begin{aligned}
&\text{Maximize or minimize} && V[u] = \int_0^T F[t, y(t), u(t)]\, dt \\
(1.9)\quad &\text{subject to} && y'(t) = f[t, y(t), u(t)] \\
& && y(0) = A \qquad (A \text{ given}) \\
&\text{and} && y(T) = Z \qquad (T, Z \text{ given})
\end{aligned}$$

Note that, in (1.9), not only does the objective functional contain u as an argument, but it has also been changed from $V[y]$ to $V[u]$. This reflects the fact that u is now the ultimate instrument of optimization. Nonetheless, this control problem is intimately related to the calculus-of-variations problem (1.8). In fact, by replacing $y'(t)$ with $u(t)$ in the integrand in (1.8), and adopting the differential equation $y'(t) = u(t)$ as the equation of motion, we immediately obtain (1.9).

The single most significant development in optimal control theory is known as *the maximum principle*. This principle is commonly associated with the Russian mathematician L. S. Pontryagin, although an American mathematician, Magnus R. Hestenes, independently produced comparable work in a Rand Corporation report in 1949.[2] The powerfulness of that principle lies in its ability to deal directly with certain constraints on the control variable. Specifically, it allows the study of problems where the admissible values of the control variable u are confined to some closed, bounded convex set $\mathscr{U}$. For instance, the set $\mathscr{U}$ may be the closed interval $[0, 1]$, requiring $0 \leq u(t) \leq 1$ throughout the planning period. If the marginal propensity to save is the control variable, for instance, then such a constraint, $0 \leq s(t) \leq 1$, may very well be appropriate. In sum, the problem

---

[2]The maximum principle is the product of the joint efforts of L. S. Pontryagin and his associates V. G. Boltyanskii, R. V. Gamkrelidze, and E. F. Mishchenko, who were jointly awarded the 1962 Lenin Prize for Science and Technology. The English translation of their work, done by K. N. Trirogoff, is *The Mathematical Theory of Optimal Processes*, Interscience, New York, 1962.

Hestenes' Rand report is titled *A General Problem in the Calculus of Variations with Applications to Paths of Least Time*. But his work was not easily available until he published a paper that extends the results of Pontryagin: "On Variational Theory and Optimal Control Theory," *Journal of SIAM, Series A, Control*, Vol. 3, 1965, pp. 23–48. The expanded version of this work is contained in his book *Calculus of Variations and Optimal Control Theory*, Wiley, New York, 1966.

Some writers prefer to call the principle *the minimum principle*, which is the more appropriate name for the principle in a slightly modified formulation of the problem. We shall use the original name, the maximum principle.

addressed by optimal control theory is (in its simple form) the same as in (1.9), except that an additional constraint,

$$u(t) \in \mathscr{U} \qquad \text{for } 0 \le t \le T$$

may be appended to it. In this light, the control problem (1.9) constitutes a special (unconstrained) case where the control set $\mathscr{U}$ is the entire real line.

The detailed discussion of optimal control theory will be undertaken in Part 3.

Dynamic Programming

Pioneered by the American mathematician Richard Bellman,[3] dynamic programming presents another approach to the control problem stated in (1.9). The most important distinguishing characteristics of this approach are two: First, it embeds the given control problem in a family of control problems, with the consequence that in solving the given problem, we are actually solving the entire family of problems. Second, for each member of this family of problems, primary attention is focused on the optimal value of the functional, V^*, rather than on the properties of the optimal state path $y^*(t)$ (as in the calculus of variations) or the optimal control path $u^*(t)$ (as in optimal control theory). In fact, an *optimal value function*—assigning an optimal value to each individual member of this family of problems—is used as a characterization of the solution.

All this is best explained with a specific discrete illustration. Referring to Fig. 1.6 (adapted from Fig. 1.1), let us first see how the "embedding" of a problem is done. Given the original problem of finding the least-cost path from point A to point Z, we consider the larger problem of finding the least-cost path from *each* point in the set $\{A, B, C, \dots, Z\}$ to the terminal point Z. There then exists a family of component problems, each of which is associated with a different initial point. This is, however, not to be confused with the variable-initial-point problem in which our task is to select one initial point as the best one. Here, we are to consider every possible point as a legitimate initial point in its own right. That is, aside from the genuine initial point A, we have adopted many pseudo initial points (B, C, etc). The problem involving the pseudo initial point Z is obviously trivial, for it does not permit any real choice or control; it is being included in the general problem for the sake of completeness and symmetry. But the component

[3] Richard E. Bellman, *Dynamic Programming*, Princeton University Press, Princeton, NJ, 1957.

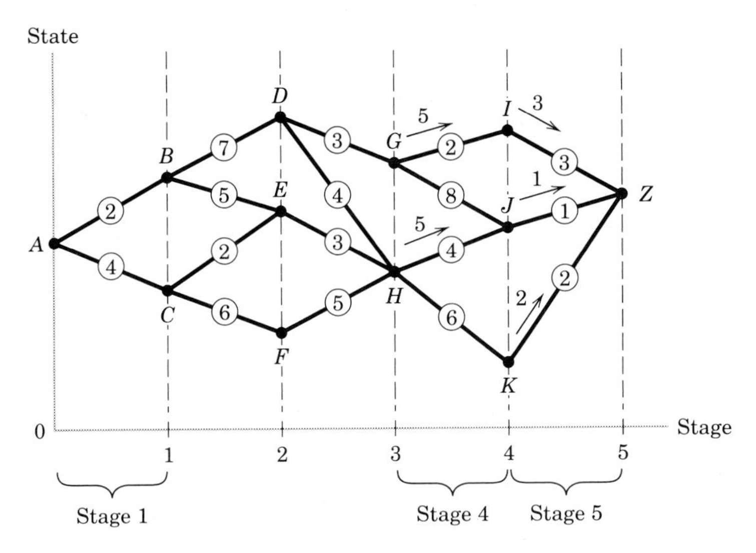

FIGURE 1.6

problems for the other pseudo initial points are not trivial. Our original problem has thus been "embedded" in a family of meaningful problems.

Since every component problem has a unique optimal path value, it is possible to write an *optimal value function*

$$V^* = V^*(i) \qquad (i = A, B, \ldots, Z)$$

which says that we can determine an optimal path value for every possible initial point. From this, we can also construct an *optimal policy function*, which will tell us how best to proceed from any specific initial point i, in order to attain $V^*(i)$ by the proper selection of a sequence of arcs leading from point i to the terminal point Z.

The purpose of the optimal value function and the optimal policy function is easy to grasp, but one may still wonder why we should go to the trouble of embedding the problem, thereby multiplying the task of solution. The answer is that the embedding process is what leads to the development of a systematic iterative procedure for solving the original problem.

Returning to Fig. 1.6, imagine that our immediate problem is merely that of determining the optimal values for stage 5, associated with the three initial points I, J, and K. The answer is easily seen to be

$$V^*(I) = 3 \qquad V^*(J) = 1 \qquad V^*(K) = 2 \tag{1.10}$$

Having found the optimal values for I, J, and K, the task of finding the least-cost values $V^*(G)$ and $V^*(H)$ becomes easier. Moving back to stage 4 and utilizing the previously obtained optimal-value information in (1.10), we can determine $V^*(G)$ as well as the optimal path GZ (from G to Z) as follows:

(1.11)

$$V^*(G) = \min\{\text{value of arc } GI + V^*(I), \text{value of arc } GJ + V^*(J)\}$$
$$= \min\{2 + 3, 8 + 1\} = 5 \qquad \text{[The optimal path } GZ \text{ is } GIZ.]$$

The fact that the optimal path from G to Z should go through I is indicated by the arrow pointing away from G; the numeral on the arrow shows the optimal path value $V^*(G)$. By the same token, we find

(1.12)

$$V^*(H) = \min\{\text{value of arc } HJ + V^*(J), \text{value of arc } HK + V^*(K)\}$$
$$= \min\{4 + 1, 6 + 2\} = 5 \qquad \text{[The optimal path } HZ \text{ is } HJZ.]$$

Note again the arrow pointing from H toward J and the numeral on it. The set of all such arrows constitutes the optimal policy function, and the set of all the numerals on the arrows constitutes the optimal value function. With the knowledge of $V^*(G)$ and $V^*(H)$, we can then move back one more stage to calculate $V^*(D)$, $V^*(E)$, and $V^*(F)$—and the optimal paths DZ, EZ, and FZ—in a similar manner. And, with two more such steps, we will be back to stage 1, where we can determine $V^*(A)$ and the optimal path AZ, that is, solve the original given problem.

The essence of the iterative solution procedure is captured in Bellman's *principle of optimality*, which states, roughly, that if you chop off the first arc from an optimal sequence of arcs, the remaining abridged sequence must still be optimal in its own right—as an optimal path from its own initial point to the terminal point. If $EHJZ$ is the optimal path from E to Z, for example, then HJZ must be the optimal path from H to Z. Conversely, if HJZ is already known to be the optimal path from H to Z, then a longer optimal path that passes through H must use the sequence HJZ at the tail end. This reasoning is behind the calculations in (1.11) and (1.12). But note that in order to apply the principle of optimality and the iterative procedure to delineate the optimal path from A to Z, we must find the optimal value associated with every possible point in Fig. 1.6. This explains why we must embed the original problem.

Even though the essence of dynamic programming is sufficiently clarified by the discrete example in Fig. 1.6, the full version of dynamic programming includes the continuous-time case. Unfortunately, the solution of continuous-time problems of dynamic programming involves the more advanced mathematical topic of partial differential equations. Besides, partial differential equations often do not yield analytical solutions. Because of this, we shall not venture further into dynamic programming in this book. The rest of the book will be focused on the methods of the calculus of variations and optimal control theory, both of which only require ordinary differential equations for their solution.

EXERCISE 1.4

1. From Fig. 1.6, find $V^*(D)$, $V^*(E)$, and $V^*(F)$. Determine the optimal paths DZ, EZ, and FZ.
2. On the basis of the preceding problem, find $V^*(B)$ and $V^*(C)$. Determine the optimal paths BZ and CZ.
3. Verify the statement in Sec. 1.1 that the minimum cost of production for the example in Fig. 1.6 (same as Fig. 1.1) is \$14, achieved on the path $ACEHJZ$.
4. Suppose that the arc values in Fig. 1.6 are profit (rather than cost) figures. For every point i in the set $\{A, B, \ldots, Z\}$, find

   (*a*) the optimal (maximum-profit) value $V^*(i)$, and
   (*b*) the optimal path from i to Z.

PART 2

THE CALCULUS OF VARIATIONS

CHAPTER 2

THE FUNDAMENTAL PROBLEM OF THE CALCULUS OF VARIATIONS

We shall begin the study of the calculus of variations with the fundamental problem:

$$\begin{aligned} &\text{Maximize or minimize} && V[y] = \int_0^T F[t, y(t), y'(t)]\, dt \\ (2.1)\quad &\text{subject to} && y(0) = A \qquad (A \text{ given}) \\ &\text{and} && y(T) = Z \qquad (T, Z \text{ given}) \end{aligned}$$

The maximization and minimization problems differ from each other in the second-order conditions, but they share the same first-order condition.

The task of variational calculus is to select from a set of admissible y paths (or *trajectories*) the one that yields an extreme value of $V[y]$. Since the calculus of variations is based on the classical methods of calculus, requiring the use of first and second derivatives, we shall restrict the set of admissible paths to those continuous curves with continuous derivatives. A smooth y path that yields an extremum (maximum or minimum) of $V[y]$ is called an *extremal*. We shall also assume that the integrand function F is twice differentiable.

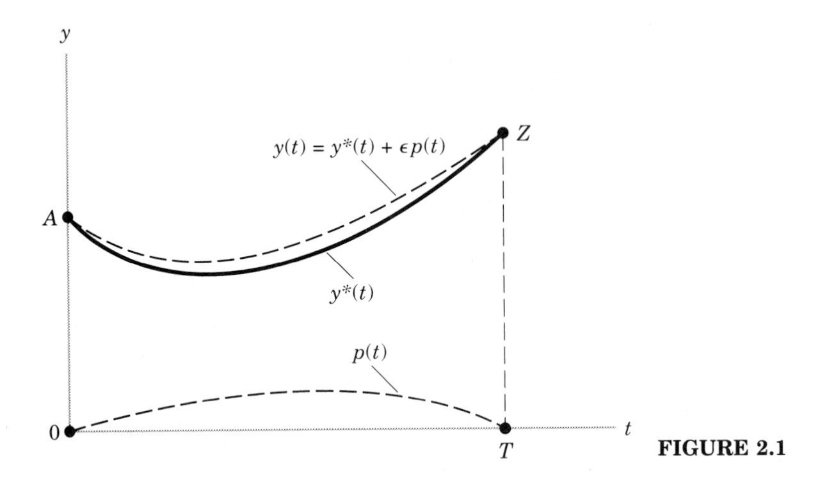

FIGURE 2.1

In locating an extremum of $V[y]$, one may be thinking of either an absolute (global) extremum or a relative (local) extremum. Since the calculus of variations is based on classical calculus methods, it can directly deal only with *relative* extrema. That is, an extremal yields an extreme value of V only in comparison with the immediately "neighboring" y paths.

2.1 THE EULER EQUATION

The basic first-order necessary condition in the calculus of variations is the *Euler equation*. Although it was formulated as early as 1744, it remains the most important result in this branch of mathematics. In view of its importance and the ingenuity of its approach, it is worthwhile to explain its rationale in some detail.

With reference to Fig. 2.1, let the solid path $y^*(t)$ be a known extremal. We seek to find some property of the extremal that is absent in the (nonextremal) neighboring paths. Such a property would constitute a necessary condition for an extremal. To do this, we need for comparison purposes a family of neighboring paths which, by specification in (2.1), must pass through the given endpoints $(0, A)$ and (T, Z). A simple way of generating such neighboring paths is by using a perturbing curve, chosen arbitrarily except for the restrictions that it be smooth and pass through the points 0 and T on the horizontal axis in Fig. 2.1, so that

$$p(0) = p(T) = 0 \tag{2.2}$$

We have chosen for illustration one with relatively small p values and small slopes throughout. By adding $\epsilon p(t)$ to $y^*(t)$, where ϵ is a small number, and by varying the magnitude of ϵ, we can perturb the $y^*(t)$ path, displacing it to various neighboring positions, thereby generating the desired neighboring paths. The latter paths can be denoted generally as

$$y(t) = y^*(t) + \epsilon p(t) \qquad [\text{implying } y'(t) = y^{*\prime}(t) + \epsilon p'(t)] \tag{2.3}$$

with the property that, as $\epsilon \to 0$, $y(t) \to y^*(t)$. To avoid clutter, only one of these neighboring paths has been drawn in Fig. 2.1.

The fact that both $y^*(t)$ and $p(t)$ are *given* curves means that each value of ϵ will determine one particular neighboring y path, and hence one particular value of $V[y]$. Consequently, instead of considering V as a *functional* of the y path, we can now consider it as a *function* of the variable ϵ—$V(\epsilon)$. This change in viewpoint enables us to apply the familiar methods of classical calculus to the function $V = V(\epsilon)$. Since, by assumption, the curve $y^*(t)$—which is associated with $\epsilon = 0$—yields an extreme value of V, we must have

$$\left.\frac{dV}{d\epsilon}\right|_{\epsilon=0} = 0 \tag{2.4}$$

This constitutes a defining property of the extremal. It follows that $dV/d\epsilon = 0$ is a necessary condition for the extremal.

As written, however, condition (2.4) is not operational because it involves the use of the arbitrary variable ϵ as well as the arbitrary perturbing function $p(t)$. What the Euler equation accomplishes is to express this necessary condition in a convenient operational form. To transform (2.4) into an operational form, however, requires a knowledge of how to take the derivatives of a definite integral.

Differentiating a Definite Integral

Consider the definition integral

$$I(x) \equiv \int_a^b F(t, x)\, dt \tag{2.5}$$

where $F(t, x)$ is assumed to have a continuous derivative $F_x(t, x)$ in the time interval $[a, b]$. Since any change in x will affect the value of the F function and hence the definite integral, we may view the integral as a function of x—$I(x)$. The effect of a change in x on the integral is given by the derivative formula:

$$\frac{dI}{dx} = \int_a^b F_x(t, x)\, dt \qquad \text{[Leibniz's rule]} \tag{2.6}$$

In words, to differentiate a definite integral with respect to a variable x which is neither the variable of integration (t) nor a limit of integration (a or b), one can simply differentiate through the integral sign with respect to x.

The intuition behind Leibniz's rule can be seen from Fig. 2.2a, where the solid curve represents $F(t, x)$, and the dotted curve represents the displaced position of $F(t, x)$ after the change in x. The vertical distance between the two curves (if the change is infinitesimal) measures the partial

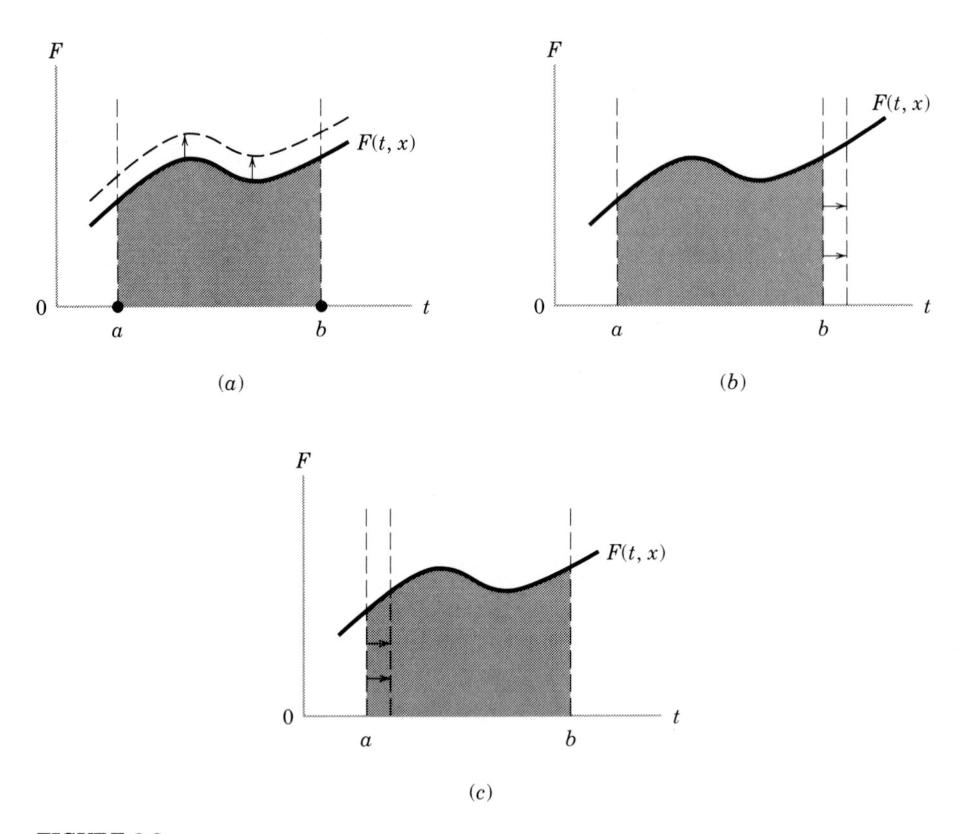

FIGURE 2.2

derivative $F_x(t, x)$ at each value of t. It follows that the effect of the change in x on the entire integral, dI/dx, corresponds to the area between the two curves, or, equivalently, the definite integral of $F_x(t, x)$ over the interval $[a, b]$. This explains the meaning of (2.6).

The value of the definite integral in (2.5) can also be affected by a change in a limit of integration. Defining the integral alternatively to be

$$J(b, a) \equiv \int_a^b F(t, x)\, dt \tag{2.7}$$

we have the following pair of derivative formulas:

$$\frac{\partial J}{\partial b} = F(t, x)\Big|_{t=b} = F(b, x) \tag{2.8}$$

$$\frac{\partial J}{\partial a} = -F(t, x)\Big|_{t=a} = -F(a, x) \tag{2.9}$$

In words, the derivative of a definite integral with respect to its upper limit of integration b is equal to the integrand evaluated at $t = b$; and the derivative with respect to its lower limit of integration a is the *negative* of the integrand evaluated at $t = a$.

In Fig. 2.2*b*, an increase in b is reflected in a rightward displacement of the right-hand boundary of the area under the curve. When the displacement is infinitesimal, the effect on the definite integral is measured by the value of the F function at the right-hand boundary—$F(b, x)$. This provides the intuition for (2.8). For an increase in the lower limit, on the other hand, the resulting displacement, as illustrated in Fig. 2.2*c*, is a rightward movement of the left-hand boundary, which *reduces* the area under the curve. This is why there is a negative sign in (2.9).

The preceding derivative formulas can also be used in combination. If, for instance, the definite integral takes the form

$$K(x) \equiv \int_a^{b(x)} F(t, x)\, dt \tag{2.10}$$

where x not only enters into the integrand function F, but also affects the upper limit of integration, then we can apply both (2.6) and (2.8), to get the total derivative

$$\frac{dK}{dx} = \int_a^{b(x)} F_x(t, x)\, dt + F[b(x), x]\, b'(x) \tag{2.11}$$

The first term on the right, an integral, follows from (2.6); the second term, representing the chain $\dfrac{dK}{db(x)} \dfrac{db(x)}{dx}$, is based on (2.8).

EXAMPLE 1 The derivative of $\int_0^2 e^{-x}\, dt$ with respect to x is, by Leibniz's rule,

$$\int_0^2 \frac{d}{dx} e^{-x}\, dt = \int_0^2 -e^{-x}\, dt = -e^{-x} t\big]_0^2 = -2e^{-x}$$

EXAMPLE 2 Similarly,

$$\frac{d}{dx} \int_2^3 x^t\, dt = \int_2^3 \frac{d}{dx} x^t\, dt = \int_2^3 t x^{t-1}\, dt$$

EXAMPLE 3 To differentiate $\int_0^{2x} 3t^2\, dt$ with respect to x which appears in the upper limit of integration, we need the chain rule as well as (2.8). The result is

$$\frac{d}{dx} \int_0^{2x} 3t^2\, dt = \left[\frac{d}{d(2x)} \int_0^{2x} 3t^2\, dt \right] \frac{d(2x)}{dx} = 3(2x)^2 \cdot 2 = 24x^2$$

Development of the Euler Equation

For ease of understanding, the Euler equation will be developed in four steps.

Step i Let us first express V in terms of ϵ, and take its derivative. Substituting (2.3) into the objective functional in (2.1), we have

$$V(\epsilon) = \int_0^T F\Big[t, \underbrace{y^*(t) + \epsilon p(t)}_{y(t)}, \underbrace{y^{*\prime}(t) + \epsilon p'(t)}_{y'(t)}\Big]\, dt \tag{2.12}$$

To obtain the derivative $dV/d\epsilon$, Leibniz's rule tells us to differentiate through the integral sign:

$$\begin{aligned} \frac{dV}{d\epsilon} &= \int_0^T \frac{\partial F}{\partial \epsilon}\, dt = \int_0^T \left(\frac{\partial F}{\partial y}\frac{dy}{d\epsilon} + \frac{\partial F}{\partial y'}\frac{dy'}{d\epsilon} \right) dt \\ &= \int_0^T \left[F_y p(t) + F_{y'} p'(t) \right] dt \qquad \text{[by (2.3)]} \end{aligned} \tag{2.13}$$

Breaking the last integral in (2.13) into two separate integrals, and setting $dV/d\epsilon = 0$, we get a more specific form of the necessary condition for an extremal as follows:

$$\int_0^T F_y p(t)\, dt + \int_0^T F_{y'} p'(t)\, dt = 0 \tag{2.14}$$

While this form of necessary condition is already free of the arbitrary variable ϵ, the arbitrary perturbing curve $p(t)$ is still present along with its derivative $p'(t)$. To make the necessary condition fully operational, we must also eliminate $p(t)$ and $p'(t)$.

Step ii To that end, we first integrate the second integral in (2.14) by parts, by using the formula:

$$\int_{t=a}^{t=b} v\, du = vu\Big|_{t=a}^{t=b} - \int_{t=a}^{t=b} u\, dv \qquad [u = u(t), v = v(t)] \tag{2.15}$$

Let $v \equiv F_{y'}$ and $u \equiv p(t)$. Then we have

$$dv \equiv \frac{dv}{dt}\, dt = \frac{dF_{y'}}{dt}\, dt \qquad \text{and} \qquad du \equiv \frac{du}{dt}\, dt = p'(t)\, dt$$

Substitution of these expressions into (2.15)—with $a = 0$ and $b = T$—

gives us

$$\int_0^T F_{y'}p'(t)\,dt = \left[F_{y'}p(t)\right]_0^T - \int_0^T p(t)\frac{d}{dt}F_{y'}\,dt \tag{2.16}$$
$$= -\int_0^T p(t)\frac{d}{dt}F_{y'}\,dt$$

since the first term to the right of the first equals sign must vanish under assumption (2.2). Applying (2.16) to (2.14) and combining the two integrals therein, we obtain another version of the necessary condition for the extremal:

$$\int_0^T p(t)\left[F_y - \frac{d}{dt}F_{y'}\right]dt = 0 \tag{2.17}$$

Step iii Although $p'(t)$ is no longer present in (2.17), the arbitrary $p(t)$ still remains. However, precisely because $p(t)$ enters in an *arbitrary* way, we may conclude that condition (2.17) can be satisfied only if the bracketed expression $[F_y - dF_{y'}/dt]$ is made to vanish for every value of t on the extremal; otherwise, the integral may not be equal to zero for some admissible perturbing curve $p(t)$. Consequently, it is a necessary condition for an extremal that

$$F_y - \frac{d}{dt}F_{y'} = 0 \qquad \text{for all } t \in [0, T] \qquad \text{[Euler equation]} \tag{2.18}$$

Note that the Euler equation is completely free of arbitrary expressions, and can thus be applied as soon as one is given a differentiable $F(t, y, y')$ function.

The Euler equation is sometimes also presented in the form

$$\int F_y\,dt = F_{y'} \tag{2.18'}$$

which is the result of integrating (2.18) with respect to t.

Step iv The nature of the Euler equation (2.18) can be made clearer when we expand the derivative $dF_{y'}/dt$ into a more explicit form. Because F is a function with three arguments (t, y, y'), the partial derivative $F_{y'}$ should also be a function of the same three arguments. The total derivative $dF_{y'}/dt$ therefore consists of three terms:

$$\frac{dF_{y'}}{dt} = \frac{\partial F_{y'}}{\partial t} + \frac{\partial F_{y'}}{\partial y}\frac{dy}{dt} + \frac{\partial F_{y'}}{\partial y'}\frac{dy'}{dt}$$
$$= F_{ty'} + F_{yy'}y'(t) + F_{y'y'}y''(t)$$

Substituting this into (2.18), multiplying through by -1, and rearranging, we arrive at a more explicit version of the Euler equation:

$$F_{y'y'}y''(t) + F_{yy'}y'(t) + F_{ty'} - F_y = 0 \qquad \text{for all } t \in [0, T] \qquad \text{[Euler equation]} \tag{2.19}$$

This expanded version reveals that the Euler equation is in general a second-order nonlinear differential equation. Its general solution will thus contain two arbitrary constants. Since our problem in (2.1) comes with two boundary conditions (one initial and one terminal), we should normally possess sufficient information to definitize the two arbitrary constants and obtain the definite solution.

EXAMPLE 4 Find the extremal of the functional

$$V[y] = \int_0^2 (12ty + y'^2)\,dt$$

with boundary conditions $y(0) = 0$ and $y(2) = 8$. Since $F = 12ty + y'^2$, we have the derivatives

$$F_y = 12t \qquad F_{y'} = 2y' \qquad F_{y'y'} = 2 \qquad \text{and} \qquad F_{yy'} = F_{ty'} = 0$$

By (2.19), the Euler equation is

$$2y''(t) - 12t = 0 \qquad \text{or} \qquad y''(t) = 6t$$

which, upon integration, yields $y'(t) = 3t^2 + c_1$, and

$$y^*(t) = t^3 + c_1 t + c_2 \qquad \text{[general solution]}$$

To definitize the arbitrary constants c_1 and c_2, we first set $t = 0$ in the general solution to get $y(0) = c_2$; from the initial condition, it follows that $c_2 = 0$. Next, setting $t = 2$ in the general solution, we get $y(2) = 8 + 2c_1$; from the terminal condition, it follows that $c_1 = 0$. The extremal is thus the cubic function

$$y^*(t) = t^3 \qquad \text{[definite solution]}$$

EXAMPLE 5 Find the extremal of the functional

$$V[y] = \int_1^5 \left[3t + (y')^{1/2}\right] dt$$

with boundary conditions $y(1) = 3$ and $y(5) = 7$. Here we have $F = 3t + (y')^{1/2}$. Thus,

$$F_y = 0 \qquad F_{y'} = \tfrac{1}{2}(y')^{-1/2} \qquad F_{y'y'} = -\tfrac{1}{4}(y')^{-3/2} \qquad \text{and} \qquad F_{yy'} = F_{ty'} = 0$$

The Euler equation (2.19) now reduces to

$$-\tfrac{1}{4}(y')^{-3/2}y''(t) = 0$$

The only way to satisfy this equation is to have a constant y', in order that $y'' = 0$. Thus, we write $y'(t) = c_1$, which integrates to the solution

$$y^*(t) = c_1 t + c_2 \qquad \text{[general solution]}$$

To definitize the arbitrary constants c_1 and c_2, we first set $t = 1$ to find $y(1) = c_1 + c_2 = 3$ (by the initial condition), and then set $t = 5$ to find $y(5) = 5c_1 + c_2 = 7$ (by the terminal condition). These two equations give us $c_1 = 1$ and $c_2 = 2$. Therefore, the extremal takes the form of the linear function

$$y^*(t) = t + 2 \qquad \text{[definite solution]}$$

EXAMPLE 6 Find the extremal of the functional

$$V[y] = \int_0^5 (t + y^2 + 3y')\, dt$$

with boundary conditions $y(0) = 0$ and $y(5) = 3$. Since $F = t + y^2 + 3y'$, we have

$$F_y = 2y \qquad \text{and} \qquad F_{y'} = 3$$

By (2.18), we may write the Euler equation as $2y = 0$, with solution

$$y^*(t) = 0$$

Note, however, that although this solution is consistent with the initial condition $y(0) = 0$, it violates the terminal condition $y(5) = 3$. Thus, we must conclude that there exists no extremal among the set of continuous curves that we consider to be admissible.

This last example is of interest because it serves to illustrate that certain variational problems with given endpoints may not have a solution. More specifically, it calls attention to one of two peculiar results that can arise when the integrand function F is *linear* in y'. One result, as illustrated in Example 6, is that no solution exists. The other possibility, shown in Example 7, is that the Euler equation is an identity, and since it is automatically satisfied, any admissible path is optimal.

EXAMPLE 7 Find the extremal of the functional

$$V[y] = \int_0^T y'\, dt$$

with boundary conditions $y(0) = \alpha$ and $y(t) = \beta$. With $F = y'$, we have

$$F_y = 0 \qquad F_{y'} = 1 \qquad \text{and} \qquad \frac{d}{dt}F_{y'} = 0$$

It follows that the Euler equation (2.18) is always satisfied. In this example, it is in fact clear from straight integration that

$$V[y] = [y(t)]_0^T = y(T) - y(0) = \beta - \alpha$$

The value of V depends only on the given terminal and initial states, regardless of the path adjoining the two given endpoints.

The reason behind these peculiarities is that when F is linear in y', $F_{y'}$ is a constant, and $F_{y'y'} = 0$, so the first term in the Euler equation (2.19) vanishes. The Euler equation then loses its status as a second-order differential equation, and will not provide two arbitrary constants in its general solution to enable us to adapt the time path to the given boundary conditions. Consequently, unless the solution path happens to pass through the fixed endpoints by coincidence, it cannot qualify as an extremal. The only circumstance under which a solution can be guaranteed for such a problem (with F linear in y' and with fixed endpoints) is when $F_y = 0$ as well, which, together with the fact that $F_{y'} =$ constant (implying $dF_{y'}/dt = 0$), would turn the Euler equation (2.18) into an identity, as in Example 7.

EXERCISE 2.1

1 In discussing the differentiation of definite integrals, no mention was made of the derivative with respect to the variable t. Is that a justifiable omission?

Find the derivatives of the following definite integrals with respect to x:

2 $I = \int_a^b x^4 \, dt$

3 $I = \int_a^b e^{-xt} \, dt$

4 $I = \int_0^{2x} e^t \, dt$

5 $I = \int_0^{2x} te^x \, dt$

Find the extremals, if any, of the following functionals:

6 $V[y] = \int_0^1 (ty + 2y'^2) \, dt$, with $y(0) = 1$ and $y(1) = 2$

7 $V[y] = \int_0^1 tyy' \, dt$, with $y(0) = 0$ and $y(1) = 1$

8 $V[y] = \int_0^2 (2ye^t + y^2 + y'^2) \, dt$, with $y(0) = 2$ and $y(2) = 2e^2 + e^{-2}$

9 $V[y] = \int_0^2 (y^2 + t^2 y') \, dt$, with $y(0) = 0$ and $y(2) = 2$

2.2 SOME SPECIAL CASES

We have written the objective functional in the general form $\int_0^T F(t, y, y')\,dt$, in which the integrand function F has three arguments: t, y, and y'. For some problems, the integrand function may not contain all three arguments. For such special cases, we can derive special versions of the Euler equation which may often (though not always) prove easier to solve.

Special case I: $F = F(t, y')$ In this special case, the F function is free of y, implying that $F_y = 0$. Hence, the Euler equation reduces to $dF_{y'}/dt = 0$, with the solution

$$F_{y'} = \text{constant} \tag{2.20}$$

It may be noted that Example 5 of the preceding section falls under this special case, although at that time we just used the regular Euler equation for its solution. It is easy to verify that the application of (2.20) will indeed lead to the same result. Here is another example of this special case.

EXAMPLE 1 Find the extremal of the functional

$$V[y] = \int_0^1 (ty' + y'^2)\,dt$$

with boundary conditions $y(0) = y(1) = 1$. Since

$$F = ty' + y'^2 \qquad \text{and} \qquad F_{y'} = t + 2y'$$

(2.20) gives us $t + 2y'(t) = \text{constant}$, or

$$y'(t) = -\tfrac{1}{2}t + c_1$$

Upon direct integration, we obtain

$$y^*(t) = -\tfrac{1}{4}t^2 + c_1 t + c_2 \qquad \text{[general solution]}$$

With the help of the boundary conditions $y(0) = y(1) = 1$, it is easily verified that $c_1 = \frac{1}{4}$ and $c_2 = 1$. The extremal is therefore the quadratic path

$$y^*(t) = -\tfrac{1}{4}t^2 + \tfrac{1}{4}t + 1 \qquad \text{[definite solution]}$$

Special case II: $F = F(y, y')$ Since F is free of t in this case, we have $F_{ty'} = 0$, so the Euler equation (2.19) simplifies to

$$F_{y'y'}y''(t) + F_{yy'}y'(t) - F_y = 0$$

The solution to this equation is by no means obvious, but it turns out that if we multiply through by y', the left-hand-side expression in the resulting new equation will be exactly the derivative $d(y'F_{y'} - F)/dt$, for

$$\begin{aligned}\frac{d}{dt}(y'F_{y'} - F) &= \frac{d}{dt}(y'F_{y'}) - \frac{d}{dt}F(y, y') \\ &= F_{y'}y'' + y'(F_{yy'}y' + F_{y'y'}y'') - (F_y y' + F_{y'}y'') \\ &= y'(F_{y'y'}y'' + F_{yy'}y' - F_y)\end{aligned}$$

Consequently, the Euler equation can be written as $d(y'F_{y'} - F)/dt = 0$, with the solution $y'F_{y'} - F =$ constant, or, what amounts to be the same thing,

$$F - y'F_{y'} = \text{constant} \tag{2.21}$$

This result—the simplified Euler equation already integrated once—is a first-order differential equation, which may under some circumstances be easier to handle than the original Euler equation (2.19). Moreover, in analytical (as against computational) applications, (2.21) may yield results that would not be discernible from (2.19), as will be illustrated in Sec. 2.4.

EXAMPLE 2 Find the extremal of the functional

$$V[y] = \int_0^{\pi/2} (y^2 - y'^2)\, dt$$

with boundary conditions $y(0) = 1$ and $y(\pi/2) = 0$. Since

$$F = y^2 - y'^2 \qquad \text{and} \qquad F_{y'} = -2y'$$

direct substitution in (2.21) gives us

$$y'^2 + y^2 = \text{constant} \qquad [\equiv a^2, \text{say}]$$

This last constant is nonnegative because the left-hand-side terms are all squares; thus we can justifiably denote it by a^2, where a is a nonnegative real number.

The reader will recognize that the equation $y'^2 + y^2 = a^2$ can be plotted as a circle, as in Fig. 2.3, with radius a and with center at the point of origin. Since y' is plotted against y in the diagram, this circle constitutes the phase line for the differential equation. Moreover, the circular loop shape of this phase line suggests that it gives rise to a cyclical time path,[1]

[1]Phase diagrams are explained in Alpha C. Chiang, *Fundamental Methods of Mathematical Economics*, 3d ed., McGraw-Hill, New York, 1984, Sec. 14.6. The phase line here is similar to phase line C in Fig. 14.3 in that section; the time path it implies is shown in Fig. 14.4c.

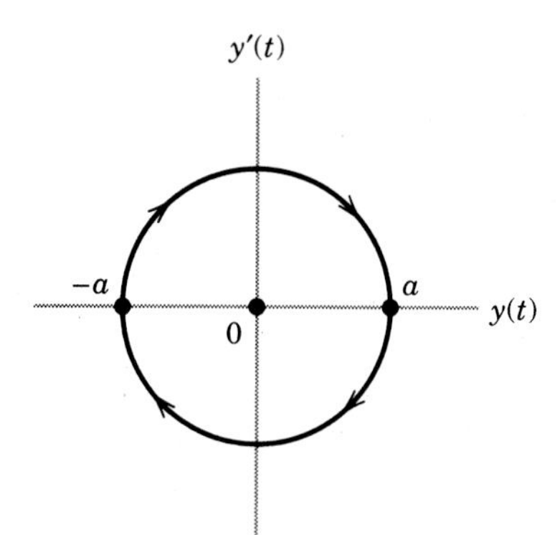

FIGURE 2.3

with the y values bounded in the closed interval $[-a, a]$, as in a cosine function with amplitude a and period 2π. Such a cosine function can be represented in general by the equation

$$y(t) = a\cos(bt + c)$$

where, aside from the amplitude parameter a, we also have two other parameters b and c which relate to the period and the phase of the function, respectively. In our case, the period should be 2π; but since the cosine function shows a period of $2\pi/b$ (obtained by setting the bt term to 2π), we infer that $b = 1$. But the values of a and c are still unknown, and they must be determined from the given boundary conditions.

At $t = 0$, we have

$$y(0) = a\cos c = 1 \qquad \text{[by the initial condition]}$$

When $t = \pi/2$, we have

$$y\left(\frac{\pi}{2}\right) = a\cos\left(\frac{\pi}{2} + c\right) = 0 \qquad \text{[by the terminal condition]}$$

To satisfy this last equation, we must have either $a = 0$ or $\cos(\pi/2 + c) = 0$. But a cannot be zero, because otherwise $a\cos c$ cannot possibly be equal to 1. Hence, $\cos(\pi/2 + c)$ must vanish, implying two possibilities: either $c = 0$, or $c = \pi$. With $c = 0$, the equation $a\cos c = 1$ becomes $a\cos 0 = 1$ or $a(1) = 1$, yielding $a = 1$. With $c = \pi$, however, the equation $a\cos c = 1$ becomes $a(-1) = 1$, giving us $a = -1$, which is inadmissible because we have restricted a to be a nonnegative number. Hence, we conclude that the boundary conditions require $a = 1$ and $c = 0$, and that the time path of y which qualifies as the extremal is

$$y^*(t) = \cos t$$

The same solution can, as we would expect, also be obtained in a straightforward manner from the original Euler equation (2.18) or (2.19).

After normalizing and rearranging, the Euler equation can be expressed as the following homogeneous second-order linear differential equation:

$$y'' + y = 0$$

Since this equation has complex characteristic roots $r_1, r_2 = \pm i$, the general solution takes the form of[2]

$$y^*(t) = \alpha \cos t + \beta \sin t$$

The boundary conditions fix the arbitrary constants α and β at the values of 1 and 0, respectively. Thus, we end up with the same definite solution:

$$y^*(t) = \cos t$$

For this example, the original Euler equation (2.18) or (2.19) turns out to be easier to use than the special one in (2.21). We illustrate the latter, not only to present it as an alternative, but also to illustrate some other techniques (such as the circular phase diagram in Fig. 2.3). The reader will have to choose the appropriate version of the Euler equation to apply to any particular problem.

EXAMPLE 3 Among the curves that pass through two fixed points A and Z, which curve will generate the smallest surface of revolution when rotated about the horizontal axis? This is a problem in physics, but it may be of interest because it is one of the earliest problems in the development of the calculus of variations. To construct the objective functional, it may be helpful to consider the curve AZ in Fig. 2.4 as a candidate for the desired extremal. When rotated about the t axis in the stipulated manner, each point on curve AZ traces out a circle parallel to the xy plane, with its center on the t axis, and with radius R equal to the value of y at that point. Since the circumference of such a circle is $2\pi R$ (in our present case, $2\pi y$), all we have to do to calculate the surface of revolution is to sum (integrate) the circumference over the entire length of the curve AZ.

An expression for the length of the curve AZ can be obtained with the help of Fig. 2.5. Let us imagine that M and N represent two points that are located very close to each other on curve AZ. Because of the extreme proximity of M and N, arc MN can be approximated by a straight line, with its length equal to the differential ds. In order to express ds in terms of the variables y and t, we resort to Pythagoras' theorem to write $(ds)^2 = (dy)^2 + (dt)^2$. It follows that $(ds)^2/(dt)^2 = (dy)^2/(dt)^2 + 1$. Taking the

[2]For an explanation of complex roots, see Alpha C. Chiang, *Fundamental Methods of Mathematical Economics*, 3d ed., McGraw-Hill, New York, 1984, Sec. 15.3. Here we have $h = 0$, and $v = 1$; thus, $e^{ht} = 1$ and $vt = t$.

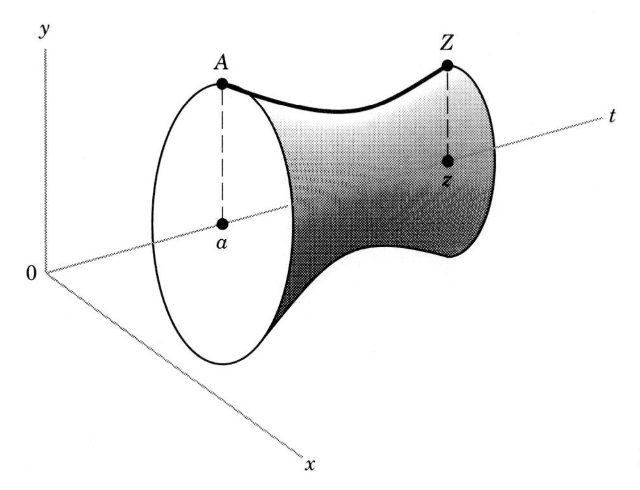

FIGURE 2.4

square root on both sides, we get

$$\frac{ds}{dt} = \sqrt{1 + \left(\frac{dy}{dt}\right)^2} = \left(1 + y'^2\right)^{1/2}$$

which yields the desired expression for ds in terms of y and t as follows:

$$ds = \left(1 + y'^2\right)^{1/2} dt \qquad \text{[arc length]} \tag{2.22}$$

The entire length of curve AZ must then be the integral of ds, namely $\int_a^z (1 + y'^2)^{1/2}\,dt$. To sum the circumference $2\pi y$ over the length of curve AZ will therefore result in the functional

$$V[y] = 2\pi \int_a^z y\left(1 + y'^2\right)^{1/2} dt$$

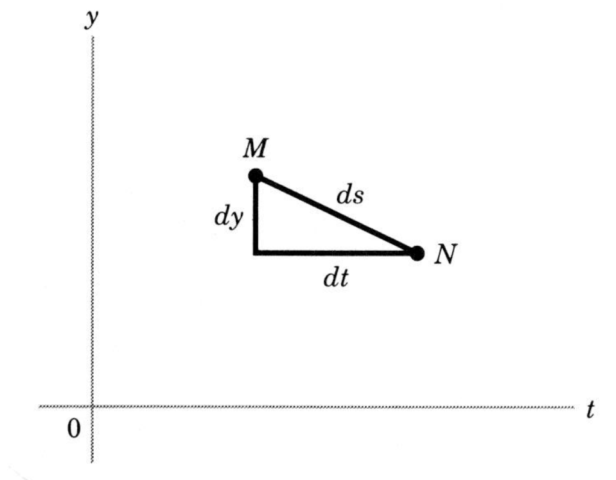

FIGURE 2.5

The reader should note that, while it is legitimate to "factor out" the "2π" (constant) part of the expression $2\pi R$, the "y" (variable) part must remain within the integral sign.

For purposes of minimization, we may actually disregard the constant 2π and take $y(1 + y'^2)^{1/2}$ to be the expression for the F function, with

$$F_{y'} = yy'(1 + y'^2)^{-1/2}$$

Since F is free of t, we may apply (2.21) to get

$$y(1 + y'^2)^{1/2} - yy'^2(1 + y'^2)^{-1/2} = c$$

This equation can be simplified by taking the following steps: (1) multiply through by $(1 + y'^2)^{1/2}$, (2) cancel out yy'^2 and $-yy'^2$ on the left-hand side, (3) square both sides and solve for y'^2 in terms of y and c, and (4) take the square root. The result is

$$y'\left(\equiv \frac{dy}{dt}\right) = \frac{1}{c}\sqrt{y^2 - c^2}$$

or

$$\frac{c\,dy}{\sqrt{y^2 - c^2}} = dt$$

In this last equation, we note that the variables y and t have been separated, so that each side can be integrated by itself. The right-hand side poses no problem, the integral of dt being in the simple form $t + k$, where k is an arbitrary constant. But the left-hand side is more complicated. In fact, to attempt to integrate it "in the raw" would take too much effort; hence, one should consult prepared tables of integration formulas to find the result:[3]

$$\int \frac{c\,dy}{\sqrt{y^2 - c^2}} = c \ln\left(\frac{y + \sqrt{y^2 - c^2}}{c}\right) + c_1$$

Equating this result with $t + k$ (and subsuming the constant c_1 under the

[3]See, for example, *CRC Standard Mathematical Tables*, 28th ed., ed. by William H. Beyer, CRC Press, Boca Raton, FL, 1987, Formula 157. The c in the denominator of the log expression on the right may be omitted without affecting the validity of the formula, because its presence or absence merely makes a difference amounting to the constant $-c \ln c$, which can be absorbed into the arbitrary constant of integration c_1, at any rate. However, by inserting the c in the denominator, our result will come out in a more symmetrical form.

constant k), we have

$$\ln\left(\frac{y + \sqrt{y^2 - c^2}}{c}\right) = \frac{t+k}{c}$$

or $$e^{(t+k)/c} = \frac{y + \sqrt{y^2 - c^2}}{c} \qquad \text{[by definition of natural log]}$$

Multiplying both sides by c, subtracting y from both sides, squaring both sides and canceling out the y^2 term, and solving for y in terms of t, we finally find the desired extremal to be

$$y^*(t) = \frac{c}{2}\left[e^{(t+k)/c} + e^{-(t+k)/c}\right] \qquad \text{[general solution]}$$

where the two constants c and k are to be definitized by using the boundary conditions.

This extremal is a modified form of the so-called *catenary* curve,

$$y = \tfrac{1}{2}(e^t + e^{-t}) \qquad \text{[catenary]} \tag{2.23}$$

whose distinguishing characteristic is that it involves the average of two exponential terms, in which the exponent of one term is the exact negative of the exponent of the other. Since the positive-exponent term gives rise to a curve that increases at an increasing rate, whereas the negative-exponent term produces an ever-decreasing curve, the average of the two has a general shape that portrays the way a flexible rope will hang on two fixed pegs. (In fact, the name *catenary* comes from the Latin word *catena*, meaning "chain.") This general shape is illustrated by curve AZ in Fig. 2.4.

Even though we have found our extremal in a curve of the catenary family, it is not certain whether the resulting surface of revolution (known as a *catenoid*) has been maximized or minimized. However, geometrical and intuitive considerations should make it clear that the surface is indeed *minimized*. With reference to Fig. 2.4, if we replace the AZ curve already drawn by, say, a new AZ curve with the opposite curvature, then a *larger* surface of revolution can be generated. Hence, the catenoid cannot possibly be a maximum.

Special case III: $F = F(y')$ When the F function depends on y' alone, many of the derivatives in (2.19) will disappear, including $F_{yy'}$, $F_{ty'}$, and F_y. In fact, only the first term remains, so the Euler equation becomes

$$F_{y'y'}y''(t) = 0 \tag{2.24}$$

To satisfy this equation, we must either have $y''(t) = 0$ or $F_{y'y'} = 0$. If $y''(t) = 0$, then obviously $y'(t) = c_1$ and $y(t) = c_1 t + c_2$, indicating that the general solution is a two-parameter family of straight lines. If, on the other hand, $F_{y'y'} = 0$, then since $F_{y'y'}$ is, like F itself, a function of y' alone, the solution of $F_{y'y'} = 0$ should appear as specific values of y'. Suppose there are one or more real solutions $y' = k_i$, then we can deduce that $y = k_i t + c$, which again represents a family of straight lines. Consequently, given an integrand function that depends on y' alone, we can always take its extremal to be a straight line.

EXAMPLE 4 Find the extremal of the functional

$$V[y] = \int_0^T (1 + y'^2)^{1/2}\, dt \tag{2.25}$$

with boundary conditions $y(0) = A$ and $y(T) = Z$. The astute reader will note that this functional has been encountered in a different guise in Example 3. Recalling the discussion of arc length leading to (2.22), we know that (2.25) measures the total length of a curve passing through two given points. The problem of finding the extremal of this functional is therefore that of finding the curve with the shortest distance between those two points.

The integrand function, $F = (1 + y'^2)^{1/2}$, is dependent on y' alone. We can therefore conclude immediately that the extremal is a straight line. But if it is desired to examine this particular example explicitly by the Euler equation, we can use (2.18). With $F_y = 0$, the Euler equation is simply $dF_{y'}/dt = 0$, and its solution is $F_{y'} =$ constant. In view of the fact that $F_{y'} = y'/(1 + y'^2)^{1/2}$, we can write (after squaring).

$$\frac{y'^2}{1 + y'^2} = c^2$$

Multiplying both sides by $(1 + y'^2)$, rearranging, and factoring out y', we can express y' in terms of c as follows: $y'^2 = c^2/(1 - c^2)$. Equivalently,

$$y' = \frac{c}{(1 - c^2)^{1/2}} = \text{constant}$$

Inasmuch as $y'(t)$, the slope of $y(t)$, is a constant, the desired extremal $y^*(t)$ must be a straight line.

Strictly speaking, we have only found an "extremal" which may either maximize or minimize the given functional. However, it is intuitively obvious that the distance between the two given points is indeed *minimized* rather than maximized by the straight-line extremal, because there is no such thing as "the longest distance" between two given points.

Special case IV: $F = F(t, y)$ In this special case, the argument y' is missing from the F function. Since we now have $F_{y'} = 0$, the Euler equation reduces simply to $F_y = 0$, or, more explicitly,

$$F_y(t, y) = 0$$

The fact that the derivative y' does not appear in this equation means that the Euler equation is not a differential equation. The problem is degenerate. Since there are no arbitrary constants in its solution to be definitized in accordance with the given boundary conditions, the extremal may not satisfy the boundary conditions except by sheer coincidence.

This situation is very much akin to the case where the F function is linear in the argument y' (Sec. 2.1, Example 6). The reason is that the function $F(t, y)$ can be considered as a special case of $F(t, y, y')$ with y' entering via a single additive term, $0y'$, with a zero coefficient. Thus, $F(t, y)$ is, in a special sense, "linear" in y'.

EXERCISE 2.2

Find the extremals of the following functionals:

1. $V[y] = \int_0^1 (t^2 + y'^2)\, dt$, with $y(0) = 0$ and $y(1) = 2$
2. $V[y] = \int_0^2 7y'^3\, dt$, with $y(0) = 9$ and $y(2) = 11$
3. $V[y] = \int_0^1 (y + yy' + y' + \frac{1}{2}y'^2)\, dt$, with $y(0) = 2$ and $y(1) = 5$
4. $V[y] = \int_a^b t^{-3} y'^2\, dt$ (Find the general solution only.)
5. $V[y] = \int_0^1 (y^2 + 4yy' + 4y'^2)\, dt$, with $y(0) = 2e^{1/2}$ and $y(1) = 1 + e$
6. $V[y] = \int_0^{\pi/2} (y^2 - y'^2)\, dt$, with $y(0) = 0$ and $y(\pi/2) = 1$

2.3 TWO GENERALIZATIONS OF THE EULER EQUATION

The previous discussion of the Euler equation is based on an integral functional with the integrand $F(t, y, y')$. Simple generalizations can be made, however, to the case of several state variables and to the case where higher-order derivatives appear as arguments in the F function.

The Case of Several State Variables

With $n > 1$ state variables in a given problem, the functional becomes

$$V[y_1, \ldots, y_n] = \int_0^T F(t, y_1, \ldots, y_n, y_1', \ldots, y_n')\, dt \tag{2.26}$$

and there will be a pair of initial condition and terminal condition for each of the n state variables.

Any extremal $y_j^*(t)$, $(j = 1, \ldots, n)$, must by definition yield the extreme path value relative to all the neighboring paths. One type of neighboring path arises from varying only one of the functions $y_j(t)$ at a time, say, $y_1(t)$, while all the other $y_j(t)$ functions are kept fixed. Then the functional depends only on the variation in $y_1(t)$ as if we are dealing with the case of a single state variable. Consequently, the Euler equation (2.18) must still hold as a necessary condition, provided we change the y symbol to y_1 to reflect the new problem. Moreover, this procedure can be similarly used to vary the other y_j functions, one at a time, to generate other Euler equations. Thus, for the case of n state variables, the single Euler equation (2.18) should be replaced by a set of n simultaneous Euler equations:

$$(2.27) \quad F_{y_j} - \frac{d}{dt}F_{y_j'} = 0 \qquad \text{for all } t \in [0, T] \qquad (j = 1, 2, \ldots, n)$$

[Euler equations]

These n equations will, along with the boundary conditions, enable us to determine the solutions $y_1^*(t), \ldots, y_n^*(t)$.

Although (2.27) is a straightforward generalization of the Euler equation (2.18)—replacing the symbol y by y_j—the same procedure cannot be used to generalize (2.19). To see why not, assume for simplicity that there are only two state variables, y and z, in our new problem. The F function will then be a function with five arguments, $F(t, y, z, y', z')$, and the partial derivatives $F_{y'}$ and $F_{z'}$, will be, too. Therefore, the total derivative of $F_{y'}(t, y, z, y', z')$ with respect to t will include *five* terms:

$$\frac{d}{dt}F_{y'}(t, y, z, y', z') = F_{ty'} + F_{yy'}y'(t) + F_{zy'}z'(t) + F_{y'y'}y''(t) + F_{z'y'}z''(t)$$

with a similar five-term expression for $dF_{z'}/dt$. The expanded version of simultaneous Euler equations corresponding to (2.19) thus looks much more complicated than (2.19) itself:

$$(2.28) \quad \begin{aligned} F_{y'y'}y''(t) + F_{z'y'}z''(t) + F_{yy'}y'(t) + F_{zy'}z'(t) + F_{ty'} - F_y &= 0 \\ F_{y'z'}y''(t) + F_{z'z'}z''(t) + F_{yz'}y'(t) + F_{zz'}z'(t) + F_{tz'} - F_z &= 0 \end{aligned}$$

for all $t \in [0, T]$

EXAMPLE 1 Find the extremal of

$$V[y, z] = \int_0^T (y + z + y'^2 + z'^2)\, dt$$

From the integrand, we find that

$$F_y = 1 \qquad F_{y'} = 2y' \qquad F_z = 1 \qquad F_{z'} = 2z'$$

Thus, by (2.27), we have two simultaneous Euler equations

$$1 - 2y'' = 0 \qquad \text{or} \qquad y'' = \tfrac{1}{2}$$
$$1 - 2z'' = 0 \qquad \text{or} \qquad z'' = \tfrac{1}{2}$$

The same result can also be obtained from (2.28).

In this particular case, it happens that each of the Euler equations contains one variable exclusively. Upon integration, the first yields $y' = \frac{1}{2}t + c_1$, and hence,

$$y^*(t) = \tfrac{1}{4}t^2 + c_1 t + c_2$$

Analogously, the extremal of z is

$$z^*(t) = \tfrac{1}{4}t^2 + c_3 t + c_4$$

The four arbitrary constants $(c_1, \ldots, c_4)$ can be definitized with the help of four boundary conditions relating to $y(0)$, $z(0)$, $y(T)$, and $z(T)$.

The Case of Higher-Order Derivatives

As another generalization, consider a functional that contains high-order derivatives of $y(t)$. Generally, this can be written as

$$V[y] = \int_0^T F(t, y, y', y'', \ldots, y^{(n)})\, dt \tag{2.29}$$

Since many derivatives are present, the boundary conditions should in this case prescribe the initial and terminal values not only of y, but also of the derivatives $y', y'', \ldots,$ up to and including the derivative $y^{(n-1)}$, making a total of $2n$ boundary conditions.

To tackle this case, we first note that an F function with a single state variable y and derivatives of y up to the nth order can be transformed into an equivalent form containing n state variables and their first-order derivatives only. In other words, the functional in (2.29) can be transformed into the form in (2.26). Consequently, the Euler equation (2.27) or (2.28) can again be applied. Moreover, the said transformation can automatically take care of the boundary conditions as well.

EXAMPLE 2 Transform the functional

$$V[y] = \int_0^T (ty^2 + yy' + y''^2)\, dt$$

with boundary conditions $y(0) = A$, $y(T) = Z$, $y'(0) = \alpha$, and $y'(T) = \beta$, into the format of (2.26). To achieve this task, we only have to introduce a new variable

$$z \equiv y' \qquad [\text{implying } z' \equiv y'']$$

Then we can rewrite the integrand as

$$F = ty^2 + yy' + y''^2 = ty^2 + yz + z'^2$$

which now contains two state variables y and z, with no derivatives higher than the first order. Substituting the new F into the functional will turn the latter into the format of (2.26).

What about the boundary conditions? For the original state variable y, the conditions $y(0) = A$ and $y(T) = Z$ can be kept intact. The other two conditions for y' can be directly rewritten as the conditions for the new state variable z: $z(0) = \alpha$ and $z(T) = \beta$. This complete the transformation.

Given the functional (2.29), it is also possible to deal with it directly instead of transforming it into the format of (2.26). By a procedure similar to that used in deriving the Euler equation, a necessary condition for an extremal can be found for (2.29). The condition, known as the *Euler-Poisson equation*, is

$$F_y - \frac{d}{dt}F_{y'} + \frac{d^2}{dt^2}F_{y''} - \cdots + (-1)^n \frac{d^n}{dt^n}F_{y^{(n)}} = 0 \tag{2.30}$$

for all $t \in [0, T]$ [Euler-Poisson equation]

This equation is in general a differential equation of order $2n$. Thus its solution will involve $2n$ arbitrary constants, which can be definitized with the help of the $2n$ boundary conditions.

EXAMPLE 3 Find an extremal of the functional in Example 2. Since we have

$$F_y = 2ty + y' \qquad F_{y'} = y \qquad F_{y''} = 2y''$$

the Euler-Poisson equation is

$$2ty + y' - \frac{dy}{dt} + \frac{d^2 2y''}{dt^2} = 0 \qquad \text{or} \qquad 2ty + 2y^{(4)} = 0$$

which is a fourth-order differential equation.

EXERCISE 2.3

1. Find the extremal of $V[y] = \int_0^1 (1 + y''^2)\,dt$, with $y(0) = 0$ and $y'(0) = y(1) = y'(1) = 1$.
2. Find the extremal of $V[y, z] = \int_a^b (y'^2 + z'^2 + y'z')\,dt$ (general solution only).

3 Find the extremal of $V[y, z] = \int_0^{\pi/2}(2yz + y'^2 + z'^2)\,dt$, with $y(0) = z(0) = 0$ and $y(\pi/2) = z(\pi/2) = 1$.

4 Example 3 of this section shows that, for the problem stated in Example 2, a necessary condition for the extremal is $2ty + 2y^{(4)} = 0$. Derive the same result by applying the definition $z \equiv y'$ and the Euler equations (2.27) to that problem.

2.4 DYNAMIC OPTIMIZATION OF A MONOPOLIST

Let us turn now to economic applications of the Euler equation. As the first example, we shall discuss the classic Evans model of a monopolistic firm, one of the earliest applications of variational calculus to economics.[4]

A Dynamic Profit Function

Consider a monopolistic firm that produces a single commodity with a quadratic total cost function[5]

$$C = \alpha Q^2 + \beta Q + \gamma \qquad (\alpha, \beta, \gamma > 0) \tag{2.31}$$

Since no inventory is considered, the output Q is always set equal to the quantity demanded. Hence, we shall use a single symbol $Q(t)$ to denote both quantities. The quantity demanded is assumed to depend not only on price $P(t)$, but also on the rate of change of price $P'(t)$:

$$Q = a - bP(t) + hP'(t) \qquad (a, b > 0;\ h \neq 0) \tag{2.32}$$

The firm's profit is thus

$$\begin{aligned}\pi &= PQ - C \\ &= P(a - bP + hP') - \alpha(a - bP + hP')^2 - \beta(a - bP + bP') - \gamma\end{aligned}$$

which is a function of P and P'. Multiplying out the above expression and collecting terms, we have the dynamic profit function

$$\begin{aligned}\pi(P, P') = &-b(1 + \alpha b)P^2 + (a + 2\alpha ab + \beta b)P \\ &- \alpha h^2 P'^2 - h(2\alpha a + \beta)P' + h(1 + 2\alpha b)PP' \\ &-(\alpha a^2 + \beta a + \gamma) \qquad \text{[profit function]}\end{aligned} \tag{2.33}$$

---

[4] G. C. Evans, "The Dynamics of Monopoly," *American Mathematical Monthly*, February 1924, pp. 77–83. Essentially the same material appears in Chapters 14 and 15 of a book by the same author, *Mathematical Introduction to Economics*, McGraw-Hill, New York, 1930, pp. 143–153.

[5] This quadratic cost function plots as a U-shaped curve. But with β positive, only the right-hand segment of the U appears in the first quadrant, giving us an upward-sloping total-cost curve for $Q \geq 0$.

The Problem

The objective of the firm is to find an optimal path of price P that maximizes the total profit Π over a finite time period $[0, T]$. This period is assumed to be sufficiently short to justify the assumption of fixed demand and cost functions, as well as the omission of a discount factor. Besides, as the first approach to the problem, both the initial price P_0 and the terminal price P_T are assumed to be given.

The objective of the monopolist is therefore to

$$\begin{aligned} &\text{Maximize} && \Pi[P] = \int_0^T \pi(P, P')\,dt \\ (2.34)\quad &\text{subject to} && P(0) = P_0 \qquad (P_0 \text{ given}) \\ &\text{and} && P(T) = P_T \qquad (T, P_T \text{ given}) \end{aligned}$$

The Optimal Price Path

Although (2.34) pertains to Special Case II, it turns out that for computation with specific functions it is simpler to use the original Euler equation (2.19), where we should obviously substitute π for F, and P for y. From the profit function (2.33), it is easily found that

$$\pi_P = -2b(1 + \alpha b)P + (a + 2\alpha ab + \beta b) + h(1 + 2\alpha b)P'$$

$$\pi_{P'} = -2\alpha h^2 P' - h(2\alpha a + \beta) + h(1 + 2\alpha b)P$$

and $\quad \pi_{P'P'} = -2\alpha h^2 \qquad \pi_{PP'} = h(1 + 2\alpha b) \qquad \pi_{tP'} = 0$

These expression turn (2.19)—after normalizing—into the specific form

$$(2.35)\quad P'' - \frac{b(1 + \alpha b)}{\alpha h^2} P = -\frac{a + 2\alpha ab + \beta b}{2\alpha h^2} \qquad \text{[Euler equation]}$$

This is a second-order linear differential equation with constant coefficients and a constant term, in the general format of

$$y'' + a_1 y' + a_2 y = a_3$$

Its general solution is known to be[6]

$$y(t) = A_1 e^{r_1 t} + A_2 e^{r_2 t} + \bar{y}$$

[6]This type of differential equation is discussed in Alpha C. Chiang, *Fundamental Methods of Mathematical Economics*, 3d ed., McGraw-Hill, New York, 1984, Sec. 15.1.

where the characteristic roots r_1 and r_2 take the values

$$r_1, r_2 = \tfrac{1}{2}\left(-a_1 \pm \sqrt{a_1^2 - 4a_2}\right)$$

and the particular integral $\bar{y}$ is

$$\bar{y} = \frac{a_3}{a_2}$$

Thus, for the Euler equation of the present model (where $a_1 = 0$), we have

$$P^*(t) = A_1 e^{r_1 t} + A_2 e^{r_2 t} + \bar{P} \tag{2.36}$$

where $$r_1, r_2 = \pm\sqrt{\frac{b(1+\alpha b)}{\alpha h^2}} \qquad \text{[characteristic roots]}$$

and $$\bar{P} = \frac{a + 2\alpha ab + \beta b}{2b(1+\alpha b)} \qquad \text{[particular integral]}$$

Note that the two characteristic roots are real and distinct under our sign specifications. Moreover, they are the exact negatives of each other. We may therefore let r denote the common absolute value of the two roots, and rewrite the solution as

$$P^*(t) = A_1 e^{rt} + A_2 e^{-rt} + \bar{P} \qquad \text{[general solution]} \tag{2.36'}$$

The two arbitrary constants A_1 and A_2 in (2.36′) can be definitized via the boundary conditions $P(0) = P_0$ and $P(T) = P_T$. When we set $t = 0$ and $t = T$, successively, in (2.36′), we get two simultaneous equations in the two variables A_1 and A_2:

$$P_0 = A_1 + A_2 + \bar{P}$$

$$P_T = A_1 e^{rT} + A_2 e^{-rT} + \bar{P}$$

with solution values

(2.37)

$$A_1 = \frac{P_0 - \bar{P} - (P_T - \bar{P})e^{rT}}{1 - e^{2rT}} \qquad A_2 = \frac{P_0 - \bar{P} - (P_T - \bar{P})e^{-rT}}{1 - e^{-2rT}}$$

The determination of these two constants completes the solution of the problem, for the entire price path $P^*(t)$ is now specified in terms of the known parameters T, P_0, P_T, α, β, γ, a, b, and h. Of these parameters, all have specified signs except h. But inasmuch as the parameter h enters into the solution path (2.36′) only through r, and only in the form of a squared term h^2, we can see that its algebraic sign will not affect our results, although its numerical value will.

At this moment, we are not equipped to discuss whether the solution path indeed maximizes (rather than minimizes) profit. Assuming that it indeed does, then a relevant question is: What can we say about the general configuration of the price path (2.36′)? Unfortunately, no simple answer can be given. If $P_T > P_0$, the monopolist's price may optimally rise steadily in the interval $[0, T]$ from P_0 to P_T or it may have to be lowered slightly before rising to the level of P_T at the end of the period, depending on parameter values. In the opposite case of $P_0 > P_T$, the optimal price path is characterized by a similar indeterminacy. Although this aspect of the problem can be pursued a little further (see Exercise 2.4, Probs. 3 and 4), specific parameter values are needed before more definite ideas about the optimal price path can be formed.

A More General View of the Problem

Evans' formulation specifies a quadratic cost function and a linear demand function. In a more general study of the dynamic-monopolist problem by Tintner, these functions are left unspecified.[7] Thus, the profit function is simply written as $\pi(P, P')$. In such a general formulation, it turns out that formula (2.21) [for Special Case II] can be used to good advantage. It directly yields a simple necessary condition

$$\pi - P'\frac{\partial \pi}{\partial P'} = c \tag{2.38}$$

that can be given a clear-cut economic interpretation.

To see this, first look into the economic meaning of the constant c. If profit π does *not* depend on the rate of change of price P'—that is, if we are dealing with the static monopoly problem as a special case of the dynamic model—then $\partial\pi/\partial P' = 0$, and (2.38) reduces to $\pi = c$. So the constant c represents the static monopoly profit. Let us therefore denote it by π_s (subscript s for static). Next, we note that if (2.38) is divided through by π, the second term on the left-hand side will involve

$$\frac{\partial \pi}{\partial P'}\frac{P'}{\pi} \equiv \epsilon_{\pi p'}$$

which represents the partial elasticity of π with respect to P'. Indeed, after performing the indicated division, the equation can be rearranged into the

[7] Gerhard Tintner, "Monopoly over Time," *Econometrica*, 1937, pp. 160–170. Tintner also tried to generalize the Evans model to the case where π depends on higher-order derivatives of P, but the economic meaning of the result is difficult to interpret.

optimization rule

$$\epsilon_{\pi p'} = 1 - \frac{\pi_s}{\pi} \tag{2.38'}$$

This rule states that the monopolistic firm should always select the price in such a way that the elasticity of profit with respect to the rate of change of price be equal to $1 - \pi_s/\pi$. The reader will note that this analytical result would not have emerged so readily if we had not resorted to the special-case formula (2.21).

It is of interest to compare this elasticity rule with a corresponding elasticity rule for the static monopolist. Since in the latter case profit depends only on P, the first-order condition for profit maximization is simply $d\pi/dP = 0$. If we multiply through by P/π, the rule becomes couched in terms of elasticity as follows: $\epsilon_{\pi p} = 0$. Thus, while the state monopolist watches the elasticity of profit with respect to price and sets it equal to zero, the dynamic monopolist must instead watch the elasticity of profit with respect to the *rate of change of price* and set it equal to $1 - \pi_s/\pi$.

The Matter of the Terminal Price

The foregoing discussion is based on the assumption that the terminal price $P(T)$ is *given*. In reality, however, the firm is likely to have discretionary control over $P(T)$ even though the terminal time T has been preset. If so, it will face the variable-terminal-point situation depicted in Fig. 1.5a, where the boundary condition $P(T) = P_T$ must be replaced by an appropriate transversality condition. We shall develop such transversality conditions in the next chapter.

EXERCISE 2.4

1. If the monopolistic firm in the Evans model faces a static linear demand ($h = 0$), what price will the firm charge for profit maximization? Call that price P_s, and check that it has the correct algebraic sign. Then compare the values of P_s and $\overline{P}$, and give the particular integral in (2.36) an economic interpretation.
2. Verify that A_1 and A_2 should indeed have the values shown in (2.37).
3. Show that the extremal $P^*(t)$ will not involve a reversal in the direction of price movement in the time interval $[0, T]$ unless there is a value $0 < t_0 < T$ such that $A_1 e^{rt_0} = A_2 e^{-rt_0}$ [i.e., satisfying the condition that the first two terms on the right-hand side of (2.36′) are equal at $t = t_0$].
4. If the coefficients A_1 and A_2 in (2.36′) are both positive, the $P^*(t)$ curve will take the shape of a catenary. Compare the location of the lowest point

on the price curve in the following three cases: (*a*) $A_1 > A_2$, (*b*) $A_1 = A_2$, and (*c*) $A_1 < A_2$. Which of these cases can possibly involve a price reversal in the interval $[0, T]$? What type of price movement characterizes the remaining cases?

5 Find the Euler equation for the Evans problem by using formula (2.18).

6 If a discount rate $\rho > 0$ is used to convert the profit $\pi(P, P')$ at any point of time to its present value, then the integrand in (2.34) will take the general form $F = \pi(P, P')e^{-\rho t}$.

(*a*) In that case, is formula (2.21) still applicable?

(*b*) Apply formula (2.19) to this F expression to derive the new Euler equation.

(*c*) Apply formula (2.18) to derive the Euler equation, and express the result as a rule regarding the rate of growth of $\pi_{P'}$.

2.5 TRADING OFF INFLATION AND UNEMPLOYMENT

The economic maladies of both inflation and unemployment inflict social losses. When a Phillips tradeoff exists between the two, what would be the best combination of inflation and unemployment over time? The answer to this question may be sought through the calculus of variations. In this section we present a simple formulation of such a problem adapted from a paper by Taylor.[8] In this formulation, the unemployment variable as such is not included; instead, it is proxied by $(Y_f - Y)$—the shortfall of current national income Y from its full-employment level Y_f.

The Social Loss Function

Let the economic ideal consist of the income level Y_f coupled with the inflation rate 0. Any deviation, positive or negative, of actual income Y from Y_f is considered undesirable, and so is any deviation of actual rate of inflation p from zero. Then we can write the social loss function as follows:

$$\lambda = (Y_f - Y)^2 + \alpha p^2 \qquad (\alpha > 0) \tag{2.39}$$

Because the deviation expressions are squared, positive and negative deviations are counted the same way (cf. Exercise 1.3, Prob. 2). However, Y deviations and p deviations do enter the loss function with different weights, in the ratio of 1 to α, reflecting the different degrees of distaste for the two types of deviations.

---

[8] Dean Taylor, "Stopping Inflation in the Dornbusch Model: Optimal Monetary Policies with Alternate Price-Adjustment Equations," *Journal of Macroeconomics*, Spring 1989, pp. 199–216. In our adapted version, we have altered the planning horizon from ∞ to a finite T.

The expectations-augmented Phillips tradeoff between $(Y_f - Y)$ and p is captured in the equation

$$p = -\beta(Y_f - Y) + \pi \qquad (\beta > 0) \tag{2.40}$$

where π, unlike its usage in the preceding section, now means the *expected* rate of inflation. The formation of inflation expectations is assumed to be adaptive:

$$\pi'\left(\equiv \frac{d\pi}{dt}\right) = j(p - \pi) \qquad (0 < j \le 1) \tag{2.41}$$

If the actual rate of inflation p exceeds the expected rate of inflation π (proving π to be an *under*estimate), then $\pi' > 0$, and π will be revised upward; if, on the other hand, the actual rate p falls short of the expected rate π (proving π to be an *over*estimate), then $\pi' < 0$, and π will be revised downward.

The last two equations together imply that

$$\pi' = -\beta j(Y_f - Y)$$

which can be rearranged as

$$Y_f - Y = \frac{-\pi'}{\beta j} \tag{2.42}$$

When substituted into (2.40), (2.42) yields

$$p = \frac{\pi'}{j} + \pi \tag{2.43}$$

And, by plugging (2.42) and (2.43) into (2.39), we are able to express the loss function entirely in terms of π and π':

$$\lambda(\pi, \pi') = \left(\frac{\pi'}{\beta j}\right)^2 + \alpha\left(\frac{\pi'}{j} + \pi\right)^2 \qquad \text{[loss function]} \tag{2.44}$$

The Problem

The problem for the government is then to find an optimal path of π that minimizes the total social loss over the time interval $[0, T]$. The initial (current) value of π is given at π_0, and the terminal value of π, a policy target, is assumed to be set at 0. To recognize the importance of the present over the future, all social losses are discounted to their present values via a positive discount rate ρ.

In view of these considerations, the objective of the policymaker is to

$$\begin{aligned} &\text{Minimize} && \Lambda[\pi] = \int_0^T \lambda(\pi, \pi')e^{-\rho t}\,dt \\ &\text{subject to} && \pi(0) = \pi_0 \qquad (\pi_0 > 0 \text{ given}) \\ &\text{and} && \pi(T) = 0 \qquad (T \text{ given}) \end{aligned} \tag{2.45}$$

The Solution Path

On the basis of the loss function (2.44), the integrand function $\lambda(\pi, \pi')e^{-\rho t}$ yields the first derivatives

$$F_\pi = 2\left(\frac{\alpha}{j}\pi' + \alpha\pi\right)e^{-\rho t}$$

$$F_{\pi'} = 2\left(\frac{1+\alpha\beta^2}{\beta^2 j^2}\pi' + \frac{\alpha}{j}\pi\right)e^{-\rho t}$$

with second derivatives

$$F_{\pi'\pi'} = 2\left(\frac{1+\alpha\beta^2}{\beta^2 j^2}\right)e^{-\rho t} \qquad F_{\pi\pi'} = \frac{2\alpha}{j}e^{-\rho t}$$

and

$$F_{t\pi'} = -2\rho\left(\frac{1+\alpha\beta^2}{\beta^2 j^2}\pi' + \frac{\alpha}{j}\pi\right)e^{-\rho t}$$

Consequently, formula (2.19) gives us (after simplification) the specific necessary condition

$$\pi'' - \rho\pi' - \Omega\pi = 0 \qquad \text{[Euler equation]} \tag{2.46}$$

$$\text{where} \quad \Omega \equiv \frac{\alpha\beta^2 j(\rho + j)}{1+\alpha\beta^2} > 0$$

Since this differential equation is homogeneous, its particular integral is zero, and its general solution is simply its complementary function:

$$\pi^*(t) = A_1 e^{r_1 t} + A_2 e^{r_2 t} \qquad \text{[general solution]} \tag{2.47}$$

$$\text{where} \quad r_1, r_2 = \tfrac{1}{2}\left(\rho \pm \sqrt{\rho^2 + 4\Omega}\right)$$

The characteristic roots are real and distinct. Moreover, inasmuch as the square root has a numerical value greater than ρ, we know that

$$r_1 > 0 \qquad \text{and} \qquad r_2 < 0 \tag{2.48}$$

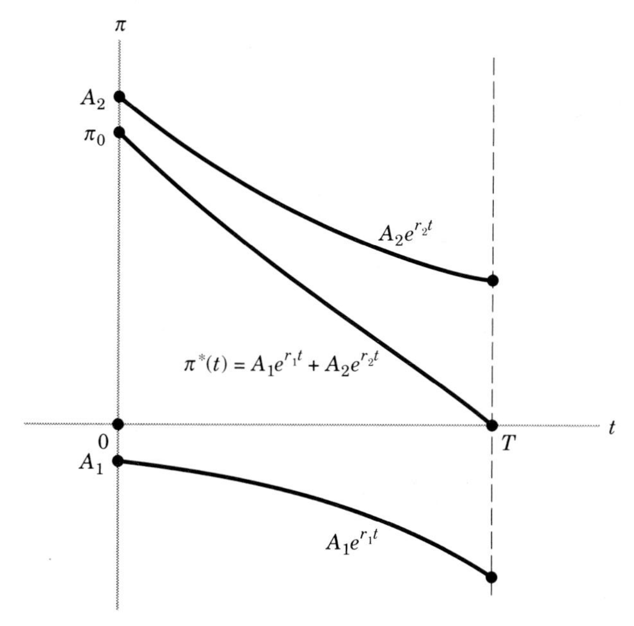

FIGURE 2.6

To definitize the arbitrary constants A_1 and A_2, we successively set $t = 0$ and $t = T$ in (2.47), and use the boundary conditions to get the pair of relations

$$A_1 + A_2 = \pi_0$$
$$A_1 e^{r_1 T} + A_2 e^{r_2 T} = 0$$

Solved simultaneously, these relations give us

$$(2.49) \qquad A_1 = \frac{-\pi_0 e^{r_2 T}}{e^{r_1 T} - e^{r_2 T}} \qquad A_2 = \frac{\pi_0 e^{r_1 T}}{e^{r_1 T} - e^{r_2 T}}$$

Because of the signs of r_1 and r_2 in (2.48), we know that

$$(2.50) \qquad A_1 < 0 \qquad A_2 > 0$$

From the sign information in (2.50) and (2.48), we can deduce that the $\pi^*(t)$ path should follow the general configuration in Fig. 2.6. With a negative A_1 and positive r_1, the $A_1 e^{r_1 t}$ component of the path emerges as the mirror image (with reference to the horizontal axis) of an exponential growth curve. On the other hand, the $A_2 e^{r_2 t}$ component is, in view of the positive A_2 and negative r_2, just a regular exponential decay curve. The π^* path, the sum of the two component curves, starts from the point $(0, \pi_0)$—where $\pi_0 = A_1 + A_2$—and descends steadily toward the point $(T, 0)$, which is vertically equidistant from the two component curves. The fact that the π^* path shows a steady downward movement can also be verified from the derivative

$$\pi^{*\prime}(t) = r_1 A_1 e^{r_1 t} + r_2 A_2 e^{r_2 t} < 0$$

Having found $\pi^*(t)$ and $\pi^{*\prime}(t)$, we can also derive some conclusions regarding p and $(Y_f - Y)$. For these, we can simply use the relations in (2.43) and (2.42), respectively.

EXERCISE 2.5

1 Verify the result in (2.46) by using Euler equation (2.18).

2 Let the objective functional in problem (2.45) be changed to $\int_0^T \frac{1}{2}\lambda(\pi, \pi')e^{-\rho t}\,dt$.

(*a*) Do you think the solution of the problem will be different?

(*b*) Can you think of any advantage in including a coefficient $\frac{1}{2}$ in the integrand?

3 Let the terminal condition in problem (2.45) be changed to $\pi(T) = \pi_T$, $0 < \pi_T < \pi_0$.

(*a*) What would be the values of A_1 and A_2?

(*b*) Can you ambiguously evaluate the signs of A_1 and A_2?

CHAPTER 3

TRANSVERSALITY CONDITIONS FOR VARIABLE-ENDPOINT PROBLEMS

The Euler equation—the basic necessary condition in the calculus of variations—is normally a second-order differential equation containing two arbitrary constants. For problems with fixed initial and terminal points, the two given boundary conditions provide sufficient information to definitize the two arbitrary constants. But if the initial or terminal point is variable (subject to discretionary choice), then a boundary condition will be missing. Such will be the case, for instance, if the dynamic monopolist of the preceding chapter faces no externally imposed price at time T, and can treat the P_T choice as an integral part of the optimization problem. In that case, the boundary condition $P(T) = P_T$ will no longer be available, and the void must be filled by a transversality condition. In this chapter we shall develop the transversality conditions appropriate to various types of variable terminal points.

3.1 THE GENERAL TRANSVERSALITY CONDITION

For expository convenience, we shall assume that only the *terminal* point is variable. Once we learn to deal with that, the technique is easily extended to the case of a variable initial point.

The Variable-Terminal-Point Problem

Our new objective is to

$$\begin{aligned} &\text{Maximize or minimize} && V[y] = \int_0^T F(t, y, y')\, dt \\ (3.1) \qquad &\text{subject to} && y(0) = A \qquad (A \text{ given}) \\ &\text{and} && y(T) = y_T \qquad (T, y_T \text{ free}) \end{aligned}$$

This differs from the previous version of the problem in that the terminal time T and terminal state y_T are now "free" in the sense that they have become a part of the optimal choice process. It is to be understood that although T is free, only positive values of T are relevant to the problem.

To develop the necessary conditions for an extremal, we shall, as before, employ a perturbing curve $p(t)$, and use the variable ϵ to generate neighboring paths for comparison with the extremal. First, suppose that T^* is the known optimal terminal time. Then any value of T in the immediate neighborhood of T^* can be expressed as

$$T = T^* + \epsilon \Delta T \tag{3.2}$$

where ϵ represents a small number, and ΔT represents an arbitrarily chosen (and fixed) small change in T. Note that since T^* is known and ΔT is a prechosen magnitude, T can be considered as a function of ϵ, $T(\epsilon)$, with derivative

$$\frac{dT}{d\epsilon} = \Delta T \tag{3.3}$$

The same ϵ is used in conjunction with the perturbing curve $p(t)$ to generate neighboring paths of the extremal $y^*(t)$:

$$y(t) = y^*(t) + \epsilon p(t) \qquad [\text{implying } y'(t) = y^{*\prime}(t) + \epsilon p'(t)] \tag{3.4}$$

However, although the $p(t)$ curve must still satisfy the condition $p(0) = 0$ [see (2.2)] to force the neighboring paths to pass through the fixed initial point, the other condition—$p(t) = 0$—should now be dropped, because y_T is free. By substituting (3.4) into the functional $V[y]$, we get a function $V(\epsilon)$ akin to (2.12), but since T is a function of ϵ by (3.2), the upper limit of integration in the V function will also vary with ϵ:

$$V(\epsilon) = \int_0^{T(\epsilon)} F\Big[t, \underbrace{y^*(t) + \epsilon p(t)}_{y(t)}, \underbrace{y^{*\prime}(t) + \epsilon p'(t)}_{y'(t)}\Big]\, dt \tag{3.5}$$

Our problem is to optimize this V function with respect to ϵ.

Deriving the General Transversality Condition

The first-order necessary condition for an extremum in V is simply $dV/d\epsilon = 0$. This latter derivative is a *total* derivative taking into account the direct effect of ϵ on V, as well as the indirect effect of ϵ on V through the upper limit of integration T.

Step i The definite integral (3.5) falls into the general form of (2.10). The derivative $dV/d\epsilon$ is therefore, by (2.11),

$$\frac{dV}{d\epsilon} = \int_0^{T(\epsilon)} \frac{\partial F}{\partial \epsilon}\, dt + F[T, y(T), y'(T)] \frac{dT}{d\epsilon} \tag{3.6}$$

The integral on the right closely resembles the one encountered in (2.13) during our earlier development of the Euler equation. In fact, much of the derivation process leading to (2.17) still applies here, with one exception. The expression $[F_{y'} p(t)]_0^T$ in (2.16) does not vanish in the present problem, but takes the value $[F_{y'} p(t)]_{t=T} = [F_{y'}]_{t=T} p(T)$, since we have assumed that $p(0) = 0$ but $p(T) \neq 0$. With this amendment to the earlier result in (2.17), we have

$$\text{First term in (3.6)} = \int_0^T p(t)\left[F_y - \frac{d}{dt}F_{y'}\right] dt + \left[F_{y'}\right]_{t=T} p(T)$$

By (3.3), we can also write

$$\text{Second term in (3.6)} = [F]_{t=T}\, \Delta T$$

Substituting these into (3.6), and setting $dV/d\epsilon = 0$, we obtain the new condition

$$\int_0^T p(t)\left[F_y - \frac{d}{dt}F_{y'}\right] dt + \left[F_{y'}\right]_{t=T} p(T) + [F]_{t=T}\, \Delta T = 0 \tag{3.7}$$

Of the three terms on the left-hand side of (3.7), each contains its own independent arbitrary element: $p(t)$ (the entire perturbing curve) in the first term, $p(T)$ (the terminal value on the perturbing curve) in the second, and ΔT (the arbitrarily chosen change in T) in the third. Thus we cannot presume any offsetting or cancellation of terms. Consequently, in order to satisfy the condition (3.7), each of the three terms must individually be set equal to zero.

It is a familiar story that when the first term in (3.7) is set equal to zero, the Euler equation emerges, as in (2.18). This establishes the fact that the Euler equation remains valid as a necessary condition in the variable-endpoint problem. The other two terms, which relate only to the terminal time T, are where we should look for the transversality condition.

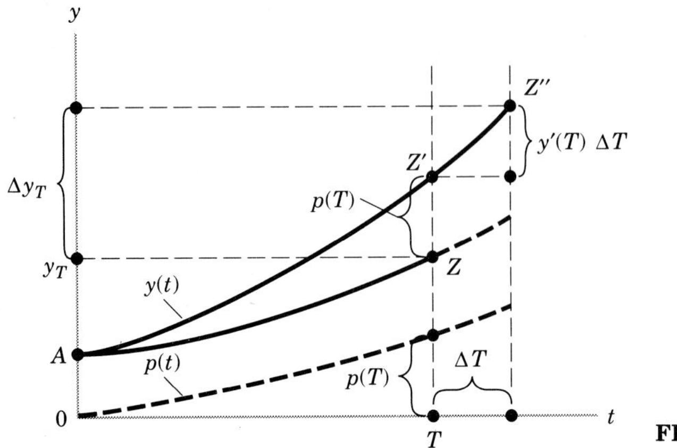

FIGURE 3.1

Step ii To this end, we first get rid of the arbitrary quantity $p(T)$ by transforming it into terms of ΔT and Δy_T—the changes in T and y_T, the two principal variables in the variable-terminal-point problem. This can be done with the help of Fig. 3.1. The AZ' curve represents a neighboring path obtained by perturbing the AZ path by $\epsilon p(t)$, with ϵ set equal to one for convenience. Note that while the two curves share the same initial point because $p(0) = 0$, they have different heights at $t = T$ because $p(T) \neq 0$ by construction of the perturbing curve.[1] The magnitude of $p(T)$, shown by the vertical distance ZZ', measures the direct change in y_T resulting from the perturbation. But inasmuch as T can also be altered by the amount $\epsilon\,\Delta T$ ($= \Delta T$ since $\epsilon = 1$), the AZ' curve should, if $\Delta T > 0$, be extended out to Z''.[2] As a result, y_T is further pushed up by the vertical distance between Z' and Z''. For a small ΔT, this second change in y_T can be approximated by $y'(T)\,\Delta T$. Hence, the total change in y_T from point Z to point Z'' is

$$\Delta y_T = p(T) + y'(T)\,\Delta T$$

Rearranging this relation allows us to express $p(T)$ in terms of Δy_T and ΔT:[3]

$$p(T) = \Delta y_T - y'(T)\,\Delta T \tag{3.8}$$

---

[1]Technically, the point T on the horizontal axis in Fig. 3.1 should be labeled T^*. We are omitting the * for simplicity. This is justifiable because the result that this discussion is leading to—(3.9)—is a transversality condition, which, as a standard practice, is stated in terms of T (without the *) anyway.

[2]Although we are initially interested only in the solid portion of the AZ path, the equation for that path should enable us also to plot the broken portion as an extension of AZ. The perturbing curve is then applied to the extended version of AZ.

[3]The result in (3.8) is valid even if ϵ is not set equal to one as in Fig. 3.1.

Step iii Using (3.8) to eliminate $p(T)$ in (3.7), and dropping the first term in (3.7), we finally arrive at the desired general transversality condition

$$[F - y'F_{y'}]_{t=T} \Delta T + [F_{y'}]_{t=T} \Delta y_T = 0 \tag{3.9}$$

This condition, unlike the Euler equation, is relevant only to one point of time, T. Its role is to take the place of the missing terminal condition in the present problem. Depending on the exact specification of the terminal line or curve, however, the general condition (3.9) can be written in various specialized forms.

3.2 SPECIALIZED TRANSVERSALITY CONDITIONS

In this section we consider five types of variable terminal points: vertical terminal line, horizontal terminal line, terminal curve, truncated vertical terminal line, and truncated horizontal terminal line.

Vertical Terminal Line (Fixed-Time-Horizon Problem)

The vertical-terminal-line case, as illustrated in Fig. 1.5*a*, involves a fixed T. Thus $\Delta T = 0$, and the first term in (3.9) drops out. But since Δy_T is arbitrary and can take either sign, the only way to make the second term in (3.9) vanish for sure is to have $F_{y'} = 0$ at $t = T$. This gives rise to the transversality condition

$$[F_{y'}]_{t=T} = 0 \tag{3.10}$$

which is sometimes referred to as the *natural boundary condition*.

Horizontal Terminal Line (Fixed-Endpoint Problem)

For the horizontal-terminal-line case, as illustrated in Fig. 1.5*b*, the situation is exactly reversed; we now have $\Delta y_T = 0$ but ΔT is arbitrary. So the second term in (3.9) automatically drops out, but the first does not. Since ΔT is arbitrary, the only way to make the first term vanish for sure is to have the bracketed expression equal to zero. Thus the transversality condition is

$$[F - y'F_{y'}]_{t=T} = 0 \tag{3.11}$$

It might be useful to give an economic interpretation to (3.10) and (3.11). To fix ideas, let us interpret $F(t, y, y')$ as a profit function, where y represents capital stock, and y' represents net investment. Net investment entails taking resources away from the current profit-making business operation, so as to build up capital which will enhance future profit. Hence, there exists a tradeoff between current profit and future profit. At any time t, with a given capital stock y, a specific investment decision—say, a decision to select the investment rate y'_0—will result in the current profit $F(t, y, y'_0)$. As to the effect of the investment decision on future profits, it enters through the intermediary of capital. The rate of capital accumulation is y'_0; if we can convert that into a value measure, then we can add it to the current profit $F(t, y, y'_0)$ to see the overall (current as well as future) profit implication of the investment decision. The imputed (or shadow) value to the firm of a unit of capital is measured by the derivative $F_{y'}$. This means that if we decide to leave (not use up) a unit of capital at the terminal time, it will entail a negative value equal to $-F_{y'}$. Thus, at $t = T$, the value measure of y'_0 is $-y'_0 F_{y'_0}$. Accordingly, the overall profit implication of the decision to choose the investment rate y'_0 is $F(t, y, y'_0) - y'_0 F_{y'_0}$. The general expression for this is $F - y'F_{y'}$, as in (3.11).

Now we can interpret the transversality condition (3.11) to mean that, in a problem with a free terminal time, the firm should select a T such that a decision to invest and accumulate capital will, at $t = T$, no longer yield any overall (current and future) profit. In other words, all the profit opportunities should have been fully taken advantage of by the optimally chosen terminal time. In addition, (3.10)—which can equivalently be written as $[-F_{y'}]_{t=T} = 0$—instructs the firm to avoid any sacrifice of profit that will be incurred by leaving a positive terminal capital. In other words, in a free-terminal-state problem, in order to maximize profit in the interval $[0, T]$ but not beyond, the firm should, at time T, us up all the capital it ever accumulated.

Terminal Curve

With a terminal curve $y_T = \phi(T)$, as illustrated in Fig. 1.5c, neither Δy_T nor ΔT is assigned a zero value, so neither term in (3.9) drops out. However, for a small arbitrary ΔT, the terminal curve implies that $\Delta y_T = \phi' \Delta T$. So it is possible to eliminate Δy_T in (3.9) and combine the two terms into the form

$$\left[F - y'F_{y'} + F_{y'}\phi'\right]_{t=T} \Delta T = 0$$

Since ΔT is arbitrary, we can deduce the transversality condition

$$\left[F + (\phi' - y')F_{y'}\right]_{t=T} = 0 \tag{3.12}$$

Unlike the last two cases, which involve a single unknown in the terminal point (either y_T or T), the terminal-curve case requires us to determine both y_T and T. Thus two relations are needed for this purpose.

The transversality condition (3.12) only provides one; the other is supplied by the equation $y_T = \phi(T)$.

Truncated Vertical Terminal Line

The usual case of vertical terminal line, with $\Delta T = 0$, specializes (3.9) to

$$[F_{y'}]_{t=T} \Delta y_T = 0 \tag{3.13}$$

When the line is truncated—restricted by the terminal condition $y_T \geq y_{\min}$ where $y_{\min}$ is a minimum permissible level of y—the optimal solution can have two possible types of outcome: $y_T^* > y_{\min}$ or $y_T^* = y_{\min}$. If $y_T^* > y_{\min}$, the terminal restriction is automatically satisfied; that is, it is nonbinding. Thus, the transversality condition is in that event the same as (3.10):

$$[F_{y'}]_{t=T} = 0 \qquad \text{for } y_T^* > y_{\min} \tag{3.14}$$

The basic reason for this is that, under this outcome, there are admissible neighboring paths with terminal values both *above* and *below* y_T^*, so that $\Delta y_T \equiv y_T - y_T^*$ can take either sign. Therefore, the only way to satisfy (3.13) is to have $F_{y'} = 0$ at $t = T$.

The other outcome, $y_T^* = y_{\min}$, on the other hand, only admits neighboring paths with terminal values $y_T \geq y_T^*$. This means that Δy_T is no longer completely arbitrary (positive or negative), but is restricted to be nonnegative. Assuming the perturbing curve to have terminal value $p(T) > 0$, as in Fig. 3.1, $\Delta y_T \geq 0$ would mean that $\epsilon \geq 0$ ($\epsilon = 1$ in Fig. 3.1). The nonnegativity of ϵ, in turn, means that the transversality condition (3.13)—which has its roots in the first-order condition $dV/d\epsilon = 0$— must be changed to an inequality as in the Kuhn-Tucker conditions.[4] For a *maximization* problem, the $\leq$ type of inequality is called for, and (3.13) should become

$$[F_{y'}]_{t=T} \Delta y_T \leq 0 \tag{3.15}$$

And since $\Delta y_T \geq 0$, (3.15) implies condition

$$[F_{y'}]_{t=T} \leq 0 \qquad \text{for } y_T^* = y_{\min} \tag{3.16}$$

Combining (3.14) and (3.16), we may write the following summary statement of the transversality condition for a maximization problem:

$$[F_{y'}]_{t=T} \leq 0 \qquad y_T^* \geq y_{\min} \qquad (y_T^* - y_{\min})[F_{y'}]_{t=T} = 0 \tag{3.17}$$

[for maximization of V]

[4] For an explanation of the Kuhn-Tucker conditions, see Alpha C. Chiang, *Fundamental Methods of Mathematical Economics*, 3d ed., McGraw-Hill, New York, 1984, Sec. 21.2.

If the problem is instead to *minimize* V, then the inequality sign in (3.15) must be reversed, and the transversality condition becomes

$$\text{(3.17')} \quad [F_{y'}]_{t=T} \geq 0 \qquad y_T^* \geq y_{\min} \qquad (y_T^* - y_{\min})[F_{y'}]_{t=T} = 0$$
$$\text{[for minimization of } V\text{]}$$

The tripartite condition (3.17) or (3.17′) may seem complicated, but its application is not. In practical problem solving, we can try the $[F_{y'}]_{t=T} = 0$ condition in (3.14) first, and check the resulting y_T^* value. If $y_T^* \geq y_{\min}$, then the terminal restriction is satisfied, and the problem solved. If $y_T^* < y_{\min}$, on the other hand, then the optimal y_T lies below the permissible range, and $[F_{y'}]_{t=T}$ fails to reach zero on the truncated terminal line. So, in order to satisfy the complementary-slackness condition in (3.17) or (3.17′), we just set $y_T^* = y_{\min}$, treating the problem as one with a fixed terminal point $(T, y_{\min})$.

Truncated Horizontal Terminal Line

The horizontal terminal line may be truncated by the restriction $T \leq T_{\max}$, where $T_{\max}$ represents a maximum permissible time for completing a task—a deadline. The analysis of such a situation is very similar to the truncated vertical terminal line just discussed. By analogous reasoning, we can derive the following transversality condition for a maximization problem:

$$\text{(3.18)} \qquad [F - y'F_{y'}]_{t=T} \geq 0 \qquad T^* \leq T_{\max}$$
$$(T^* - T_{\max})[F - y'F_{y'}]_{t=T} = 0 \qquad \text{[for maximization of } V\text{]}$$

If the problem is to *minimize* V, the first inequality in (3.18) must be changed, and the transversality condition is

$$\text{(3.18')} \qquad [F - y'F_{y'}]_{t=T} \leq 0 \qquad T^* \leq T_{\max}$$
$$(T^* - T_{\max})[F - y'F_{y'}]_{t=T} = 0 \qquad \text{[for minimization of } V\text{]}$$

EXAMPLE 1 Find the curve with the shortest distance between the point (0, 1) and the line $y = 2 - t$. Referring to Fig. 3.2, we see that this is a problem with a terminal curve $y(T) = 2 - T$, but otherwise it is similar to the shortest-distance problem in Example 4 of Sec. 2.2. Here, the problem is to

$$\begin{aligned} &\text{Minimize} && V[y] = \int_0^T (1 + y'^2)^{1/2}\, dt \\ &\text{subject to} && y(0) = 1 \\ &\text{and} && y(T) = 2 - T \end{aligned}$$

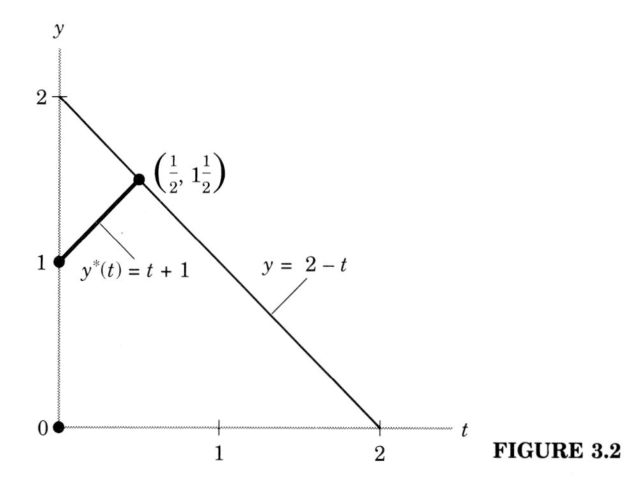

FIGURE 3.2

It has previously been shown that with the given integrand, the Euler equation yields a straight-line extremal, say,

$$y^* = at + b$$

with two arbitrary constants a and b. From the initial condition $y(0) = 1$, we can readily determine that $b = 1$. To determine the other constant, a, we resort to the transversality condition (3.12). Since we have

$$F = \left(1 + y'^2\right)^{1/2} \qquad \phi' = -1 \qquad F_{y'} = y'\left(1 + y'^2\right)^{-1/2}$$

the transversality condition can be written as

$$\left(1 + y'^2\right)^{1/2} + (-1 - y')y'\left(1 + y'^2\right)^{-1/2} = 0 \qquad (\text{at } t = T)$$

Multiplying through by $(1 + y'^2)^{1/2}$ and simplifying, we can reduce this equation to the form $y' = 1$ (at $t = T$). But the extremal actually has a constant slope, $y^{*\prime} = a$, at all values of t. Thus, we must have $a = 1$. And the extremal is therefore

$$y^*(t) = t + 1$$

As shown in Fig. 3.2, the extremal is a straight line that meets the terminal line at the point $(\frac{1}{2}, 1\frac{1}{2})$. Moreover, we note that the slopes of the terminal line and the extremal are, respectively, -1 and $+1$. Thus the two lines are *orthogonal* (perpendicular) to each other. What the transversality condition does in this case is to require the extremal to be orthogonal to the terminal line.

EXAMPLE 2 Find the extremal of the functional

$$V[y] = \int_0^T \left(ty' + y'^2\right) dt$$

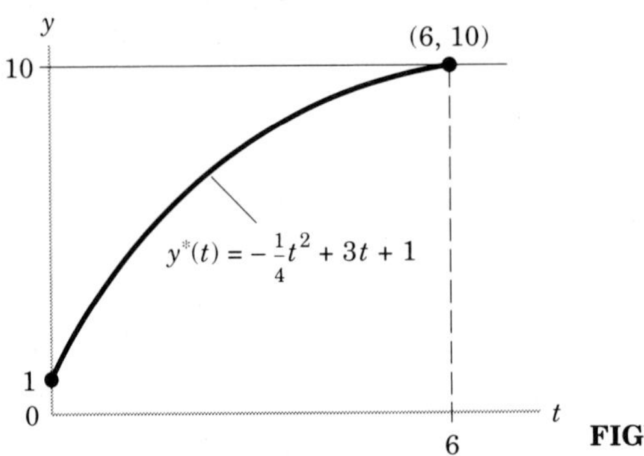

FIGURE 3.3

with boundary conditions $y(0) = 1$, $y_T = 10$, and T free. Here we have a horizontal terminal line as depicted in Fig. 3.3.

The general solution of the Euler equation for this problem has previously been found to be a quadratic path (Sec. 2.2, Example 1):

$$y^*(t) = -\tfrac{1}{4}t^2 + c_1 t + c_2$$

Since $y(0) = 1$, we have $c_2 = 1$. But the other constant c_1 must be definitized with the help of the transversality condition (3.11). Since $F = ty' + y'^2$, so that $F_{y'} = t + 2y'$, that condition becomes

$$ty' + y'^2 - y'(t + 2y') = 0 \qquad (\text{at } t = T)$$

which reduces to $-y'^2 = 0$ or $y' = 0$ at $t = T$. That is, the extremal is required to have a zero slope at the terminal time. To meet this condition, we differentiate the general solution to get $y^{*\prime}(t)$, then set $t = T$, and let $y^{*\prime}(T) = 0$. The result is the equation $-T/2 + c_1 = 0$; it follows that $c_1 = T/2$. However, we still cannot determine the specific value of c_1 without knowing the value of T.

To find T, we make use of the information that $y_T = 10$ (horizontal terminal line). Substituting $c_1 = T/2$ and $c_2 = 1$ into the general solution, setting $t = T$, and letting the resulting expression take the stipulated value 10, we obtain the equation

$$-\tfrac{1}{4}T^2 + \tfrac{1}{2}T^2 + 1 = 10$$

The solution values of T are therefore ± 6. Rejecting the negative value, we end up with $T^* = 6$. Then it follows that $c_1 = 3$, and the extremal is

$$y^*(t) = -\tfrac{1}{4}t^2 + 3t + 1$$

This time path is shown in Fig. 3.3. As required by the transversality condition, the extremal indeed attains a zero slope at the terminal point (6, 10). Unlike in Example 1, where the transversality condition translates into an orthogonality requirement, the transversality condition in the present example dictates that the extremal share a common slope with the given horizontal terminal line at $t = T$.

Application to the Dynamic Monopoly Model

Let us now consider the Evans model of a dynamic monopolist as a problem with a fixed terminal time T, but a variable terminal state $P_T \geq P_{\min}$. With the vertical terminal line thus truncated, the appropriate transversality condition is (3.17). For simplicity, we shall in this discussion assign specific numerical values to the parameters, for otherwise the solution expressions will become too unwieldy.

Let the cost function and demand function be

$$C = \tfrac{1}{10}Q^2 + 1000 \qquad \left[\text{i.e., } \alpha = \tfrac{1}{10},\ \beta = 0,\ \gamma = 1000\right]$$

$$Q = 160 - 8P + 100P' \qquad [\text{i.e., } a = 160,\ b = 8,\ h = 100]$$

Then the profit function becomes

$$\pi \equiv PQ - C = 416P - 14.4P^2 + 260PP' - 1000P'^2 - 3200P' - 3560$$

which implies that

$$\pi_{P'} = 260P - 2000P' - 3200 \tag{3.19}$$

This is the derivative needed in the transversality condition.

Since the postulated parameter values yield the characteristic roots and particular integral [see (2.36)]:

$$r_1, r_2 = \pm 0.12 \qquad \overline{P} = 14\tfrac{4}{9}$$

the general solution of the Euler equation is

$$P^*(t) = A_1 e^{0.12t} + A_2 e^{-0.12t} + 14\tfrac{4}{9} \tag{3.20}$$

If we further assume the boundary conditions to be

$$P_0 = 11\tfrac{4}{9} \qquad P_T = 15\tfrac{4}{9} \qquad \text{and} \qquad T = 2$$

then, according to (2.37), the constants A_1 and A_2 should, in the fixed-terminal-point problem, have the values (after rounding):

$$A_1 = 6.933 \qquad A_2 = -9.933$$

The reader can verify that substitution of these two constants into (3.20) does (apart from rounding errors) produce the terminal price $P^*(2) = 15\frac{4}{9}$ as required.

Now adopt a variable terminal state $P_T \geq 10$. Using the transversality condition (3.17), we first set the $\pi_{p'}$ expression in (3.19) equal to zero at $t = T = 2$. Upon normalizing, this yields the condition

$$P'(T) - 0.13P(T) = -1.6 \tag{3.21}$$

The $P(T)$ term here refers to the general solution $P^*(t)$ in (3.20) evaluated at $t = T = 2$. And the $P'(T)$ is the derivative of (3.20) evaluated at the same point of time. That is,

$$P(T) = A_1 e^{0.24} + A_2 e^{-0.24} + 14\tfrac{4}{9}$$
$$P'(T) = 0.12 A_1 e^{0.24} - 0.12 A_2 e^{-0.24}$$

Thus, (3.21) can be rewritten more specifically as

$$-0.01 A_1 e^{0.24} - 0.25 A_2 e^{-0.24} = 0.2778$$

To solve for A_1 and A_2, we need to couple this condition with the condition that governs the initial point,

$$A_1 + A_2 = -3$$

obtained from (3.20) by setting $t = 0$ and equating the result to $P_0 = 11\frac{4}{9}$. The solution values of A_1 and A_2 turn out to be (after rounding):

$$A_1 = 4.716 \qquad \text{and} \qquad A_2 = -7.716$$

giving us the definite solution

$$P^*(t) = 4.716 e^{0.12t} - 7.716 e^{-0.12t} + 14\tfrac{4}{9} \tag{3.22}$$

It remains to check whether this solution satisfies the terminal specification $P_T \geq 10$. Setting $t = T = 2$ in (3.22), we find that $P^*(2) = 14.37$. Since this does meet the stipulated restriction, the problem is solved.[5]

The Inflation-Unemployment Tradeoff Again

The inflation-unemployment problem in Sec. 2.5 has a fixed terminal point that requires the expected rate of inflation, π, to attain a value of zero at the given terminal time T: $\pi(T) = 0$. It may be interesting to ask: What will happen if the terminal condition comes as a vertical terminal line instead?

---

[5] While Evan's original treatment does include a discussion of the variable terminal price, it neither makes use of the transversality condition nor gives a completely determined price path for that case. For his alternative method of treatment, see his *Mathematical Introduction to Economics*, McGraw-Hill, New York, 1930, pp. 150–153.

From (2.47), the general solution of the Euler equation is

$$\pi^*(t) = A_1 e^{r_1 t} + A_2 e^{r_2 t} \tag{3.23}$$

The initial condition, still assumed to be $\pi(0) = \pi_0 > 0$, requires (by setting $t = 0$ in the general solution) that

$$A_1 + A_2 = \pi_0 \tag{3.24}$$

With a vertical terminal line at the given T, however, we now must use the transversality condition $[F_{y'}]_{t=T} = 0$. In the tradeoff model, this takes the form

$$F_{\pi'} = 2\left(\frac{1+\alpha\beta^2}{\beta^2 j^2}\pi' + \frac{\alpha}{j}\pi\right)e^{-\rho t} = 0 \qquad (\text{at } t = T) \tag{3.25}$$

which can be satisfied if and only if the expression in the parentheses is zero. Thus, the transversality condition can be written (after simplification) as

$$\pi'(T) + \sigma\pi(T) = 0 \qquad \text{where} \quad \sigma \equiv \frac{\alpha\beta^2 j}{1+\alpha\beta^2} \tag{3.25'}$$

The $\pi(T)$ and $\pi'(T)$ expressions are, of course, to be obtained from the general solution $\pi^*(t)$ and its derivative, both evaluated at $t = T$. Using (3.23), we can thus give condition (3.25′) the specific form

$$(r_1 + \sigma)A_1 e^{r_1 T} + (r_2 + \sigma)A_2 e^{r_2 T} = 0 \tag{3.25''}$$

When (3.24) and (3.25″) are solved simultaneously, we finally find the definite values of the constants A_1 and A_2:

$$A_1 = \frac{-\pi_0(r_2+\sigma)e^{r_2 T}}{(r_1+\sigma)e^{r_1 T} - (r_2+\sigma)e^{r_2 T}} \qquad A_2 = \frac{\pi_0(r_1+\sigma)e^{r_1 T}}{(r_1+\sigma)e^{r_1 T} - (r_2+\sigma)e^{r_2 T}} \tag{3.26}$$

These can be used to turn the general solution into the definite solution.

Now that the terminal state is endogenously determined, what value of $\pi^*(T)$ emerges from the solution? To see this, we substitute (3.26) into the general solution (3.23), and evaluate the result at $t = T$. The answer turns out to be

$$\pi^*(T) = 0$$

In view of the given loss function, this requirement—driving the rate of expected inflation down to zero at the terminal time—is not surprising. And since the earlier version of the problem in Sec. 2.5 already has the target value of π set at $\pi_T = 0$, the switch to the vertical-terminal-line formulation does not modify the result.

EXERCISE 3.2

1 For the functional $V[y] = \int_0^T (t^2 + y'^2)\, dt$, the general solution to the Euler equation is $y^*(t) = c_1 t + c_2$ (see Exercise 2.2, Prob. 1).
  (*a*) Find the extremal if the initial condition is $y(0) = 4$ and the terminal condition is $T = 2$, y_T free.
  (*b*) Sketch a diagram showing the initial point, the terminal point, and the extremal.

2 How will the answer to the preceding problem change, if the terminal condition is altered to: $T = 2$, $y_T \geq 3$?

3 Let the terminal condition in Prob. 1 be changed to: $y_T = 5$, T free.
  (*a*) Find the new extremal. What is the optimal terminal time T^*?
  (*b*) Sketch a diagram showing the initial point, the terminal line, and the extremal.

4 (*a*) For the functional $V[y] = \int_0^T (y'^2/t^3)\, dt$, the general solution to the Euler equation is $y^*(t) = c_1 t^4 + c_2$ (see Exercise 2.2, Prob. 4). Find the extremal(s) if the initial condition is $y(0) = 0$, and the terminal condition is $y_T = 3 - T$.
  (*b*) Sketch a diagram to show the terminal curve and the extremal(s).

5 The discussion leading to condition (3.16) for a truncated vertical terminal line is based on the assumption that $p(T) > 0$. Show that if the perturbing curve is such that $p(T) < 0$ instead, the same condition (3.16) will emerge.

6 For the truncated-horizontal-terminal-line problem, use the same line of reasoning employed for the truncated vertical terminal line to derive transversality condition (3.18).

3.3 THREE GENERALIZATIONS

What we have learned about the variable terminal point can be generalized in three directions.

A Variable Initial Point

If the initial point is variable, then the boundary condition $y(0) = A$ no longer holds, and an initial transversality condition is needed to fill the void. Since the transversality conditions developed in the preceding section can be applied, *mutatis mutandis*, to the case of variable initial points, there is no need for further discussion. If a problem has *both* the initial and terminal points variable, then two transversality conditions have to be used to definitize the two arbitrary constants that arise from the Euler equation.

The Case of Several State Variables

When several state variables appear in the objective functional, so that the integrand is

$$F(t, y_1, \ldots, y_n, y'_1, \ldots, y'_n)$$

the general (terminal) transversality condition (3.9) must be modified into

$$\begin{aligned}(3.27)\qquad & \left[F - (y_1'F_{y_1'} + \cdots + y_n'F_{y_n'})\right]_{t=T} \Delta T \\ & + \left[F_{y_1'}\right]_{t=T} \Delta y_{1T} + \cdots + \left[F_{y_n'}\right]_{t=T} \Delta y_{nT} = 0\end{aligned}$$

It should be clear that (3.9) constitutes a special case of (3.27) with $n = 1$.

Given a fixed terminal time, the first term in (3.27) drops out because $\Delta T = 0$. Similarly, if any variable y_j has a fixed terminal value, then $\Delta y_{jT} = 0$ and the jth term in (3.27) drops out. For the terms that remain, however, we may expect all the Δ expressions to represent independently determined arbitrary quantities. Thus, there cannot be any presumption that the terms in (3.27) can cancel out one another. Consequently, each term that remains will give rise to a separate transversality condition.

The following examples illustrate the application of (3.27) when $n = 2$, with the state variables denoted by y and z. The general transversality condition for $n = 2$ is

$$(3.27')\qquad \left[F - (y'F_{y'} + z'F_{z'})\right]_{t=T} \Delta T + \left[F_{y'}\right]_{t=T} \Delta y_T + \left[F_{z'}\right]_{t=T} \Delta z_T = 0$$

EXAMPLE 1 Assume that T is fixed, but y_T and z_T are both free. Then we have $\Delta T = 0$, but Δy_T and Δz_T are arbitrary. Eliminating the first term in (3.27′) and setting the other two terms individually equal to zero, we get the transversality conditions

$$\left[F_{y'}\right]_{t=T} = 0 \qquad \text{and} \qquad \left[F_{z'}\right]_{t=T} = 0$$

which the reader should compare with (3.10).

EXAMPLE 2 Suppose that the terminal values of y and z are required to satisfy the restrictions

$$y_T = \phi(T) \qquad \text{and} \qquad z_T = \psi(T)$$

Then we have a pair of terminal curves. For a small ΔT, we may expect the following to hold:

$$\Delta y_T = \phi' \Delta T \qquad \text{and} \qquad \Delta z_T = \psi' \Delta T$$

Using these to eliminate Δy_T and Δz_T in (3.27′), we obtain

$$\left[F + (\phi' - y')F_{y'} + (\psi' - z')F_{z'}\right]_{t=T} \Delta T = 0$$

Because ΔT is arbitrary, the transversality condition emerges as

$$\left[F + (\phi' - y')F_{y'} + (\psi' - z')F_{z'}\right]_{t=T} = 0$$

which the reader should compare with (3.12).

With this transversality condition, the terminal curves $y_T = \phi(T)$ and $z_T = \psi(T)$, and the initial conditions $y(0) = y_0$ and $z(0) = z_0$, we now have five equations to determine T^* as well as the four arbitrary constants in the two Euler equations.

The Case of Higher Derivatives

When the functional $V[y]$ has the integrand $F(t, y, y', \dots, y^{(n)})$, the (terminal) transversality condition again requires modification. In view of the rarity of high-order derivatives in economic applications, we shall state here the general transversality condition for the case of $F(t, y, y', y'')$ only:

$$\left[F - y'F_{y'} - y''F_{y''} + y'\frac{d}{dt}F_{y''}\right]_{t=T} \Delta T + \left[F_{y'} - \frac{d}{dt}F_{y''}\right]_{t=T} \Delta y_T + \left[F_{y''}\right]_{t=T} \Delta y'_T = 0 \tag{3.28}$$

The new symbol appearing in this condition, $\Delta y'_T$, means the change in the terminal slope of the y path when it is perturbed. In terms of Fig. 3.1, $\Delta y'_T$ would mean the difference between the slope of the AZ'' path at Z'' and the slope of the AZ path at Z. If the problem specifies that the terminal slope must remain unchanged, then $\Delta y'_T = 0$, and the last term in (3.28) drops out. If the terminal slope is free to vary, then the last term will call for the condition $F_{y''} = 0$ at $t = T$.

EXERCISE 3.3

1 For the functional $V[y] = \int_0^T (y + yy' + y' + \frac{1}{2}y'^2)\,dt$, the general solution of the Euler equation is $y^*(t) = \frac{1}{2}t^2 + c_1 t + c_2$ (see Exercise 2.2, Prob. 3). If we have a vertical initial line at $t = 0$ and a vertical terminal line at $t = 1$, write out the transversality conditions, and use them to definitize the constants in the general solution.

2 Let the vertical initial line in the preceding problem be truncated with the restriction $y^*(0) \geq 1$, but keep the terminal line unchanged.

(*a*) Is the original solution still acceptable? Why?

(*b*) Find the new extremal.

3 In a problem with the functional $\int_0^T F(t, y, z, y', z')\,dt$, suppose that $y(0) = A$, $z(0) = B$, $y_T = C$, $z_T = D$, T free (A, B, C, D are constants).

(a) How many transversality condition(s) does the problem require? Why?

(b) Write out the transversality condition(s).

4 In the preceding problem, suppose that we have instead $y(0) = A$, $z(0) = B$, $y_T = C$, and $z_T = \psi(T)$, T free (A, B, C are constants).

(a) How many transversality condition(s) does the problem require? Why?

(b) Write out the transversality condition(s).

3.4 THE OPTIMAL ADJUSTMENT OF LABOR DEMAND

Consider a firm that has decided to raise its labor input from L_0 to a yet undetermined level L_T after encountering a wage reduction at $t = 0$. The adjustment of labor input is assumed to entail a cost C that varies with $L'(t)$, the rate of change of L. Thus the firm has to decide on the best speed of adjustment toward L_T as well as the magnitude of L_T itself. This is the essence of the labor adjustment problem discussed in a paper by Hamermesh.[6]

The Problem

For simplicity, let the profit of the firm be expressed by the general function $\pi(L)$, with $\pi''(L) < 0$, as illustrated in Fig. 3.4. The labor input is taken to be the sole determinant of profit because we have subsumed all aspects of production and demand in the profit function. The cost of adjusting L is assumed to be

$$C(L') = bL'^2 + k \qquad (b > 0, \text{ and } k > 0 \text{ when } L' \neq 0) \tag{3.29}$$

Thus the net profit at any time is $\pi(L) - C(L')$.

The problem of the firm is to maximize the total net profit Π over time during the process of changing the labor input. Inasmuch as it must choose not only the optimal L_T, but also an optimal time T^* for completing the adjustment, we have both the terminal state and terminal time free. Another feature to note about the problem is that Π should include not only the net profit from $t = 0$ to $t = T^*$ (a definite integral), but also the capitalized value of the profit in the post-T^* period, which is affected by the choice of L_T and T, too. Since the profit rate at time T is $\pi(L_T)$, its present value is $\pi(L_T)e^{-\rho t}$, where ρ is the given discount rate and T is to be set equal to T^*. So the capitalized value of that present value is,

---

[6] Daniel S. Hamermesh, "Labor Demand and the Structure of Adjustment Costs," *American Economic Review*, September 1989, pp. 674–689.

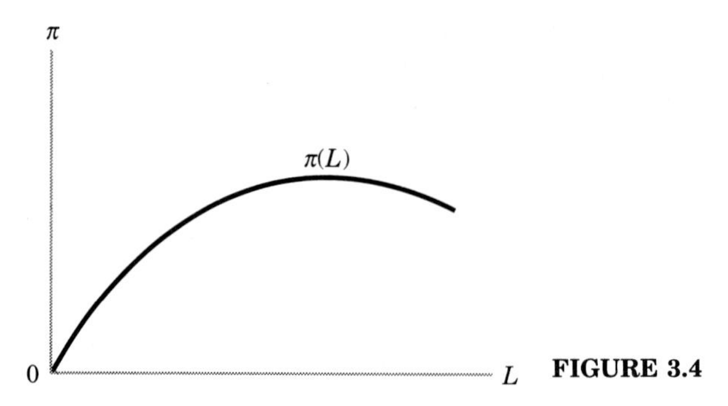

FIGURE 3.4

according to the standard capitalization formula, $\pi(L_T)e^{-\rho T}/\rho$. The full statement of the problem is therefore

$$\text{Maximize} \quad \Pi[L] = \int_0^T \left[\pi(L) - bL'^2 - k\right] e^{-\rho t}\, dt + \frac{1}{\rho}\pi(L_T)e^{-\rho T}$$

$$(3.30) \quad \text{subject to} \quad L(0) = L_0 \qquad (L_0 \text{ given})$$

$$\text{and} \quad L(T) = L_T \qquad (L_T > L_0 \text{ free, } T \text{ free})$$

If the last term in the functional were a constant, we could ignore it in the optimization process, because the same solution path would emerge either way. But since that term—call it $z(T)$—varies with the optimal choice of L_T and T, we must explicitly take it into account. From our earlier discussion of the problem of Mayer and the problem of Bolza, we have learned that

$$(3.31) \qquad z(T) = \int_0^T z'(t)\, dt + z(0) \qquad [\text{see } (1.7)]$$

Thus, by defining

$$z(t) = \frac{1}{\rho}\pi(L)e^{-\rho t} \qquad \text{so that } z'(t) = \left[-\pi(L) + \frac{1}{\rho}\pi'(L)L'\right]e^{-\rho t}$$

we can write the post-T^* term in (3.30) as

(3.31′)

$$\frac{1}{\rho}\pi(L_T)e^{-\rho T} = \int_0^T \left[-\pi(L) + \frac{1}{\rho}\pi'(L)L'\right] e^{-\rho t}\, dt + \frac{1}{\rho}\pi(L_0)$$

Substitution of (3.31′) into the functional in (3.30) yields, after combining the two integrals, the new but equivalent function

$$(3.32) \quad \Pi(L) = \int_0^T \left[-bL'^2 - k + \frac{1}{\rho}\pi'(L)L'\right] e^{-\rho t}\, dt + \frac{1}{\rho}\pi(L_0)$$

While this functional *still* contains an extra term outside of the integral, that term is a constant. So it affects only the optimal Π value, but not the optimal L path, nor the optimal values of L_T and T.

The Solution

To find the Euler equation, we see from (3.32) that

$$F = \left[-bL'^2 - k + \frac{1}{\rho}\pi'(L)L' \right] e^{-\rho t}$$

$$F_L = \frac{1}{\rho}\pi''(L)L'e^{-\rho t} \qquad F_{L'} = \left[-2bL' + \frac{1}{\rho}\pi'(L) \right] e^{-\rho t}$$

Thus, from formula (2.18), we get the condition

$$L'' - \rho L' + \frac{\pi'(L)}{2b} = 0 \qquad \text{[Euler equation]} \tag{3.33}$$

The transversality condition in this problem is twofold. Both L_T and T being free, we must satisfy both transversality conditions (3.10) and (3.11).[7] The condition $[F_{L'}]_{t=T} = 0$ means that

$$L' - \frac{\pi'(L)}{2\rho b} = 0 \qquad (\text{at } t = T) \tag{3.34}$$

And the condition $[F - L'F_{L'}]_{t=T} = 0$ means (after simplification) that

$$L'^2 = \frac{k}{b} \qquad \text{or} \qquad L' = \sqrt{\frac{k}{b}} \qquad (\text{at } t = T) \tag{3.35}$$

where we take the positive square root because L is supposed to *increase* from L_0 to L_T. The two transversality conditions, plus the initial condition $L(0) = L_0$, can provide the information needed for definitizing the arbitrary constants in the solution path, as well as for determining the optimal L_T and T. In using (3.34) and (3.35), it is understood that the L' symbol refers to the derivative of the general solution of the Euler equation, evaluated at $t = T$.

To obtain a specific quantitative solution, it is necessary to specify the form of the profit function $\pi(L)$. Since our primary purpose of citing this example is to illustrate a case with both the terminal state and the terminal

[7]Technically, condition (3.17) for a truncated vertical terminal line should be used in lieu of (3.10). However, since L_T should be strictly greater than L_0, the complementary-slackness requirement would reduce (3.17) to (3.10).

time free, and to illustrate the problem of Bolza, we shall not delve into specific quantitative solution.

EXERCISE 3.4

1 (*a*) From the two transversality conditions (3.34) and (3.35), deduce the location of the optimal terminal state L_T^* with reference to Fig. 3.4.
(*b*) How would an increase in ρ affect the location of L_T^*? How would an increase in b or k affect L_T^*?
(*c*) Interpret the economic meaning of your result in (*b*).

2 In the preceding problem, let the profit function be

$$\pi(L) = 2mL - nL^2 \qquad (0 < n < m)$$

(*a*) Find the value of L_T^*.
(*b*) At what L value does π reach a peak?
(*c*) In light of (*b*), what can you say about the location of L_T^* in relation to the $\pi(L)$ curve?

3 (*a*) From the transversality condition (3.35), deduce the location of the optimal terminal time T^* with reference to a graph of the solution path $L^*(t)$.
(*b*) How would an increase in k affect the location of T^*? How would an increase in b affect T^*?

CHAPTER

4

SECOND-ORDER CONDITIONS

Our discussion has so far concentrated on the identification of the extremal(s) of a problem, without attention to whether they maximize or minimize the functional $V[y]$. The time has come to look at the latter aspect of the problem. This involves checking the second-order conditions. But we shall also discuss a sufficient condition based on the concept of concavity and convexity. A simple but useful test known as the Legendre necessary conditions for maximum and minimum will also be introduced.

4.1 SECOND-ORDER CONDITIONS

By treating the functional $V[y]$ as a function $V(\epsilon)$ and setting the first derivative $dV/d\epsilon$ equal to zero, we have derived the Euler equation and transversality conditions as first-order necessary conditions for an extremal. To distinguish between maximization and minimization problems, we can take the second derivative $d^2V/d\epsilon^2$, and use the following standard second-order conditions in calculus:

Second-order necessary conditions:

$$\frac{d^2V}{d\epsilon^2} \le 0 \qquad [\text{for maximization of } V]$$

$$\frac{d^2V}{d\epsilon^2} \ge 0 \qquad [\text{for minimization of } V]$$

Second-order sufficient conditions:

$$\frac{d^2V}{d\epsilon^2} < 0 \qquad [\text{for maximization of } V]$$

$$\frac{d^2V}{d\epsilon^2} > 0 \qquad [\text{for minimization of } V]$$

The Second Derivative of V

To find $d^2V/d\epsilon^2$, we differentiate the $dV/d\epsilon$ expression in (2.13) with respect to ϵ, bearing in mind that (1) all the partial derivatives of $F(t, y, y')$ are, like F itself, functions of t, y, and y', and (2) y and y' are, in turn, both functions of ϵ, with derivatives

$$\frac{dy}{d\epsilon} = p(t) \quad \text{and} \quad \frac{dy'}{d\epsilon} = p'(t) \qquad [\text{by } (2.3)] \tag{4.1}$$

Thus, we have

$$\frac{d^2V}{d\epsilon^2} = \frac{d}{d\epsilon}\left(\frac{dV}{d\epsilon}\right) = \frac{d}{d\epsilon}\int_0^T \left[F_y p(t) + F_{y'} p'(t)\right] dt \tag{4.2}$$

$$= \int_0^T \left[p(t)\frac{d}{d\epsilon}F_y + p'(t)\frac{d}{d\epsilon}F_{y'}\right] dt \qquad [\text{by Leibniz's rule}]$$

In view of the fact that

$$\frac{d}{d\epsilon}F_y = F_{yy}\frac{dy}{d\epsilon} + F_{y'y}\frac{dy'}{d\epsilon} = F_{yy}p(t) + F_{y'y}p'(t) \qquad [\text{by } (4.1)]$$

and, similarly,

$$\frac{d}{d\epsilon}F_{y'} = F_{yy'}p(t) + F_{y'y'}p'(t)$$

the second derivative (4.2) emerges (after simplification) as

$$\frac{d^2V}{d\epsilon^2} = \int_0^T \left[F_{yy}p^2(t) + 2F_{yy'}p(t)p'(t) + F_{y'y'}p'^2(t)\right] dt \tag{4.2'}$$

The Quadratic-Form Test

The second derivative in (4.2′) is a definite integral with a quadratic form as its integrand. Since t spans the interval $[0, T]$, we have, of course, not one, but an infinite number of quadratic forms in the integral. Nevertheless, if it can be established that the quadratic form—with F_{yy}, $F_{yy'}$, and $F_{y'y'}$ evaluated on the extremal—is negative definite for every t, then $d^2V/d\epsilon^2 < 0$, and the extremal maximizes V. Similarly, positive definiteness of the quadratic form for every t is sufficient for minimization of V. Even if we can only establish sign *semi*definiteness, we can at least have the second-order necessary conditions checked.

For some reason, however, the quadratic-form test was totally ignored in the historical development of the classical calculus of variations. In a more recent development (see the next section), however, the concavity/convexity of the integrand function F is used in a sufficient condition. While concavity/convexity does not per se require differentiability, it is true that if the F function does possess continuous second derivatives, then its concavity/convexity can be checked by means of the sign semidefiniteness of the second-order total differential of F. So the quadratic-form test definitely has a role to play in the calculus of variations.

4.2 THE CONCAVITY/CONVEXITY SUFFICIENT CONDITION

A Sufficiency Theorem for Fixed-Endpoint Problems

Just as a concave (convex) objective function in a static optimization problem is sufficient to identify an extremum as an absolute maximum (minimum), a similar sufficiency theorem holds in the calculus of variations:

(4.3) For the fixed-endpoint problem (2.1), if the integrand function $F(t, y, y')$ is concave in the variables (y, y'), then the Euler equation is sufficient for an absolute maximum of $V[y]$. Similarly, if $F(t, y, y')$ is convex in (y, y'), then the Euler equation is sufficient for an absolute minimum of $V[y]$.

It should be pointed out that concavity/convexity in (y, y') means concavity/convexity in the two variables y and y' *jointly*, not in each variable separately.

We shall give here the proof of this theorem for the concave case.[1] Central to the proof is the defining property of a differentiable concave function: The function $F(t, y, y')$ is concave in (y, y') if, and only if, for any

[1]This proof is due to Akira Takayama. See his *Mathematical Economics*, 2d ed., Cambridge University Press, Cambridge, 1985, pp. 429–430.

pair of distinct points in the domain, $(t, y^*, y^{*\prime})$ and (t, y, y'), we have

$$(4.4) \quad F(t,y,y') - F(t,y^*,y^{*\prime}) \le F_y(t,y^*,y^{*\prime})(y-y^*) + F_{y'}(t,y^*,y^{*\prime})(y'-y^{*\prime}) = F_y(t,y^*,y^{*\prime})\epsilon p(t) + F_{y'}(t,y^*,y^{*\prime})\epsilon p'(t) \qquad \text{[by (2.3)]}$$

Here $y^*(t)$ denotes the optimal path, and $y(t)$ denotes any other path. By integrating both sides of (4.4) with respect to t over the interval $[0, T]$, we obtain

$$(4.5) \quad V[y] - V[y^*] \le \epsilon \int_0^T \left[F_y(t,y^*,y^{*\prime})p(t) + F_{y'}(t,y^*,y^{*\prime})p'(t)\right] dt$$

$$= \epsilon \int_0^T p(t)\left[F_y(t,y^*,y^{*\prime}) - \frac{d}{dt}F_{y'}(t,y^*,y^{*\prime})\right] dt$$

[integration of the $F_{y'}(t,y^*,y^{*\prime})p'(t)$ term by parts as in (2.16)]

$= 0$ [since $y^*(t)$ satisfies the Euler equation (2.18)]

In other words, $V[y] \le V[y^*]$, where $y(t)$ can refer to any other path. We have thus identified $y^*(t)$ as a V-maximizing path, and at the same time demonstrated that the Euler equation is a sufficient condition, given the assumption of a concave F function. The opposite case of a convex F function for minimizing V can be proved analogously.

Note that if the F function is *strictly concave* in (y, y'), then the weak inequality $\le$ in (4.4) and (4.5) will become the strict inequality $<$. The result, $V[y] < V[y^*]$, will then establish $V[y^*]$ to be a *unique* absolute maximum of V. By the same token, a *strictly convex* F will make $V[y^*]$ a *unique* absolute minimum.

Generalization to Variable Terminal Point

The proof of the sufficiency theorem (4.3) is based on the assumption of fixed endpoints. But it can easily be generalized to problems with a vertical terminal line or truncated vertical terminal line.

To show this, first recall that the integration-by-parts process in (2.16) originally produced an extra term $[F_{y'}p(t)]_0^T$, which later drops out because it reduces to zero. The reason is that, with fixed endpoints, the perturbing curve $p(t)$ is characterized by $p(0) = p(T) = 0$. When we switch to the problem with a variable terminal point, with T fixed but $y(T)$ free, $p(T)$ is no longer required to be zero. For this reason, we must admit an extra term

$$\epsilon\left[F_{y'}p(t)\right]_{t=T} = \left[F_{y'}(y-y^*)\right]_{t=T} \qquad \text{[by (2.3)]}$$

on the right-hand side of the second and the third lines of (4.5). Upon rearrangement, (4.5) now becomes

$$V[y] \le V[y^*] + \left[F_{y'}(y - y^*)\right]_{t=T}$$

where $F_{y'}$ is to be evaluated along the optimal path, and $(y - y^*)$ represents the deviation of any admissible neighboring path $y(t)$ from the optimal path $y^*(t)$.

If the last term in the last inequality is zero, then obviously the original conclusion—that $V[y^*]$ is an absolute maximum—still stands. Moreover, if the said term is negative, we can again be sure that $V[y^*]$ constitutes an absolute maximum. It is only when $[F_{y'}(y - y^*)]_{t=T}$ is positive that we are thrown into doubt. In short, the concavity condition on $F(t, y, y')$ in (4.3) only needs to be supplemented in the present case by a nonpositivity condition on the expression $[F_{y'}(y - y^*)]_{t=T}$.

But this supplementary condition is automatically met when the transversality condition is satisfied for the vertical-terminal-line problem, namely, $[F_{y'}]_{t=T} = 0$. As for the truncated case, the transversality condition calls for either $[F_{y'}]_{t=T} = 0$ (when the minimum acceptable terminal value y_{min} is nonbinding), or $y^* = y_{min}$ (when that terminal value is binding, thereby in effect turning the problem into one with a fixed terminal point). Either way, the supplementary condition is met. Thus, if the integrand function F is concave (convex) in the variables (y, y') in a problem with a vertical terminal line or a truncated vertical terminal line, then the Euler equation *plus* the transversality condition are sufficient for an absolute maximum (minimum) of $V[y]$.

Checking Concavity/Convexity

For any function f, whether differentiable or not, the concavity feature can be checked via the basic defining property that, for two distinct points u and v in the domain, and for $0 < \theta < 1$, F is concave if and only if

$$\theta f(u) + (1 - \theta) f(v) \le f[\theta u + (1 - \theta)v]$$

For convexity, the inequality $\le$ is reversed to $\ge$. Checking this property can, however, be very involved and tedious. For our purposes, since $F(t, y, y')$ is already assumed to have continuous second derivatives, we can take the simpler alternative of checking the sign definiteness or semidefiniteness of the quadratic form

$$q = F_{yy}\, dy^2 + 2F_{yy'}\, dy\, dy' + F_{y'y'}\, dy'^2 \tag{4.6}$$

This quadratic form can, of course, be equivalently rewritten as

$$q = F_{y'y'}\, dy'^2 + 2F_{y'y}\, dy'\, dy + F_{yy}\, dy^2 \tag{4.6'}$$

which would fit in better with the ensuing discussion of the Legendre condition. Once the sign of the quadratic form is ascertained, inferences can readily be drawn regarding concavity/convexity as follows: The $F(t, y, y')$ function is concave (convex) in (y, y') if, and only if, the quadratic form q is everywhere negative (positive) semidefinite; and the F function is strictly concave (strictly convex) if (but *not* only if) q is everywhere negative (positive) definite.

It should be noted that concavity/convexity is a *global* concept. This is why it is related to absolute extrema. This is also why the quadratic form q is required to be *everywhere* negative (positive) semidefinite for concavity (convexity) of F, meaning that the sign semidefiniteness of q should hold for all points in the domain (in the yy' space) for all t. This is a stronger condition than the earlier-mentioned quadratic-form criterion relating to (4.2′), because in the latter, the second derivatives of F are to be evaluated on the extremal only. From another point of view, however, the concavity/convexity sufficient condition is less strong; it only calls for sign *semi*definiteness, whereas, for (4.2′), a sufficient condition would require a definite sign.

The sign definiteness and semidefiniteness of a quadratic form can be checked with determinantal and characteristic-root tests. Since these tests are applicable not only in the present context of the calculus of variations, but also in that of optimal control theory later, it may prove worthwhile to present here a self-contained explanation of these tests.

The Determinantal Test for Sign Definiteness

The determinantal test for the sign definiteness of the quadratic form q is the simplest to use. We first write the discriminant of q,

$$|D| \equiv \begin{vmatrix} F_{y'y'} & F_{y'y} \\ F_{yy'} & F_{yy} \end{vmatrix} \qquad \text{[from (4.6')]} \tag{4.7}$$

and then define the following two principal minors:

$$|D_1| \equiv |F_{y'y'}| = F_{y'y'} \qquad \text{and} \qquad |D_2| \equiv \begin{vmatrix} F_{y'y'} & F_{y'y} \\ F_{yy'} & F_{yy} \end{vmatrix} \tag{4.8}$$

The test is

$$\begin{aligned} &\text{Negative definiteness of } q && \Leftrightarrow && |D_1| < 0,\ |D_2| > 0 \\ &\text{Positive definiteness of } q && \Leftrightarrow && |D_1| > 0,\ |D_2| > 0 \end{aligned} \tag{4.9}$$

(everywhere in the domain)

Though easy to apply, this test may be overly stringent. It is intended for *strict* concavity/convexity, whereas the sufficient condition (4.3) merely stipulates weak concavity/convexity.

The Determinantal Test for Sign Semidefiniteness

The determinantal test for the sign *semi*definiteness of q requires a larger number of determinants to be examined, because under this test we must consider all the possible ordering sequences in which the variables of the quadratic form can be arranged. For the present two-variable case, there are two possible ordering sequences for the two variables—(y', y) and (y, y'). Therefore, we need to consider only one additional discriminant besides (4.7), namely,

$$|D^0| \equiv \begin{vmatrix} F_{yy} & F_{yy'} \\ F_{y'y} & F_{y'y'} \end{vmatrix} \qquad \text{[from (4.6)]} \tag{4.10}$$

whose principal minors are

$$|D^0{}_1| \equiv |F_{yy}| = F_{yy} \qquad \text{and} \qquad |D^0{}_2| \equiv \begin{vmatrix} F_{yy} & F_{yy'} \\ F_{y'y} & F_{y'y'} \end{vmatrix} \tag{4.11}$$

For notational convenience, let us refer to the 1×1 principal minors $|D_1|$ and $|D^0{}_1|$ collectively as $|\tilde{D}_1|$, and the 2×2 principal minors $|D_2|$ and $|D^0{}_2|$ collectively as $|\tilde{D}_2|$. Further, let us use the notation $|\tilde{D}_i| \geq 0$ to mean that *each* member of $|\tilde{D}_i|$ is ≥ 0. Then the test for sign semidefiniteness is as follows:

$$\begin{array}{lcl} \text{Negative semidefiniteness of } q & \Leftrightarrow & |\tilde{D}_1| \leq 0,\ |\tilde{D}_2| \geq 0 \\ \text{Positive semidefiniteness of } q & \Leftrightarrow & |\tilde{D}_1| \geq 0,\ |\tilde{D}_2| \geq 0 \end{array} \tag{4.12}$$

(everywhere in the domain)

The Characteristic-Root Test

The alternative to the use of determinants is to apply the characteristic-root test, which can reveal sign semidefiniteness as well as sign definiteness all at once. To use this test, first write the characteristic equation

$$\begin{vmatrix} F_{y'y'} - r & F_{y'y} \\ F_{yy'} & F_{yy} - r \end{vmatrix} = 0 \qquad \text{[from (4.7)]} \tag{4.13}$$

Then solve for the characteristic roots r_1 and r_2. The test revolves around the signs of these roots as follows:

$$(4.14)\quad \begin{array}{lcl} \text{Negative definiteness of } q & \Leftrightarrow & r_1 < 0,\ r_2 < 0 \\ \text{Positive definiteness of } q & \Leftrightarrow & r_1 > 0,\ r_2 > 0 \\ \text{Negative semidefiniteness of } q & \Leftrightarrow & r_1 \le 0,\ r_2 \le 0 \\ & & (\text{at least one root} = 0) \\ \text{Positive semidefiniteness of } q & \Leftrightarrow & r_1 \ge 0,\ r_2 \ge 0 \\ & & (\text{at least one root} = 0) \end{array}$$

For this test, it is useful to remember that the two roots are tied to the determinant $|D_2|$ in (4.8) by the following two relations:[2]

$$(4.15)\quad r_1 + r_2 = \text{sum of principal-diagonal elements} = F_{y'y'} + F_{yy}$$

$$(4.16)\quad r_1 r_2 = |D_2|$$

For one thing, these two relations provide a means of double-checking the calculation of r_1 and r_2. More importantly, they allow us to infer that, if $|D_2| < 0$, then r_1 and r_2 must be opposite in sign, so that q must be indefinite in sign. Consequently, once $|D_2|$ is found to be negative, there is no need to go through the steps in (4.13) and (4.14).

EXAMPLE 1 The shortest-distance problem (Example 4 of Sec. 2.2) has the integrand function $F = (1 + y'^2)^{1/2}$. Since there is only one variable, y', in this function, it is easy to check the second-order condition. In so doing, bear in mind that the square root in the F expression is a *positive* square root because it represents a distance, which cannot be negative. This point becomes important when we evaluate the second derivative.

The first derivative of F is, by the chain rule,

$$F_{y'} = \tfrac{1}{2}\left(1 + y'^2\right)^{-1/2} 2y' = \left(1 + y'^2\right)^{-1/2} y'$$

Further differentiation yields

$$\begin{aligned} F_{y'y'} &= -\left(1 + y'^2\right)^{-3/2} y'^2 + \left(1 + y'^2\right)^{-1/2} \\ &= \left(1 + y'^2\right)^{-3/2}\left[-y'^2 + \left(1 + y'^2\right)\right] \\ &= \left(1 + y'^2\right)^{-3/2} > 0 \qquad [\text{positive square root}] \end{aligned}$$

[2]See the discussion relating to local stability analysis in Sec. 18.6 of Alpha C. Chiang, *Fundamental Methods of Mathematical Economics*, 3d ed., McGraw-Hill, New York, 1984.

The positive sign for $F_{y'y'}$—positive for all y' values and for all t—means that F is strictly convex in the only variable y' everywhere. We can thus conclude that the extremal found from the Euler equation yields a unique minimum distance between the two given points.

EXAMPLE 2 Is the Euler equation sufficient for maximization or minimization if the integrand of the objective functional is $F(t, y, y') = 4y^2 + 4yy' + y'^2$?

To check the concavity/convexity of F, we first find the second derivatives of F. Since

$$F_y = 8y + 4y' \qquad \text{and} \qquad F_{y'} = 4y + 2y'$$

the required second derivatives are

$$F_{y'y'} = 2 \qquad F_{y'y} = F_{yy'} = 4 \qquad F_{yy} = 8$$

Thus, with reference to (4.8), we have

$$|D_1| = 2 \qquad |D_2| = \begin{vmatrix} 2 & 4 \\ 4 & 8 \end{vmatrix} = 0$$

The quadratic form q is not positive definite.

However, with reference to (4.11), we find that

$$|D^0{}_1| = 8 \qquad |D^0{}_2| = 0$$

So the condition for positive semidefiniteness in (4.12) is satisfied.

To illustrate the characteristic-root test, we first write the characteristic equation

$$\begin{vmatrix} 2-r & 4 \\ 4 & 8-r \end{vmatrix} = r^2 - 10r = 0$$

Its two roots are $r_1 = 10$ and $r_2 = 0$. [This result, incidentally, confirms (4.15) and (4.16).] Since the roots are unconditionally nonnegative, q is everywhere positive semidefinite according to (4.14). It follows that the F function is convex, and the Euler equation is indeed sufficient for the minimization of $V[y]$.

EXAMPLE 3 Let us now check whether the Euler equation is sufficient for profit maximization in the dynamic monopoly problem of Sec. 2.4. The profit function (2.33) gives rise to the second derivatives

$$\pi_{P'P'} = -2\alpha h^2 \qquad \pi_{P'P} = \pi_{PP'} = h(1 + 2\alpha b) \qquad \pi_{PP} = -2b(1 + \alpha b)$$

Using the determinantal test (4.9), we see that even though $|D_1|$ ($\pi_{P'P'}$ here) is negative as required for the negative definiteness of q, $|D_2|$ is also negative:

$$\begin{vmatrix} \pi_{P'P'} & \pi_{P'P} \\ \pi_{PP'} & \pi_{PP} \end{vmatrix} = 4\alpha bh^2(1 + \alpha b) - \left[h(1 + 2\alpha b)\right]^2 = -h^2$$

Hence, by (4.16), the characteristic roots r_1 and r_2 have opposite signs. The F function is not concave, and the Euler equation is not sufficient for profit maximization.

As we shall see in the next section (Example 3), however, the problem does satisfy a second-order necessary condition for a maximum.

Extension to n-Variable Problems

When the problem contains n state variables, the concavity and convexity sufficiency conditions are still applicable. But, in that case, the F function must be concave/convex (as the case may be) in all the n variables and their derivatives $(y_1, \ldots, y_n, y_1', \ldots, y_n')$, jointly. For the characteristic-root test, this would mean a higher-degree polynomial equation to solve. Once the roots are found, however, we need only to subject all of them to sign restrictions similar to those in (4.14). In the determinantal test for sign definiteness, there would be a larger discriminant with more principal minors to be checked. But the technique is well known.[3]

With regard to the determinantal test for sign *semi*definiteness, the procedure for the n-variable case would become more complex. This is because the number of principal minors generated by different ordering sequences of the variables will quickly multiply. Even with only two state variables, say, (y, z), concavity/convexity must now be checked with regard to as many as *four* variables (y, z, y', z'). For notational simplicity, let us number these four variables, respectively, as the first, second, third, and fourth. Then we can arrange the second-order derivatives of F into the following 4×4 discriminant:

$$(4.17) \qquad |D| \equiv \begin{vmatrix} F_{11} & F_{12} & F_{13} & F_{14} \\ F_{21} & F_{22} & F_{23} & F_{24} \\ F_{31} & F_{32} & F_{33} & F_{34} \\ F_{41} & F_{42} & F_{43} & F_{44} \end{vmatrix} \equiv \begin{vmatrix} F_{yy} & F_{yz} & F_{yy'} & F_{yz'} \\ F_{zy} & F_{zz} & F_{zy'} & F_{zz'} \\ F_{y'y} & F_{y'z} & F_{y'y'} & F_{y'z'} \\ F_{z'y} & F_{z'z} & F_{z'y'} & F_{z'z'} \end{vmatrix}$$

Since each of the variables can be taken as the "first" variable in a different

[3]See, for example, Alpha C. Chiang, *Fundamental Methods of Mathematical Economics*, 3d ed., McGraw-Hill, New York, 1984, Sec. 11.3.

ordering sequence, there exist four possible 1×1 principal minors:

$$F_{yy} \quad F_{zz} \quad F_{y'y'} \quad F_{z'z'} \tag{4.18}$$

We shall denote these 1×1 principal minors collectively by $|\tilde{D}_1|$.

The total number of possible 2×2 principal minors is 12, but only six distinct values can emerge from them. These values can be calculated from the following six principal minors:

$$\begin{vmatrix} F_{yy} & F_{yz} \\ F_{zy} & F_{zz} \end{vmatrix} \quad \begin{vmatrix} F_{yy} & F_{yy'} \\ F_{y'y} & F_{y'y'} \end{vmatrix} \quad \begin{vmatrix} F_{yy} & F_{yz'} \\ F_{z'y} & F_{z'z'} \end{vmatrix}$$
$$\begin{vmatrix} F_{zz} & F_{zy'} \\ F_{y'z} & F_{y'y'} \end{vmatrix} \quad \begin{vmatrix} F_{zz} & F_{zz'} \\ F_{z'z} & F_{z'z'} \end{vmatrix} \quad \begin{vmatrix} F_{y'y'} & F_{y'z'} \\ F_{z'y'} & F_{z'z'} \end{vmatrix} \tag{4.19}$$

The reason why the other six can be ignored is that the principal minor with (F_{ii}, F_{jj}) in the diagonal is always equal in value to the one with the two variables in reverse order, (F_{jj}, F_{ii}), in the diagonal. The 2×2 principal minors in (4.19) will be collectively referred to as $|\tilde{D}_2|$.

As to 3×3 principal minors, the total number available is 24. From these, however, only four distinct values can arise, and they can be calculated from the following principal minors:

$$\begin{vmatrix} F_{yy} & F_{yz} & F_{yy'} \\ F_{zy} & F_{zz} & F_{zy'} \\ F_{y'y} & F_{y'z} & F_{y'y'} \end{vmatrix} \quad \begin{vmatrix} F_{yy} & F_{yz} & F_{yz'} \\ F_{zy} & F_{zz} & F_{zz'} \\ F_{z'y} & F_{z'z} & F_{z'z'} \end{vmatrix}$$
$$\begin{vmatrix} F_{yy} & F_{yy'} & F_{yz'} \\ F_{y'y} & F_{y'y'} & F_{y'z'} \\ F_{z'y} & F_{z'y'} & F_{z'z'} \end{vmatrix} \quad \begin{vmatrix} F_{zz} & F_{zy'} & F_{zz'} \\ F_{y'z} & F_{y'y'} & F_{y'z'} \\ F_{z'z} & F_{z'y'} & F_{z'z'} \end{vmatrix} \tag{4.20}$$

We can ignore the others because the principal minor with (F_{ii}, F_{jj}, F_{kk}) in the diagonal always has the same value as those with F_{ii}, F_{jj}, F_{kk} arranged in any other order in the diagonal. We shall refer to the principal minors in (4.20) collectively as $|\tilde{D}_3|$.

Finally, we need to consider only one 4×4 principal minor, $|\tilde{D}_4|$, which is identical in value with $|D|$ in (4.17).

One question naturally arises: How does one ascertain the number of distinct values that can arise from principal minors of various dimensions? The answer is that we can just apply the formula for the number of combinations of n objects taken r at a time, denoted by C_r^n:

$$C_r^n = \frac{n!}{r!(n-r)!} \tag{4.21}$$

For (4.18), we have four variables taken one at a time; thus the number of combinations is

$$C_1^4 = \frac{4!}{1!\,3!} = 4$$

For 2×2 principal minors, we have

$$C_2^4 = \frac{4!}{2!\,2!} = 6$$

And for 3×3 principal minors, the formula yields

$$C_3^4 = \frac{4!}{3!\,1!} = 4$$

Finally, for 4×4 principal minors, a unique value exists, because

$$C_4^4 = \frac{4!}{4!\,0!} = 1$$

Once all the principal minors have been calculated, the test for sign semidefiniteness is similar to (4.12). Using the notation $|\tilde{D}_i| \geq 0$ to mean that *each* member of $|\tilde{D}_i|$ is ≥ 0, we have the following:

$$\begin{aligned} &\text{Negative semidefiniteness of } q \\ &\Leftrightarrow \quad |\tilde{D}_1| \leq 0,\ |\tilde{D}_2| \geq 0,\ |\tilde{D}_3| \leq 0,\ |\tilde{D}_4| \geq 0 \\ &\text{Positive semidefiniteness of } q \\ &\Leftrightarrow \quad |\tilde{D}_1| \geq 0,\ |\tilde{D}_2| \geq 0,\ |\tilde{D}_3| \geq 0,\ |\tilde{D}_4| \geq 0 \\ &\qquad\qquad\qquad \text{(everywhere in the domain)} \end{aligned} \tag{4.22}$$

EXERCISE 4.2

1 For Prob. 1 of Exercise 2.2:

(*a*) Check whether the F function is strictly concave/convex in (y, y') by the determinantal test (4.9).

(*b*) If that test fails, check for concavity/convexity by the determinantal test (4.12) or the characteristic-root test.

(*c*) Is the sufficient condition for a maximum/minimum satisfied?

2 Perform the tests mentioned in Prob. 1 above on Prob. 3 of Exercise 2.2.

3 Perform the tests mentioned in Prob. 1 above on Prob. 5 of Exercise 2.2.

4 (*a*) Show that, in the inflation-unemployment tradeoff model of Sec. 2.5, the integrand function is strictly convex.
(*b*) What specifically can you conclude from that fact?

4.3 THE LEGENDRE NECESSARY CONDITION

The concavity feature described in (4.4) is a *global* concept. When that feature is present in the integrand function F, the extremal is guaranteed to maximize $V[y]$. But when F is not globally concave, as may often happen (such as in the dynamic monopoly problem), we would have to settle for some weaker conditions. The same is true for convexity. In this section, we introduce a second-order necessary condition known as the *Legendre condition*, which is based on *local* concavity/convexity. Though not as powerful as a sufficient condition, it is very useful and indeed is used frequently.

The Legendre Condition

The great merit of the Legendre condition lies in its extreme simplicity, for it involves nothing but the sign of $F_{y'y'}$. Legendre, the mathematician, thought at one time that he had discovered a remarkably neat sufficient condition: $F_{y'y'} < 0$ for a maximum of V, and $F_{y'y'} > 0$ for a minimum of V. But, unfortunately, he was mistaken. Nonetheless, the weak-inequality version of this condition is indeed correct as a necessary condition:

$$\begin{aligned} &\text{Maximization of } V[y] \quad \Rightarrow \quad F_{y'y'} \le 0 \quad \text{for all } t \in [0, T] \\ &\text{Minimization of } V[y] \quad \Rightarrow \quad F_{y'y'} \ge 0 \quad \text{for all } t \in [0, T] \end{aligned} \tag{4.23}$$

[Legendre necessary condition]

The $F_{y'y'}$ derivative is to be evaluated along the extremal.

The Rationale

To understand the rationale behind (4.23), let us first transform the middle term in the integrand of (4.2′) into a squared term by integration by parts. Let $v \equiv F_{yy'}$ and $u \equiv p^2(t)$, so that

$$dv = \frac{dF_{yy'}}{dt}\,dt \qquad \text{and} \qquad du = 2p(t)p'(t)\,dt$$

Then the middle term in (4.2′) can be rewritten as

$$(4.24)\qquad \int_0^T v\,du = uv\Big|_0^T - \int_0^T u\,dv$$
$$= F_{yy'}p^2(t)\Big|_0^T - \int_0^T p^2(t)\frac{dF_{yy'}}{dt}\,dt$$
$$= 0 - \int_0^T p^2(t)\frac{dF_{yy'}}{dt}\,dt \qquad [p(0) = p(T) = 0 \text{ assumed}]$$

where all the limits of integration refer to values of t (not u or v). Plugging this into (4.2′) yields

$$(4.25)\qquad \frac{d^2V}{d\epsilon^2} = \int_0^T \left[\left(F_{yy} - \frac{dF_{yy'}}{dt}\right)p^2(t) + F_{y'y'}p'^2(t)\right]dt$$

where the integrand now consists of two squared terms $p^2(t)$ and $p'^2(t)$.

It turns out the $p'^2(t)$ term will dominate the other one over the interval $[0, T]$, and thus the nonpositivity (nonnegativity) of its coefficient $F_{y'y'}$ is necessary for the nonpositivity (nonnegativity) of $d^2V/d\epsilon^2$ for maximization (minimization) of V, as indicated in (4.23). But because of the presence of the other term in (4.25), the condition is not sufficient. That the $p'^2(t)$ term will dominate can be given an intuitive explanation with the help of the illustrative perturbing curves in Fig. 4.1. Figure 4.1*a* shows that $p'(t)$, the slope of $p(t)$, can take large absolute values even if $p(t)$ maintain small values throughout; Fig. 4.1*b* shows that in order to attain large $p(t)$ values, $p'(t)$ would in general have to take large absolute values, too. These considerations suggest that the $p'^2(t)$ term tends to dominate.

Note that, to reach the result in (4.25), we have made the assumption that $p(0) = p(T) = 0$ on the perturbing curve. What will happen if the terminal point is variable and $p(T)$ is not required to be zero? The answer

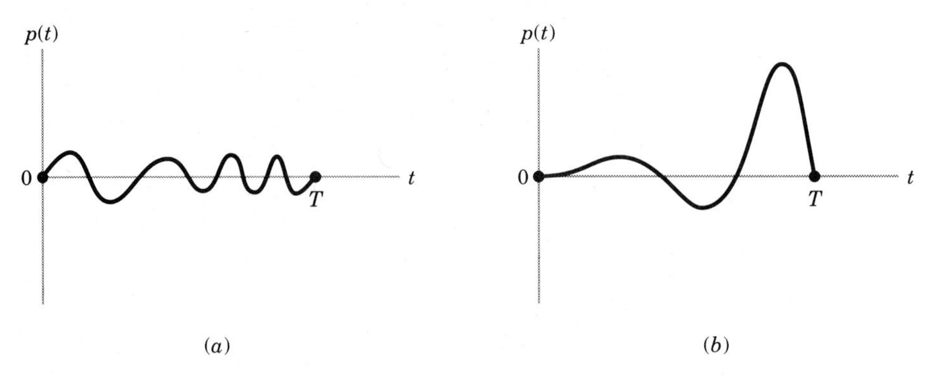

FIGURE 4.1

is that the zero term on the last line of (4.24) will then be replaced by the new expression $[F_{yy'}p^2(t)]_{t=T}$. But even then, it is possible for some admissible neighboring path to have its terminal point located exactly where $y^*(T)$ is, thereby duplicating the result of $p(T) = 0$, and reducing that new expression to zero. This means that the eventuality of (4.25) is still possible, and the Legendre condition is still needed. Hence, the Legendre necessary condition is valid whether the terminal point is fixed or variable.

EXAMPLE 1 In the shortest-distance problem (Example 1, preceding section), we have already found that

$$F_{y'y'} = (1 + y'^2)^{-3/2} > 0$$

Thus, by (4.23), the Legendre necessary condition for a minimum is satisfied. This is only to be expected, since that problem has earlier been shown to satisfy the second-order *sufficient* condition for minimization.

EXAMPLE 2 In Example 3 of Sec. 2.2, we seek to find a curve between two fixed points that will generate the smallest surface of revolution. Does the extremal (pictured in Fig. 2.4) minimize the surface of revolution? Since that problem has

$$F_{y'} = yy'(1 + y'^2)^{-1/2}$$

we can, after differentiating and simplifying, get

$$F_{y'y'} = y(1 + y'^2)^{-3/2}$$

The $(1 + y'^2)^{-3/2}$ expression is positive, and y—represented by the height of the AZ curve in Fig. 2.4—is positive along the entire path. Therefore, $F_{y'y'}$ is unconditionally positive for all t in the interval $[a, z]$, and the Legendre condition for a minimum is satisfied.

To make certain that $y(t)$ is indeed positive for all t, we see from the general solution of the problem

$$y^*(t) = \frac{c}{2}[e^{(t+k)/c} + e^{-(t+k)/c}]$$

that $y^*(t)$ takes the sign of the constant c, because the two terms in parentheses are always positive. This means that $y^*(t)$ can only have a *single* sign throughout. Given that points A and Z show positive y values, $y^*(t)$ must be positive for *all* t in the interval $[a, z]$.

Unlike the shortest-distance problem, however, the minimum-surface-of-revolution problem does not satisfy the sufficient condition.

EXAMPLE 3 For the dynamic monopoly problem, we have found in the preceding section that the integrand function $\pi(P, P')$ is not concave, so it does not satisfy the sufficient condition (4.3). However, since

$$\pi_{p'p'} = -2\alpha h^2 < 0$$

the Legendre necessary condition for a maximum is indeed satisfied.

The n-Variable Case

The Legendre necessary condition can, with proper modification, also be applied to the problem with n state variables $(y_1, \ldots, y_n)$. But instead of checking whether $F_{y'y'} \leq 0$ or $F_{y'y'} \geq 0$, as in (4.23), we must now check whether the $n \times n$ matrix $[F_{y_i'y_j'}]$—or, alternatively, the quadratic form whose coefficients are $F_{y_i'y_j'}$—is negative semidefinite (for maximization of V) or positive semidefinite (for minimization of V). For this purpose, first define

$$|\Delta| = \begin{vmatrix} F_{y_1'y_1'} & \cdots & F_{y_1'y_n'} \\ F_{y_n'y_1'} & \cdots & F_{y_n'y_n'} \end{vmatrix} \tag{4.26}$$

where the second-order derivatives are to be evaluated along the extremals. Then write all the 1×1 principal minors $|\tilde{\Delta}_1|$, all the 2×2 principal minors $|\tilde{\Delta}_2|$, and so forth, as explained in the text following (4.17). The Legendre second-order necessary condition is

$$\begin{aligned} &\text{Maximization of } V \quad \Rightarrow \quad |\tilde{\Delta}_1| \leq 0,\ |\tilde{\Delta}_2| \geq 0, \ldots, (-1)^n |\tilde{\Delta}_n| \geq 0 \\ &\qquad\qquad\qquad\qquad\qquad \text{for all } t \in [0, T] \\ &\text{Minimization of } V \quad \Rightarrow \quad |\tilde{\Delta}_1| \geq 0,\ |\tilde{\Delta}_2| \geq 0, \ldots, |\tilde{\Delta}_n| \geq 0 \\ &\qquad\qquad\qquad\qquad\qquad \text{for all } t \in [0, T] \end{aligned} \tag{4.27}$$

[Legendre necessary condition]

It is clear that (4.27) includes (4.23) as a special case.

Note that the Legendre condition in (4.27), though apparently similar to the determinantal test for sign semidefiniteness in (4.22), differs from the latter in two essential respects. First, the sign-semidefiniteness test in (4.22) involves derivatives of the F_{yy} and $F_{yy'}$ types as well as $F_{y'y'}$, but the Legendre condition is based exclusively on the $F_{y'y'}$ type. This is why we employ the symbol $|\Delta|$ in (4.26) and (4.27), so as to keep it distinct from the $|D|$ symbol used earlier in (4.22). Second, unlike the global concavity/convexity characterization, the Legendre condition is local in nature; so the second-order derivatives in (4.26) are to be evaluated along the extremals only.

EXAMPLE 4 In Exercise 2.3, Prob. 2, two state variables appear in the integrand function

$$F = y'^2 + z'^2 + y'z'$$

Since

$$F_{y'} = 2y' + z' \qquad \text{and} \qquad F_{z'} = 2z' + y'$$

we find that

$$F_{y'y'} = 2 \qquad F_{y'z'} = F_{z'y'} = 1 \qquad F_{z'z'} = 2$$

Substitution of these into (4.26) results in

$$|\Delta| = \begin{vmatrix} 2 & 1 \\ 1 & 2 \end{vmatrix}$$

In this case, we find that

$$|\tilde{\Delta}_1| = 2 \qquad \text{and} \qquad |\tilde{\Delta}_2| = \begin{vmatrix} 2 & 1 \\ 1 & 2 \end{vmatrix} = 3$$

Thus, by (4.27), the Legendre condition for a minimum is satisfied.

EXERCISE 4.3

1. Check whether the extremals obtained in Probs. 6, 8, and 9 of Exercise 2.1 satisfy the Legendre necessary condition for a maximum/minimum.
2. Check whether Prob. 3 of Exercise 2.3, a two-variable problem, satisfies the Legendre condition for a maximum/minimum.
3. Does the inflation-unemployment problem of Sec. 2.5 satisfy the Legendre condition for minimization of total social loss?

4.4 FIRST AND SECOND VARIATIONS

Our discussion of first-order and second-order conditions has hitherto been based on the concept of the first derivative $dV/d\epsilon$ and the second derivative $d^2V/d\epsilon^2$. An alternative way of viewing the problems centers around the concepts of *first variation* and *second variation*, which are directly tied to the name *calculus of variations* itself.

The calculus of variations involves a comparison of path values $V[y]$ and $V[y^*]$. The deviation of $V[y]$ from $V[y^*]$ is

$$\Delta V \equiv V[y] - V[y^*] = \int_0^T F(t, y, y')\, dt - \int_0^T F(t, y^*, y^{*\prime})\, dt \tag{4.28}$$

When the integrand of the first integral is expanded into a Taylor series around the point $(t, y^*, y^{*\prime})$, it would contain a term $F(t, y^*, y^{*\prime})$, which would allow us to cancel out the second integral in (4.28). That Taylor series is

$$(4.29)\quad \begin{aligned} F(t,y,y') = {} & F(t,y^*,y^{*\prime}) \\ & + \left[F_t(t-t) + F_y(y-y^*) + F_{y'}(y'-y^{*\prime})\right] \\ & + \frac{1}{2!}\Big[F_{tt}(t-t)^2 + F_{yy}(y-y^*)^2 + F_{y'y'}(y'-y^{*\prime})^2 \\ & \quad + 2F_{ty}(t-t)(y-y^*) + 2F_{ty'}(t-t)(y'-y^{*\prime}) \\ & \quad + 2F_{yy'}(y-y^*)(y'-y^{*\prime})\Big] + \cdots + R_n \end{aligned}$$

where all the partial derivatives of F are to be evaluated at $(t, y^*, y^{*\prime})$. We have, for the sake of completeness, included the terms that involve $(t - t)$, but, of course, these would all drop out. Recalling (2.3), we can substitute $y - y^* = \epsilon p$, and $y' - y^{*\prime} = \epsilon p'$, to get

$$(4.29')\quad \begin{aligned} F(t,y,y') = {} & F(t,y^*,y^{*\prime}) + F_y \epsilon p + F_{y'}\epsilon p' \\ & + \frac{1}{2!}\left[F_{yy}(\epsilon p)^2 + F_{y'y'}(\epsilon p')^2 + 2F_{yy'}(\epsilon p)(\epsilon p')\right] \\ & + \cdots + R_n \end{aligned}$$

Using this result in (4.28), we can then transform the latter into

$$(4.30)\quad \begin{aligned} \Delta V = {} & \epsilon \int_0^T (F_y p + F_{y'} p')\, dt \\ & + \frac{\epsilon^2}{2}\int_0^T \left(F_{yy}p^2 + F_{y'y'}p'^2 + 2F_{yy'}pp'\right) dt \\ & + \text{h.o.t. (higher-order-terms)} \end{aligned}$$

In the calculus-of-variations literature, the first integral in (4.30) is referred to as the *first variation*, denoted by δV:

$$(4.31)\quad \delta V \equiv \int_0^T (F_y p + F_{y'}p')\, dt = \frac{dV}{d\epsilon} \qquad \text{[by (2.13)]}$$

Similarly, the second integral in (4.30) is known as the *second variation*, denoted by $\delta^2 V$:

$$(4.32)\quad \delta^2 V \equiv \int_0^T \left(F_{yy}p^2 + F_{y'y'}p'^2 + 2F_{yy'}pp'\right) dt = \frac{d^2V}{d\epsilon^2} \qquad \text{[by (4.2')]}$$

In a maximization problem, where $\Delta V \leq 0$, it is clear from (4.30) that it is necessary to have $\delta V = 0$, since ϵ can take either sign at every point of time. This is equivalent to the first-order condition $dV/d\epsilon = 0$, which in our earlier discussion has led to the Euler equation. Once the $\delta V = 0$ condition is met, it is further necessary to have $\delta^2 V \leq 0$, since the coefficient $\epsilon^2/2$ is nonnegative at every point of time. This condition is equivalent to $d^2V/d\epsilon^2 \leq 0$, the second-order necessary condition of Legendre.

Analogous reasoning would show that, for a minimization problem, it is necessary to have $\delta V = 0$ and $\delta^2 V \geq 0$. In sum, the first-order and second-order necessary conditions can be derived from either the derivative route or the variation route. We have chosen to follow the derivative route because it provides more of an intuitive grasp of the underlying reasoning process.

CHAPTER 5

INFINITE PLANNING HORIZON

For an individual, it is generally adequate to plan for a finite time interval $[0, T]$, for even the most far-sighted person probably would not plan too far beyond his or her expected lifetime. But for society as a whole, or even for a corporation, there may be good reasons to expect or assume its existence to be permanent. It may therefore be desirable to extend its planning horizon indefinitely into the future, and change the interval of integration in the objective functional from $[0, T]$ to $[0, \infty]$. Such an extension of horizon has the advantage of rendering the optimization framework more comprehensive. Unfortunately, it also has the offsetting drawback of compromising the plausibility of the often-made assumption that all parameter values in a model will remain constant throughout the planning period. More importantly, the infinite planning horizon entails some methodological complications.

5.1 METHODOLOGICAL ISSUES OF INFINITE HORIZON

With an infinite horizon, at least two major methodological issues must be addressed. One has to do with the convergence of the objective functional, and the other concerns the matter of transversality conditions.

The Convergence of the Objective Functional

The convergence problem arises because the objective functional, now in the form of $\int_0^\infty F(t, y, y')\,dt$, is an improper integral which may or may not have a finite value. In the case where the integral diverges, there may exist more than one $y(t)$ path that yields an infinite value for the objective functional and it would be difficult to determine which among these paths is optimal.[1] It is true that even in the divergent situation, various ways have been proposed to make an appropriate selection of a time path from among all the paths with an infinite integral value. But that topic is complicated and, besides, it does not pertain to the calculus of variations as such, so we shall not delve into it here.[2] Instead, as a preliminary to the discussion of economic applications in the two ensuing sections, we shall make some comments on certain conditions that are (or allegedly are) sufficient for convergence.

Condition I Given the improper integral $\int_0^\infty F(t, y, y')\,dt$, if the integrand F is finite throughout the interval of integration, and if F attains a zero value at some finite point of time, say, t_0, and remains at zero for all $t > t_0$, then the integral will converge.

This is in the nature of a *sufficient* condition. Although the integral nominally has an infinite horizon, the effective upper limit of integration is a finite value, t_0. Thus, the given improper integral reduces in effect to a proper one, with assurance that it will integrate to a finite value.

Condition II (False) Given the improper integral $\int_0^\infty F(t, y, y')\,dt$, if $F \to 0$ as $t \to \infty$, then the integral will converge.

This condition is often taken to be a sufficient condition, but it is not. To see this, consider the two integrals

$$I_1 = \int_0^\infty \frac{1}{(t+1)^2}\,dt \qquad \text{and} \qquad I_2 = \int_0^\infty \frac{1}{t+1}\,dt \tag{5.1}$$

Each of these has an integrand that tends to zero as $t \to \infty$. But while I_1

[1] For a detailed discussion of the problem of convergence, see S. Chakravarty, "The Existence of an Optimum Savings Program," *Econometrica*, January 1962, pp. 178–187.

[2] The various optimality criteria in the divergent case are concisely summarized in Atle Seierstad and Knut Sydsæter, *Optimal Control Theory with Economic Applications*, Elsevier, New York, 1987, pp. 231–237.

converges, I_2 does not:

$$(5.2) \qquad I_1 = \lim_{b\to\infty}\left[\frac{-1}{t+1}\right]_0^b = 1 \qquad I_2 = \lim_{b\to\infty}\left[\ln(t+1)\right]_0^b = \infty$$

What accounts for this sharp difference in results? The answer lies in the speed at which the integrand falls toward zero. In the case of I_1, where the denominator of the integrand is a squared term, the fraction falls with a sufficiently high speed as t takes increasingly larger values, resulting in convergence. In the case of I_2, on the other hand, the speed of decline is not great enough, and the integral diverges.[3] The upshot is that, in and of itself, the condition "$F \to 0$ as $t \to \infty$" does not guarantee convergence.

It is of interest also to ask the converse question: Is the condition "$F \to 0$ as $t \to \infty$" a necessary condition for $\int_0^\infty F(t, y, y')\,dt$ to converge? Such a necessary condition seems intuitively plausible, and it is in fact commonly accepted as such.[4] But counterexamples can be produced to show that an improper integral can converge even though F does not tend to zero, so the condition is not necessary. For instance, if

$$F(t) = \begin{cases} 1 & \text{if } n \le t \le n + \dfrac{1}{n^2} \qquad (n = 1, 2, \dots) \\ 0 & \text{otherwise} \end{cases}$$

then $F(t)$ has no limit as t becomes infinite; yet the integral $\int_0^\infty F(t)\,dt$ converges to the magnitude $\sum_{k=1}^\infty (1/k^2)$.[5] As another example, the integral $\int_0^\infty \sin t^2\,dt$ can be shown to converge to the value $\frac{1}{2}\sqrt{\pi/2}$ even though the sine-function integrand does not have zero as its limit.[6]

Note, however, that these counterexamples involve either an unusual type of discontinuous integrand or an integrand that periodically changes sign, thereby allowing the contributions to the integral from neighboring time intervals to cancel out one another. Such functions are not usually

[3]For an integral $\int_0^\infty F(t)\,dt$ whose integrand maintains a single sign (say, positive), the following criterion can be used: The integral *converges* if $F(t)$ vanishes at infinity to a higher order than the first, that is, if there exists a number $\alpha > 1$ such that, for all values of t (no matter how large), $0 < F(t) \le M/t^\alpha$ holds, where M is a constant. The integral *diverges* if $F(t)$ vanishes at infinity to an order not higher than the first, that is, if there is a positive constant N such that $tF(t) \ge N > 0$. See R. Courant, *Differential and Integral Calculus*, translated from German by E. J. McShane, 2d ed., Interscience, New York, 1937, Vol. 1, pp. 249–250.

[4]See, for example, S. Chakravarty, "The Existence of an Optimum Savings Program," *Econometrica*, January 1962, p. 182, and G. Hadley and M. C. Kemp, *Variational Methods in Economics*, North-Holland, Amsterdam, 1971, pp. 52, 63.

[5]See Watson Fulks, *Advanced Calculus: An Introduction to Analysis*, 3d ed., Wiley, New York, 1978, pp. 570–571.

[6]See R. Courant, *Differential and Integral Calculus*, Interscience, New York, Vol. 1, p. 253.

used in the objective functionals in economic models. Typically, the integrand in an economic model—representing a profit function, utility function, and the like—is assumed to be continuous and nonnegative. For such functions, the counterexamples would be irrelevant.

Condition III In the integral $\int_0^\infty F(t, y, y')\,dt$, if the integrand takes the form of $G(t, y, y')e^{-\rho t}$, where ρ is a positive rate of discount, and the G function is bounded, then the integral will converge.

A distinguishing feature of this integral is the presence of the discount factor $e^{-\rho t}$ which, *ceteris paribus*, provides a dynamic force to drive the integrand down toward zero over time at a good speed. When the $G(t, y, y')$ component of the integrand is positive (as in most economic applications) and has an upper bound, say, $\hat{G}$, then the downward force of $e^{-\rho t}$ is sufficient to make the integral converge. More formally, since the value of the G function can never exceed the value of the constant $\hat{G}$, we can write

$$\int_0^\infty G(t, y, y')e^{-\rho t}\,dt \le \int_0^\infty \hat{G}e^{-\rho t}\,dt = \frac{\hat{G}}{\rho} \tag{5.3}$$

The last equality, based on the formula for the present value of a perpetual constant flow, shows that the second integral in (5.3), with the upper bound $\hat{G}$ in the integrand, is convergent. It follows that the first integral, whose integrand is $G \le \hat{G}$, must also be convergent. We have here another *sufficient* condition for convergence.

There also exist other sufficient conditions, but they are not simple to apply, and will not be discussed here. In the following, if the F function is given in the general form, we shall always assume that the integral is convergent. This would imply, for normal economic applications, that the integrand function F tends to zero as t tends to infinity. When specific functions are used, on the other hand, convergence is something to be explicitly checked.

The Matter of Transversality Conditions

Transversality conditions enter into the picture when either the terminal time or the terminal state, or both, are variable. When the planning horizon is infinite, there is no longer a specific terminal T value for us to adhere to. And the terminal state may also be left open. Thus transversality conditions are needed.

To develop the transversality conditions, we can use the same procedure as in the finite-horizon problem of Chap. 3. Essentially, these conditions would emerge naturally as a by-product of the process of deriving the Euler equation, as in (3.9), which emanates from (3.7). In the present context, (3.9) must be modified to

$$(5.4) \qquad \left[F - y'F_{y'}\right]_{t\to\infty} \Delta T + \left[F_{y'}\right]_{t\to\infty} \Delta y_T = 0$$

where each of the two terms must individually vanish.

Since there is no fixed T in the present context, ΔT is perforce nonzero, and this necessitates the condition

$$(5.5) \qquad \lim_{t\to\infty} \left(F - y'F_{y'}\right) = 0 \qquad \text{[transversality condition for infinite horizon]}$$

This condition has the same economic interpretation as transversality condition (3.11) in the finite-horizon framework. If the problem pertains to a profit-maximizing firm, for instance, the condition requires the firm to take full advantage of all profit opportunities.

As to the second term in (5.4), if an asymptotic terminal state is specified in the problem:

$$(5.6) \qquad \lim_{t\to\infty} y(t) = y_\infty = \text{a given constant}$$

[which is the infinite-horizon counterpart of $y(T) = Z$ in the finite-horizon problem], then the second term in (5.4) will vanish on its own ($\Delta y_T = 0$) and no transversality condition is needed. But if the terminal state is free, then we should impose the additional condition

$$(5.7) \qquad \lim_{t\to\infty} F_{y'} = 0 \qquad \text{[transversality condition for free terminal state]}$$

This condition can be given the same economic interpretation as transversality condition (3.10) for the finite-horizon problem. If the problem is that of a profit-maximizing firm with the state variable y representing capital stock, the message of (5.7) is for the firm to use up all its capital as $t \to \infty$.

It may be added that if the free terminal state is subject to a restriction such as $y_\infty \geq y_{\min}$, then the Kuhn-Tucker conditions should be used [cf. (3.17) and (3.17′)]. In practical applications, however, we can always apply (5.7) first. If the restriction $y_\infty \geq y_{\min}$ is satisfied by the solution, then we are done with the problem. Otherwise, we have to use $y_{\min}$ as a given terminal state.

Although these transversality conditions are intuitively reasonable, their validity is sometimes called into question. This is because in the field of optimal control theory, to be discussed in Part 3, alleged counterexamples have been presented to show that standard transversality conditions are not necessarily applicable in infinite-horizon problems. And the doubts have spread to the calculus of variations. When we come to the topic of optimal control, however, we shall argue that these counterexamples are not really valid counterexamples. Nevertheless, it is only fair to warn the reader at this point that there exists a controversy surrounding this aspect of infinite-horizon problems.

One way out of the dilemma is to avoid the transversality conditions and use plain economic reasoning to determine what the terminal state should be, as $t \to \infty$. Such an approach is especially feasible in problems where the integrand function F either does not contain an explicit t argument (an "autonomous" problem in the mathematical sense or where the t argument appears explicitly only in the discount factor $e^{-\rho t}$ (an "autonomous" problem in economists' usage). In a problem of this type, there is usually an implicit terminal state y_∞—a steady state—toward which the system would gravitate in order to fulfill the objective of the problem. If we can determine y_∞, then we can simply use the terminal condition (5.6) in place of a transversality condition.

5.2 THE OPTIMAL INVESTMENT PATH OF A FIRM

The gross investment of a firm, I_g, has two components: net investment, $I \equiv dK/dt$, and replacement investment, assumed equal to δK, where δ is the depreciation rate of capital K. Since both components of investment are intimately tied to capital, the determination of an optimal investment path, $I_g^*(t)$, is understandably contingent upon the determination of an optimal capital path, $K^*(t)$. Assuming that investment plans are always realizable, then, once $K^*(t)$ has been found, the optimal investment path is simply

$$I_g^*(t) = K^{*\prime}(t) + \delta K^*(t) \tag{5.8}$$

But if, for some reason, there exist obstacles to the execution of investment plans, then some other criterion must be used to determine $I_g^*(t)$ from $K^*(t)$. In the present section, we present two models of investment that illustrate both of these eventualities.

The Jorgenson Model[7]

In the Jorgenson neoclassical theory of investment, the firm is assumed to produce its output with capital K and labor L with the "neoclassical production function" $Q = Q(K, L)$ that allows substitution between the two inputs. This feature distinguishes it from the acceleration theory of investment in which capital is tied to output in a fixed ratio. The neoclassical production function is usually accompanied by the assumptions of positive but diminishing marginal products ($Q_K, Q_L > 0$; $Q_{KK}, Q_{LL} < 0$) and constant returns to scale (linear homogeneity).

[7] Dale W. Jorgenson, "Capital Theory and Investment Behavior," *American Economic Review*, May 1963, pp. 247–259. The version given here omits certain less essential features of the original model.

The firm's cash revenue at any point of time is PQ, where P is the given product price. Its cash outlay at any time consists of its wage bill, WL (W denotes the money wage rate), and its expenditure on new capital, mI_g (m denotes the price of "machine"). Thus the net revenue at any point of time is

$$PQ(K, L) - WL - m(K' + \delta K)$$

Applying the discount factor $e^{-\rho t}$ to this expression and summing over time, we can express the present-value net worth N of the firm as

$$N[K, L] = \int_0^{\infty} [PQ(K, L) - WL - m(K' + \delta K)]e^{-\rho t}\, dt \tag{5.9}$$

The firm's objective is to maximize its net worth N by choosing an optimal K path and an optimal L path.

The objective functional in (5.9) is an improper integral. In view of the presence of the discount factor $e^{-\rho t}$, however, the integral will converge, according to Condition III of the preceding section, if the bracketed net-revenue expression has an upper bound. Such would be the case if we can rule out an infinite value for K', that is, if K is not allowed to make a vertical jump.

There are two state variables, K and L, in the objective functional; the other symbols denote parameters. There will be two Euler equations yielding two optimal paths, $K^*(t)$ and $L^*(t)$, and the $K^*(t)$ path can then lead us to the $I_g^*(t)$ path. We may expect the firm to have a given initial capital, K_0, but the terminal capital is left open.

Optimal Capital Stock

In applying the Euler equations

$$F_K - \frac{d}{dt}F_{K'} = 0 \qquad F_L - \frac{d}{dt}F_{L'} = 0$$

to the present model, we first observe that the integrand in (5.9) is linear in both K' and L'. (The L' term is altogether absent; i.e., it has a zero coefficient.) Thus, in line with our discussion in Sec. 2.1, the Euler equations will not be differential equations, and the problem is degenerate. The partial derivatives of the integrand are

$$F_K = (PQ_K - m\delta)e^{-\rho t} \qquad F_{K'} = -me^{-\rho t}$$
$$F_L = (PQ_L - W)e^{-\rho t} \qquad F_{L'} = 0$$

and the Euler equations indeed emerge as the twin conditions

$$Q_K = \frac{m(\delta + \rho)}{P} \qquad \text{and} \qquad Q_L = \frac{W}{P} \qquad \text{for all } t \geq 0 \tag{5.10}$$

with no derivatives of t in them. In fact, they are nothing but the standard "marginal physical product = real marginal resource cost" conditions[8] that arise in static contexts, but are now required to hold at every t.

The important point about (5.10) is that it implies *constant* K^* and L^* solutions. Taking a generalized Cobb-Douglas production function $Q = K^\alpha L^\beta$, $(\alpha + \beta \neq 1)$, for instance, we have $Q_K = \alpha K^{\alpha-1} L^\beta$ and $Q_L = \beta K^\alpha L^{\beta-1}$. When these are substituted into (5.10), we obtain (after eliminating L)

$$(5.11) \quad K^* = \left[\frac{m(\delta + \rho)}{\alpha P}\right]^{(\beta-1)/(1-\alpha-\beta)} \left(\frac{W}{\beta P}\right)^{-\beta/(1-\alpha-\beta)} = \text{constant}$$

And substitution of this K^* back into (5.10) can give us a similar constant expression for L^*. The constant K^* value represents a first-order necessary condition to be satisfied at all points of time, including $t = 0$. Therefore, unless the firm's initial capital, K_0, happens to be equal to K^*, or the firm is a newly instituted enterprise free to pick its initial capital (variable initial state), the condition cannot be satisfied.

The Flexible Accelerator

If it is not feasible to force a jump in the state variable K from K_0 to K^*, the alternative is a gradual move to K^*. Jorgenson adopts what amounts to the so-called *flexible-accelerator mechanism* to effect a gradual adjustment in net investment. The essence of the flexible accelerator is to eliminate in a systematic way the existing discrepancy between a target capital, $\bar{K}$, and the actual capital at time t, $K(t)$:[9]

$$(5.12) \qquad I(t) = j\left[\bar{K} - K(t)\right] \qquad (0 < j < 1)$$

The gradual nature of the adjustment in the capital level would presumably enable the firm to alleviate the difficulties that might be associated with abrupt and wholesale changes. While the particular construct in (5.12) was originally introduced by economists merely as a convenient postulate, a theoretical justification has been provided for it in a classic paper by Eisner and Strotz.

---

[8] This fact is immediately obvious from the second equation in (5.10). As for the first equation, $m\delta$ means the marginal depreciation cost, and $m\rho$ means the interest-earning opportunity cost of holding capital, and their sum constitutes the marginal *user cost* of capital, which is the exact counterpart of W on the labor side.

[9] The correct application of the flexible accelerator calls for setting the target $\bar{K}$ at the level in (5.11), but the Jorgenson paper actually uses a substitute target that varies with time, which is suboptimal. For a discussion of this, see John P. Gould, "The Use of Endogenous Variables in Dynamic Models of Investment," *Quarterly Journal of Economics*, November 1969, pp. 580–599.

The Eisner-Strotz Model[10]

The Eisner-Strotz model focuses on net investment as a process that expands a firm's plant size. Thus replacement investment is ignored. Assuming that the firm has knowledge of the profit rate π associated with each plant size, as measured by the capital stock K, we have a profit function $\pi(K)$—which, incidentally, provides an interesting contrast to the profit function $\pi(L)$ in the labor-adjustment model of Sec. 3.4. To expand the plant, an adjustment cost C is incurred whose magnitude varies positively with the speed of expansion $K'(t)$. So we have an increasing function $C = C(K')$, through which the firm's internal difficulties of plant adjustment as well as the external obstacles to investment (such as pressure on the capital-goods industry supply) can be explicitly taken into account in the optimization problem of the firm. If the C function adequately reflects these adjustment difficulties, then once the $K^*(t)$ path is found, we may just take its derivative $K^{*\prime}(t)$ as the optimal net investment path without having to use some ad hoc expedient like the flexible-accelerator mechanism.

The objective of the firm is to choose a $K^*(t)$ path that maximizes the total present value of its net profit over time:

$$\text{(5.13)} \qquad \begin{aligned} &\text{Maximize} \qquad \Pi[K] = \int_0^\infty [\pi(K) - C(K')]e^{-\rho t}\,dt \\ &\text{subject to} \qquad K(0) = K_0 \qquad (K_0 \text{ given}) \end{aligned}$$

This functional is again an improper integral, but inasmuch as the net return can be expected to be bounded from above, convergence should not be a problem by Condition III of the preceding section. Note that although we have specified a fixed initial capital stock, the terminal capital stock is left open. Note also that this problem is autonomous.

The Quadratic Case

It is possible to apply the Euler equation to the general-function formulation in (5.13). For more specific results, however, let us assume that both the π and C function are quadratic, as graphed in Fig. 5.1:

$$\pi = \alpha K - \beta K^2 \qquad (\alpha, \beta > 0) \tag{5.14}$$

$$C = aK'^2 + bK' \qquad (a, b > 0) \tag{5.15}$$

[10] Robert Eisner and Robert H. Strotz, "Determinants of Business Investment," in *Impacts of Monetary Policy*, A Series of Research Studies Prepared for the Commission on Money and Credit, Prentice-Hall, Englewood Cliffs, NJ, 1963, pp. 60–233. The theoretical model discussed here is contained in Sec. 1 of their paper; see also their App. A.

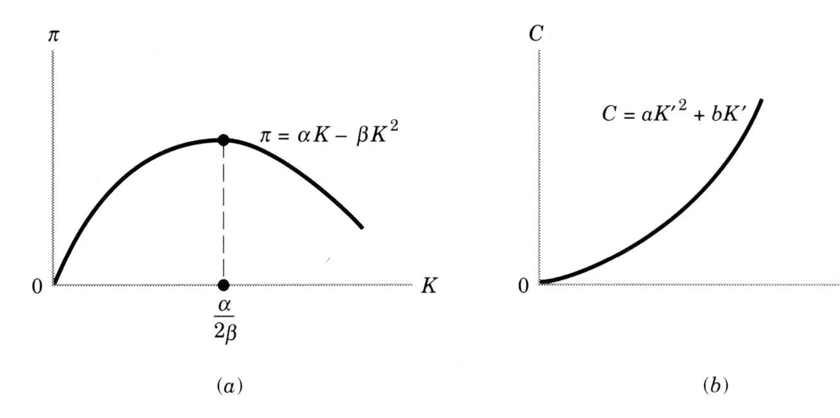

FIGURE 5.1

The profit rate has a maximum, attained at $K = \alpha/2\beta$. The C curve has been drawn only in the first quadrant, where $K' > 0$, because we are confining our attention to the case of plant expansion. Considered together, these two curves should provide an upper bound for the net return $\pi - C$, to guarantee convergence of the improper integral in (5.13).

In this quadratic model, we have

$$F = \left(\alpha K - \beta K^2 - aK'^2 - bK'\right)e^{-\rho t}$$

with derivatives:

$$\begin{aligned} &F_K = (\alpha - 2\beta K)e^{-\rho t} \qquad F_{K'} = -(2aK' + b)e^{-\rho t} \\ (5.16) \qquad &F_{K'K'} = -2ae^{-\rho t} \qquad F_{KK'} = 0 \qquad F_{KK} = -2\beta e^{-\rho t} \\ &F_{tK'} = \rho(2aK' + b)e^{-\rho t} \end{aligned}$$

Thus, by (2.19), we have the Euler equation

$$K'' - \rho K' - \frac{\beta}{a}K = \frac{b\rho - \alpha}{2a} \tag{5.17}$$

with general solution[11]

$$K^*(t) = A_1 e^{r_1 t} + A_2 e^{r_2 t} + \overline{K} \tag{5.18}$$

[11]The Euler equation given by Eisner and Strotz contains a minor error—the coefficient of the K term in (5.17) is shown as $-\beta/2a$ instead of $-\beta/a$. As a result, their characteristic roots and particular integral do not coincide with (5.18). This fact, however, does not affect their qualitative conclusions.

where $$r_1, r_2 = \frac{1}{2}\left(\rho \pm \sqrt{\rho^2 + \frac{4\beta}{a}}\right) \qquad \text{[characteristic roots]}$$

and $$\overline{K} = \frac{\alpha - b\rho}{2\beta} \qquad \text{[particular integral]}$$

The characteristic roots r_1 and r_2 are both real, because the expression under the square-root sign is positive. Moreover, since the square root itself exceeds ρ, it follows that $r_1 > \rho > 0$, but $r_2 < 0$. As to the particular integral $\overline{K}$, it is uncertain in sign. Economic considerations would dictate, however, that it be positive, because it represents the intertemporal equilibrium level of K. For the problem to be sensible, therefore, we must stipulate that

$$\alpha > b\rho \tag{5.19}$$

The Definite Solution

To get the definite solution, we first make use of the initial condition $K(0) = K_0$, which enables us—after setting $t = 0$ in (5.18)—to write

$$K_0 = A_1 + A_2 + \overline{K} \tag{5.20}$$

We would like to see whether we can apply the terminal condition (5.6). Even though no explicit terminal state is given in the model, the autonomous nature of the problem suggests that it has an implicit terminal state—the ultimate plant size (capital stock) to which the firm is to expand. As Fig. 5.1a indicates, the highest attainable profit rate occurs at plant size $K = \alpha/2\beta$. Accordingly, we would not expect the firm to want to expand beyond that size. After the cost function C is taken into account, moreover, the firm may very well wish to select an ultimate size even smaller. In fact, from our knowledge of differential equations, the particular integral $\overline{K}$ would be the logical candidate for the ultimate plant size K_∞. Referring to (5.18), however, we see that while the second exponential term (where $r_2 < 0$) tends to zero as $t \to \infty$, the first exponential term (where $r_1 > 0$) tends to $\pm\infty$ if $A_1 \gtrless 0$. Since neither $+\infty$ nor $-\infty$ is acceptable as the terminal value for K on economic grounds, the only way out is to set $A_1 = 0$. Coupling this information with (5.20), we then have $A_2 = K_0 - \overline{K}$, so that the optimal K path is

$$K^*(t) = \left(K_0 - \overline{K}\right)e^{r_2 t} + \overline{K} \tag{5.21}$$

A few things may be noted about this result. First, since $r_2 < 0$, $K^*(t)$ converges to the particular integral $\overline{K}$; thus $\overline{K}$ indeed emerges as the optimal terminal plant size K_∞. Moreover, given that b and ρ are positive, $\overline{K} = (\alpha - b\rho)/2\beta$ is less than $\alpha/2\beta$. Thus, after explicitly taking into account the adjustment cost, the firm indeed ends up with a plant size

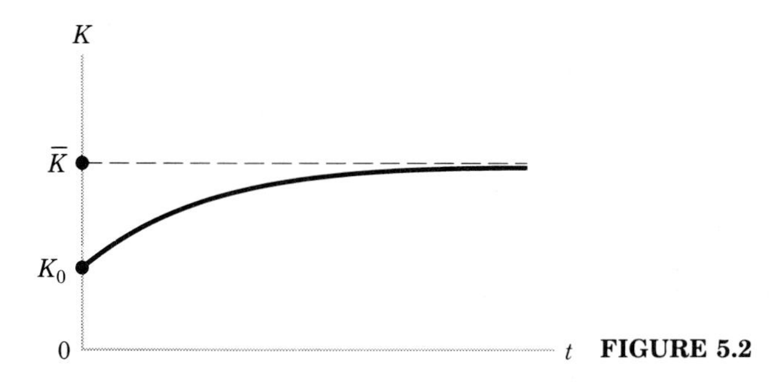

FIGURE 5.2

smaller than the π-maximizing one shown in Fig. 5.1a. Finally, the discrepancy between the initial and terminal sizes, $K_0 - \overline{K}$, is subject to exponential decay. The $K^*(t)$ path should therefore have the general shape illustrated in Fig. 5.2.

The Transversality Conditions

Although we have been able to definitize the arbitrary constants without using any transversality conditions, it would be interesting to see where the transversality condition (5.5) would lead us. Using the information in (5.16), we know that

$$F - K'F_{K'} = \left(\alpha K - \beta K^2 + aK'^2\right)e^{-\rho t} \tag{5.22}$$

For the K and K' terms, we should use the K^* in (5.18) and its derivative, namely,

$$K^* = A_1 e^{r_1 t} + A_2 e^{r_2 t} + \overline{K}$$

$$K^{*\prime} = A_1 r_1 e^{r_1 t} + A_2 r_2 e^{r_2 t}$$

Plugging these into (5.22), we get the complicated result

$$\begin{aligned}(5.22')\qquad F^* - K^{*\prime}F_{K^{*\prime}} &= A_1(\alpha - 2\beta\overline{K})e^{(r_1-\rho)t} + A_1{}^2\left(ar_1{}^2 - \beta\right)e^{(2r_1-\rho)t}\\ &\quad + A_1A_2(2ar_1r_2 - 2\beta)e^{(r_1+r_2-\rho)t} + A_2(\alpha - 2\beta\overline{K})e^{(r_2-\rho)t}\\ &\quad + A_2{}^2\left(ar_2{}^2 - \beta\right)e^{(2r_2-\rho)t} + \left(\alpha\overline{K} - \beta\overline{K}^2\right)e^{-\rho t}\end{aligned}$$

The transversality condition is that (5.22′) vanish as $t \to \infty$. Since r_2 is negative, the last three terms cause no difficulty. But the exponential in the third term reduces to $e^0 = 1$, because $r_1 + r_2 = \rho$ [see (5.18)], and that term will not tend to zero. Worse, the first two terms are explosive since $r_1 > \rho$. In order to force the first three terms to go to zero, the only way is to set

$A_1 = 0$. Thus, the transversality condition leads us to precisely what we have concluded via economic reasoning.

As to transversality condition (5.7), it is not really needed in this problem. But if we try to apply it, it will also tell us to set $A_1 = 0$ in order to tame an explosive exponential expression.

The Optimal Investment Path and the Flexible Accelerator

If the adjustment-cost function $C = C(K')$ fully takes into account the various internal and external difficulties of adjustment, then the optimal investment path is simply the derivative of the definite solution (5.21):

$$I^*(t) = K^{*\prime}(t) = r_2(K_0 - \overline{K})e^{r_2 t} \tag{5.23}$$

But since (5.21) implies that

$$(K_0 - \overline{K})e^{r_2 t} = K^*(t) - \overline{K}$$

the investment path (5.23) can be simplified to

$$I^*(t) = r_2[K^*(t) - \overline{K}] = -r_2[\overline{K} - K^*(t)] \tag{5.23'}$$

The remarkable thing about (5.23′) is that it exactly depicts the flexible-accelerator mechanism in (5.12). The discrepancy between $\overline{K}$, the target plant size, and $K^*(t)$, the actual plant size on the extremal at any point of time, is systematically eliminated through the positive coefficient $-r_2$. The only unsettled matter is that, in the flexible accelerator, $-r_2$ must be less than one. This requirement is met if we impose the following additional restriction on the model:

$$\frac{\beta}{a} < 1 + \rho \tag{5.24}$$

What makes the flexible accelerator in (5.23′) fundamentally different from (5.12) is that the former is no longer a postulated behavior pattern, but an optimization rule that emerges from the solution procedure. The Eisner-Strotz model, in its quadratic version, has thus contributed a much needed theoretical justification for the popularly used flexible-accelerator mechanism.

EXERCISE 5.2

1 Let the profit function and the adjustment-cost function (5.14) and (5.15) take the specific forms

$$\pi = 50K - K^2 \qquad C = K'^2 + 2K'$$

but leave the discount rate ρ in the parametric form. Find the optimal path of K (definite solution) by directly applying the Euler equation (2.19).

2 Use the solution formula (5.21) to check your answer to the preceding problem.

3 Find the Euler equation for the general formulation of the Eisner-Strotz model as given in (5.13).

4 Let the profit function be $\pi = 50K - K^2$ as in Prob. 1, but change the adjustment-cost function to $C = K'^2$. Use the general Euler equation you derived in the preceding problem for (5.13) to get the optimal K path.

5 Demonstrate that the parameter restriction (5.24) is necessary to make $-r_2$ a positive fraction.

5.3 THE OPTIMAL SOCIAL SAVING BEHAVIOR

Among the very first applications of the calculus of variations to economics is the classic paper by Frank Ramsey on the optimal social saving behavior.[12] This paper has exerted an enormous if delayed influence on the current literature on optimal economic growth, and is well worth a careful review.

The Ramsey Model

The central question addressed by Ramsey is that of intertemporal resource allocation: How much of the national output at any point of time should be for current consumption to yield current utility, and how much should be saved (and invested) so as to enhance future production and consumption, and hence yield future utility?

The output is assumed to be produced with two inputs, capital K and labor L. The production function, $Q = Q(K, L)$, is time invariant, since no technological progress is assumed. Other simplifying assumptions include the absence of depreciation for capital and a stationary population. However, the amount of labor services rendered can still vary. The output can either be consumed or saved, but what is saved always results in investment and capital accumulation. Then, using the standard symbols, we have $Q = C + S = C + K'$, or

$$C = Q(K, L) - K' \tag{5.25}$$

Consumption contributes to social welfare via a social utility (index) function $U(C)$, with nonincreasing marginal utility, $U''(C) \le 0$. This speci-

---

[12] Frank P. Ramsey, "A Mathematical Theory of Saving," *Economic Journal*, December 1928, pp. 543–559.

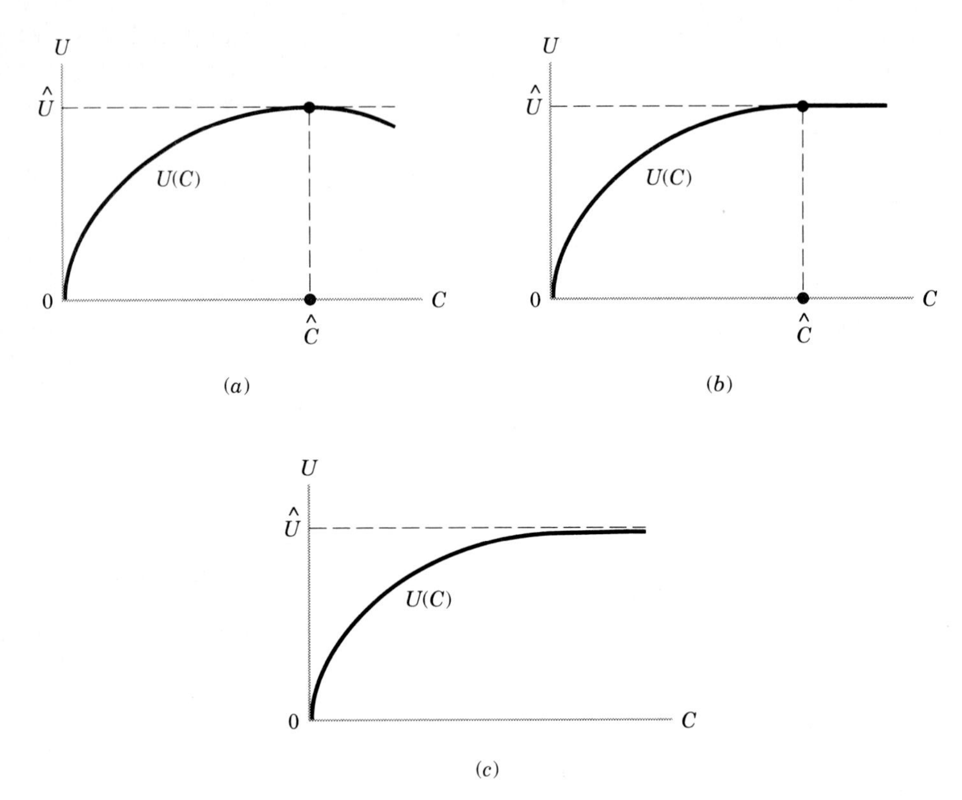

FIGURE 5.3

fication is consistent with all three utility functions illustrated in Fig. 5.3. In every case, there is an upper bound for utility, $\hat{U}$. That level of utility is attainable at a finite consumption level $\hat{C}$ in Figs. 5.3*a* and *b*, but can only be approached asymptotically in Fig. 5.3*c*.

In order to produce its consumption goods, society incurs disutility of labor $D(L)$, with nondecreasing marginal disutility, $D''(L) \geq 0$. The net social utility is therefore $U(C) - D(L)$, where C and L—as K and Q—are functions of time. The economic planner's problem is to maximize the social utility for the current generation as well as for all generations to come:

$$\text{Maximize} \int_0^\infty [U(C) - D(L)]\,dt \tag{5.26}$$

In the integrand, U depends solely on C, and, by (5.25), C depends on K, L, and K', whereas D depends only on L. Thus, this problem has two state variables, K and L. Because there is no L' term in the integrand, however, the problem is degenerate on the L side, and in general it would be inappropriate to stipulate an initial condition on L.

The Question of Convergence

Unlike the functional of the Eisner-Strotz model, the improper integral in (5.26) does not contain a discount factor. This omission is not the result of neglect; it stems from Ramsey's view that it is "ethically indefensible" for the (current-generation) planner to discount the utility of future generations. While this may be commendable on moral grounds, the absence of a discount factor unfortunately forfeits the opportunity to take advantage of Condition III of Sec. 5.1 to establish convergence, even if the integrand has an upper bound. In fact, since the net utility is expected to be positive as t becomes infinite, the integral is likely to diverge.

To handle this difficulty, Ramsey replaces (5.26) with the following substitute problem:

$$(5.26') \qquad \begin{aligned} &\text{Minimize} && \int_0^\infty [B - U(C) + D(L)]\, dt \\ &\text{subject to} && K(0) = K_0 \qquad (K_0 \text{ given}) \end{aligned}$$

where B (for Bliss) is a postulated maximum attainable level of net utility. Since the new functional measures the amount by which the net utility $U(C) - D(L)$ falls short of Bliss, it is to be *minimized* rather than maximized. Intuitively, an optimal allocation plan should either take society to Bliss, or lead it to approach Bliss asymptotically. If so, the integrand in (5.26′) will fall steadily to the zero level, or approach zero as $t \to \infty$. As far as Ramsey is concerned, the convergence problem is thereby resolved. The substitution of (5.26) by (5.26′), referred to as "the Ramsey device," is widely accepted as sufficient for convergence. Our earlier discussion of Condition I in Sec. 5.1 would confirm that if the integrand attains zero at some finite time and remains at zero thenceforth, then convergence indeed is ensured. But, as underscored in Condition II, the mere fact that the integrand tends to zero as t becomes infinite does not in and of itself guarantee convergence. The integrand must also fall sufficiently fast over time.

Even though the convergence problem is not clearly resolved by Ramsey, we shall now proceed on the assumption that the general-function integrand in (5.26′) does converge.

The Solution of the Model

The problem as stated in (5.26′) is an autonomous problem in the state variables K and L. From the integrand

$$F = B - U(C) + D(L) \qquad \text{where } C = Q(K, L) - K'$$

we can get, on the L side, the derivatives

$$F_L = -U'(C)\frac{\partial C}{\partial L} + D'(L) \equiv -\mu Q_L + D'(L) \qquad [\mu \equiv U'(C)]$$

$$F_{L'} = 0$$

For notational simplicity, we are using μ to denote marginal utility. (The abbreviation of marginal utility is "mu.") Like $U(C)$, μ is a function of C and hence indirectly a function of K, L, and K'. On the K side, we can obtain the derivatives

$$F_K = -U'(C)\frac{\partial C}{\partial K} = -\mu Q_K$$

$$F_{K'} = -U'(C)\frac{\partial C}{\partial K'} = -U'(C)(-1) = U'(C) = \mu$$

These derivatives will enable us to apply the Euler equations (2.27).

We note first that since $F_{L'} = 0$, the Euler equation for the L variable, $F_L - dF_{L'}/dt = 0$, reduces simply to $F_L = 0$, or

$$D'(L) = \mu Q_L \qquad \text{for all } t \geq 0 \tag{5.27}$$

The marginal disutility of labor must, at each point of time, be equated to the product of the marginal utility of consumption and the marginal product of labor. On the K side, the Euler equation $F_K - dF_{K'}/dt = 0$ gives us the condition $-\mu Q_K - d\mu/dt = 0$, or

$$\frac{d\mu/dt}{\mu} = -Q_K \qquad \text{for all } t \geq 0 \tag{5.28}$$

This result prescribes a rule about consumption: μ, the marginal utility of consumption, must at every point of time have a growth rate equal to the negative of the marginal product of capital. Following this rule, we can map out the optimal path for μ. Once the optimal μ path is found, (5.27) can be used to determine an optimal L path. As mentioned earlier, however, the problem is degenerate in L, so it is not appropriate to preset an initial level $L(0)$.

The Optimal Investment and Capital Paths

Of greater interest to us is the optimal K path and the related investment (and savings) path K'. Instead of deducing these from the previous results, let us find this information by taking advantage of the fact that the present problem falls under Special Case II of Sec. 2.2. Without t as an explicit argument in the integrand function, the Euler equation for K is, according to (2.21), $F - K'F_{K'}$ = constant, or

$$B - U(C) + D(L) - K'\mu = \text{constant} \qquad \text{for all } t \geq 0 \tag{5.29}$$

This equation can be solved for K' as soon as the (arbitrary) constant on the right-hand side can be assigned a specific value. To find this value, we note that this constant is to hold for all t, including $t \to \infty$. We can thus make use of the fact that, as $t \to \infty$, the economic objective of the model is to have $U(C) - D(L)$ tend to Bliss. This means that U must tend to $\hat{U}$ and μ must tend to zero as t becomes infinite. It follows that the arbitrary constant in (5.29) has to be zero. If so, the optimal path of K' is

$$K^{*\prime} = \frac{B - U(C) + D(L)}{\mu} \tag{5.30}$$

or, with the time argument explicitly written out,

$$K^{*\prime}(t) = \frac{B - U[C(t)] + D[L(t)]}{\mu(t)} \tag{5.30'}$$

This result is known as the *Ramsey rule*. It stipulates that, optimally, the rate of capital accumulation must at any point of time be equal to the ratio of the shortfall of net utility from Bliss to the marginal utility of consumption. On the surface, this rule is curiously independent of the production function. And this led Ramsey to conclude (page 548) that the production function would matter only insofar as it may affect the determination of Bliss. But this conclusion is incorrect. According to (5.28), μ is to be optimally chosen by reference to Q_K. This means that the denominator of (5.30′) does depend crucially on the production function.

A Look at the Transversality Conditions

We shall now show that the use of the infinite-horizon transversality condition would also dictate that the constant in (5.29) be set equal to zero. When condition (5.5) is applied to the two state variables in the present problem, it requires that

$$\lim_{t \to \infty} (F - L'F_{L'}) = 0 \qquad \text{and} \qquad \lim_{t \to \infty} (F - K'F_{K'}) = 0$$

In view of the fact that $F_{L'} = 0$, the first of these conditions reduces to the condition that $F \to 0$ as $t \to \infty$. This would mean the net utility $U(C) - D(L)$ must tend to Bliss. Note that, by itself, this condition still leaves the constant in (5.29) uncertain. However, the other condition will fix the constant at zero because $F - K'F_{K'}$ is nothing but the left-hand-side expression in (5.29).

The present problem implicitly specifies the terminal state at Bliss. Consequently, the transversality condition (5.7) is not needed.

From the Ramsey rule, it is possible to go one step further to find the $K^*(t)$ path by integrating (5.30′). For that, however, we need specific forms of $U(C)$ and $D(L)$ functions. The general solution of (5.30′) will contain one arbitrary constant, which can be definitized by the initial condition $K(0) = K_0$. And that would complete the solution of the model. We shall not attempt to illustrate the model with specific functions; rather, we shall show the application of phase-diagram analysis to it in the next section.

EXERCISE 5.3

1 Let the production function and the utility function in the Ramsey model take the specific forms:

$$Q(K) = rK \qquad (r > 0)$$

$$U = \hat{U} - \frac{1}{b}C^{-b} \qquad (b > 0)$$

with initial capital K_0.

(*a*) Apply the condition (5.28) to derive a $C^*(t)$ path. Discuss the characteristics of this path.

(*b*) Use the Ramsey rule (5.30′) to find $K^{*\prime}(t)$. Discuss the characteristics of this path, and compare it with $C^*(t)$. How specifically are the two paths related to each other? [*Hint:* Since the $D(L)$ term is absent in the problem, we have $B = \hat{U}$.]

(*c*) Bearing in mind that $Q_0 = rK_0$ (production of Q_0), and $Q_0 = C^*_0 + K^{*\prime}(0)$ (allocation of Q_0), and using the relationship between $K^{*\prime}(0)$ and C^*_0 implied by the result in (*b*), find an expression for C^*_0 in terms of K_0. Write out the definite solution for $K^{*\prime}(t)$.

(*d*) Integrate $K^{*\prime}(t)$ to obtain the $K^*(t)$ path. What is the rate of growth of K^*? How does that rate vary over time?

2 Find the $K^*(t)$ path for the preceding problem by the alternative procedure outlined below:

(*a*) Express the integrand of the functional $[B - U(C)]$ in terms of K and K', that is, as $F(K, K')$.

(*b*) Apply the Euler equation (2.18) or (2.19) to get a second-order differential equation. Find the general solution $K^*(t)$, with arbitrary constants A_1 and A_2. Then derive the optimal savings path $K^{*\prime}(t)$; retain the arbitrary constants.

(*c*) Find the savings ratio $K^{*\prime}(t)/Q^*(t) = K^{*\prime}(t)/rK^*(t)$. Show that this ratio approaches one as $t \to \infty$, unless one of the arbitrary constants is zero.

(*d*) Is a unitary limit for the savings ratio economically acceptable? Use your answer to this question to definitize one of the constants. Then definitize the other constant by the initial condition $K(0) = K_0$. Compare the resulting definite solution with the one obtained in the preceding problem.

5.4 PHASE-DIAGRAM ANALYSIS

In models of dynamic optimization, two-variable phase diagrams are prevalently employed to obtain qualitative analytical results.[13] This is particularly true when general functions are used in the model, and the time horizon is infinite. For a calculus-of-variations problem with a single state variable, the Euler equation comes as a single second-order differential equation. But it is usually a simple matter to convert that equation into a system of two simultaneous first-order equations in two variables. The two-variable phase diagram can then be applied to the problem in a straightforward manner. We shall illustrate the technique in this section with the Eisner-Strotz model and the Ramsey model. We shall then adapt the phase diagram to a finite-horizon context.

The Eisner-Strotz Model Once Again

In their model of a firm's optimal investment in its plant, discussed in Sec. 5.2, Eisner and Strotz are able to offer a quantitative solution to the problem when specific quadratic profit and cost-of-adjustment functions are assumed. Writing these functions as

$$\pi = \alpha K - \beta K^2 \qquad (\alpha, \beta > 0) \qquad [\text{from (5.14)}]$$

$$C = aK'^2 + bK' \qquad (a, b > 0) \qquad [\text{from (5.15)}]$$

they derive the Euler equation

$$K'' - \rho K' - \frac{\beta}{a} K = \frac{b\rho - \alpha}{2a} \qquad [\text{from (5.17)}] \tag{5.31}$$

where ρ is a positive rate of discount. From the quantitative solution,

$$K^*(t) = \left(K_0 - \overline{K}\right)e^{r_2 t} + \overline{K} \qquad (r_2 < 0) \qquad [\text{from (5.21)}] \tag{5.32}$$

$$\text{where} \quad \overline{K} = \frac{\alpha - b\rho}{2\beta}$$

---

[13]The method of two-variable phase diagrams is explained in Alpha C. Chiang, *Fundamental Methods of Mathematical Economics*, 3d ed., McGraw-Hill, New York, 1984, Sec. 18.5.

we see that the capital stock K (the plant size) should optimally approach its specific target value $\overline{K}$ by following a steady time path that exponentially closes the gap between the initial value and the target value of K.

Now we shall show how the Euler equation can be analyzed by means of a phase diagram. We begin by introducing a variable I (net investment)

(5.33)
$$I(t) \equiv K'(t) \qquad [\text{implying } I'(t) \equiv K''(t)]$$

which would enable us to rewrite the Euler equation (5.31) as a system of two first-order differential equations:

(5.34)
$$I' = \rho I + \frac{\beta}{a}K + \frac{b\rho - \alpha}{2a} \qquad [\text{i.e., } I' = f(I, K)]$$
$$K' = I \qquad [\text{i.e., } K' = g(I, K)]$$

This is an especially simple system, since both the f and g functions are linear in the two variables I and K, and in fact K is even absent in the g function.

The Phase Diagram

To construct the phase diagram, our first order of business is to draw two demarcation curves, $I' = 0$ and $K' = 0$. Individually, each of these curves serves to delineate the subset of points in the IK space where the variable in question can be stationary ($dI/dt = 0$ *or* $dK/dt = 0$). Jointly, the two curves determine at their intersection the intertemporal equilibrium—steady state—of the entire system ($dI/dt = 0$ *and* $dK/dt = 0$).

Setting $I' = 0$ in (5.34) and solving for K, we get

$$K = \frac{\alpha - b\rho}{2\beta} - \frac{a\rho}{\beta}I \qquad [\text{equation for } I' = 0 \text{ curve}] \tag{5.35}$$

Similarly, by setting $K' = 0$, we get

$$I = 0 \qquad [\text{equation for } K' = 0 \text{ curve}] \tag{5.36}$$

These two equations both plot as straight lines in the phase space, as shown in Fig. 5.4, with the $I' = 0$ curve sloping downward, and the $K' = 0$ curve coinciding with the vertical axis. The unique intertemporal equilibrium occurs at the vertical intercept of the $I' = 0$ curve, which, by (5.35), is $K = (\alpha - b\rho)/2\beta$, a positive magnitude by (5.19). Note that this equilibrium coincides exactly with the particular integral $\overline{K}$ (target plant size) found in (5.18). This is of course only to be expected.

By definition, all points on the $I' = 0$ curve are characterized by stationarity in the variable I. We have therefore drawn a couple of vertical "sketching bars" on the curve, to indicate that there should not be any east-west movement for points on the curve. Similarly, horizontal "sketch-

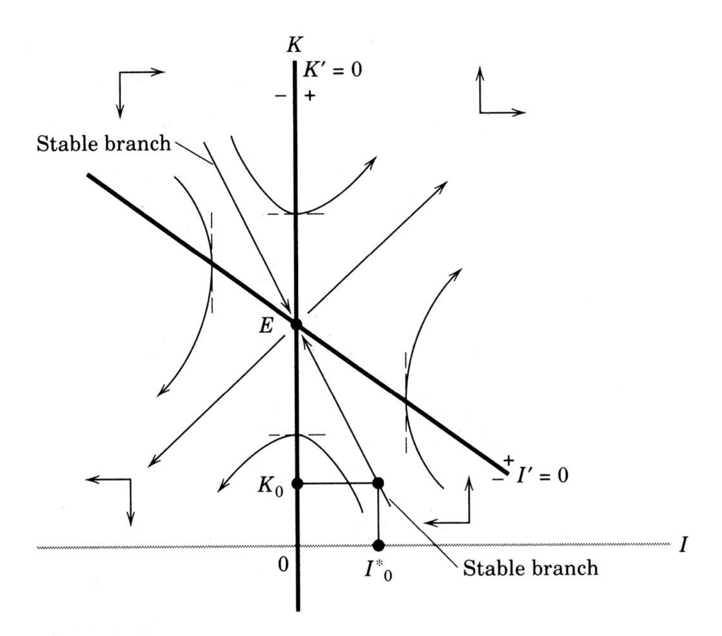

FIGURE 5.4

ing bars" have been attached to the $K' = 0$ curve to indicate that the stationarity of K forbids any north-south movement on the curve.

Points off the demarcation curves are, on the other hand, very much involved in dynamic motion. The direction of movement depends on the signs of the derivatives I' and K' at a particular point in the IK space; the speed of movement depends on the magnitudes of those derivatives. From (5.34), we find by differentiation that

$$\frac{\partial I'}{\partial I} = \rho > 0 \qquad \text{and} \qquad \frac{\partial K'}{\partial I} = 1 > 0 \tag{5.37}$$

The positive sign of $\partial I'/\partial I$ implies that, moving from west to east (with I increasing), I' should be going from a negative region, through zero, to a positive region (with I' increasing). To record this $(-, 0, +)$ sign sequence, a minus sign has been added to the left of the $I' = 0$ label, and a plus sign has been added to the right of the $I' = 0$ label. Accordingly, we have also drawn leftward I-arrowheads $(I' < 0)$ to the left, and rightward I-arrowheads $(I' > 0)$ to the right, of the $I' = 0$ curve. For K', the positive sign of $\partial K'/\partial I$ similarly implies that, going from west to east, K' should be moving from a negative region, through zero, to a positive region. Hence, the sign sequence is again $(-, 0, +)$, and this explains the minus sign to the left, and the plus sign to the right, of the $K' = 0$ label. The K-arrowheads are downward $(K' < 0)$ to the left, and upward $(K' > 0)$ to the right, of the $K' = 0$ curve.

Following the directional restrictions imposed by the IK-arrowheads and sketching bars, we can draw a family of streamlines, or trajectories, to portray the dynamics of the system from any conceivable initial point. Every point should be on some streamline, and there should be an infinite number of streamlines. But usually we only draw a few. The reader should verify in Fig. 5.4 that the streamlines conform to the IK-arrowheads, and, in particular, that they are consistent with the sketching bars which require them to cross the $I' = 0$ curve with infinite slope and cross the $K' = 0$ curve with zero slope.

The Saddle-Point Equilibrium

The intertemporal equilibrium of the system occurs at point E, the intersection of the two demarcation curves. The way the streamlines are structured, we see two *stable branches* leading toward E (one from the southeast direction and the other from the northwest), two *unstable branches* leading away from E (one toward the northeast and the other toward the southwest), and the other streamlines heading toward E at first but then turning away from it. As a result, we have a so-called *saddle-point equilibrium*.

It is clear that the only way to reach the target level of capital at E is to get onto the stable branches. Given the initial capital K_0, for instance, it is mandatory that we select I^*_0 as the initial rate of investment, for only that choice will place us on the "yellow brick road" to the equilibrium. A different initial capital will, of course, call for a different I^*_0. The restriction on the choice of I^*_0 is thus the surrogate for a transversality condition. And this serves to underscore the fact that the transversality condition constitutes an integral part of the optimization rule. While all the streamlines, being based on (5.34), satisfy the Euler equation, only the one that also satisfies the transversality condition or its equivalent—one that lies along a stable branch—qualifies as an optimal path.

Traveling toward the northwest on the stable branch in Fig. 5.4, we find a steady climb in the level of K, implying a steady narrowing of the discrepancy between K_0 and the target capital level. This is precisely what (5.32) tells us. Accompanying the increase in K is a steady decline in the rate of investment I. Again, this is consistent with the earlier quantitative solution, for by differentiating (5.32) twice with respect to t, we do find

$$I^{*\prime}(t) = K^{*\prime\prime}(t) = r_2^2(K_0 - \overline{K})e^{r_2 t} < 0$$

since, in the present example, $K_0 < \overline{K}$.

A Simplified Ramsey Model

To simplify the analysis of the Ramsey model, we shall assume that the labor input is constant, thereby reducing the production function to $Q(K)$.

Then we will have only one state variable and one Euler equation with which to contend. Another consequence of omitting L and $D(L)$ is to make Bliss identical with the upper bound of the utility function, $\hat{U}$.

The Euler equation for the original Ramsey model can be written as

$$-\mu Q_K - \frac{d\mu}{dt} = 0 \qquad \text{[from (5.28)]} \tag{5.38}$$

or, making explicit the dependence of the marginal product on K,

$$-\mu Q'(K) - \mu' = 0 \tag{5.38'}$$

This is a first-order differential equation in the variable μ. However, since the variable K also appears in the equation, we need a differential equation in the variable K to close the dynamic system. This need is satisfied by (5.25), $C = Q(K, L) - K'$, which, when adapted to the present simplified context, becomes

$$K' = Q(K) - C(\mu) \tag{5.39}$$

The replacement of $Q(K, L)$ by $Q(K)$ is due to the elimination of L. The substitution of C by the function $C(\mu)$ reflects the fact that in the usual marginal-utility function, μ is a monotonically decreasing function of C, so that C can be taken as an inverse function of marginal utility μ.

Combining (5.38′) and (5.39), we get a two-equation system

$$\begin{aligned} K' &= Q(K) - C(\mu) \qquad &[\text{i.e., } K' = f(K, \mu)] \\ \mu' &= -\mu Q'(K) \qquad &[\text{i.e., } \mu' = g(K, \mu)] \end{aligned} \tag{5.40}$$

containing two general functions, $Q(K)$ and $C(\mu)$.

Before we proceed, certain qualitative restrictions should be placed on the shapes of these functions. For the time being, let us assume that the $U(C)$ function is shaped as in the top diagram of Fig. 5.5, with B as the Bliss level of utility and C_B the Bliss level of consumption. There exists *consumption saturation* at C_B, and further consumption beyond that level entails a reduction in U. From this function, we can derive the marginal-utility function $\mu(C)$ in the diagram directly beneath, with $\mu = 0$ at $C = C_B$. Since $\mu(C)$ is monotonic, the inverse function $C(\mu)$ exists, with $C'(\mu) < 0$. The diagram in the lower-left corner, on the other hand, sets forth the assumed shape of $Q(K)$, the other general function, here plotted—for a good reason—with Q on the horizontal rather than the vertical axis. The underlying assumption for this curve is that the marginal product of capital $Q'(K)$ is positive throughout—there is no *capital saturation*—but the law of diminishing returns prevails throughout.

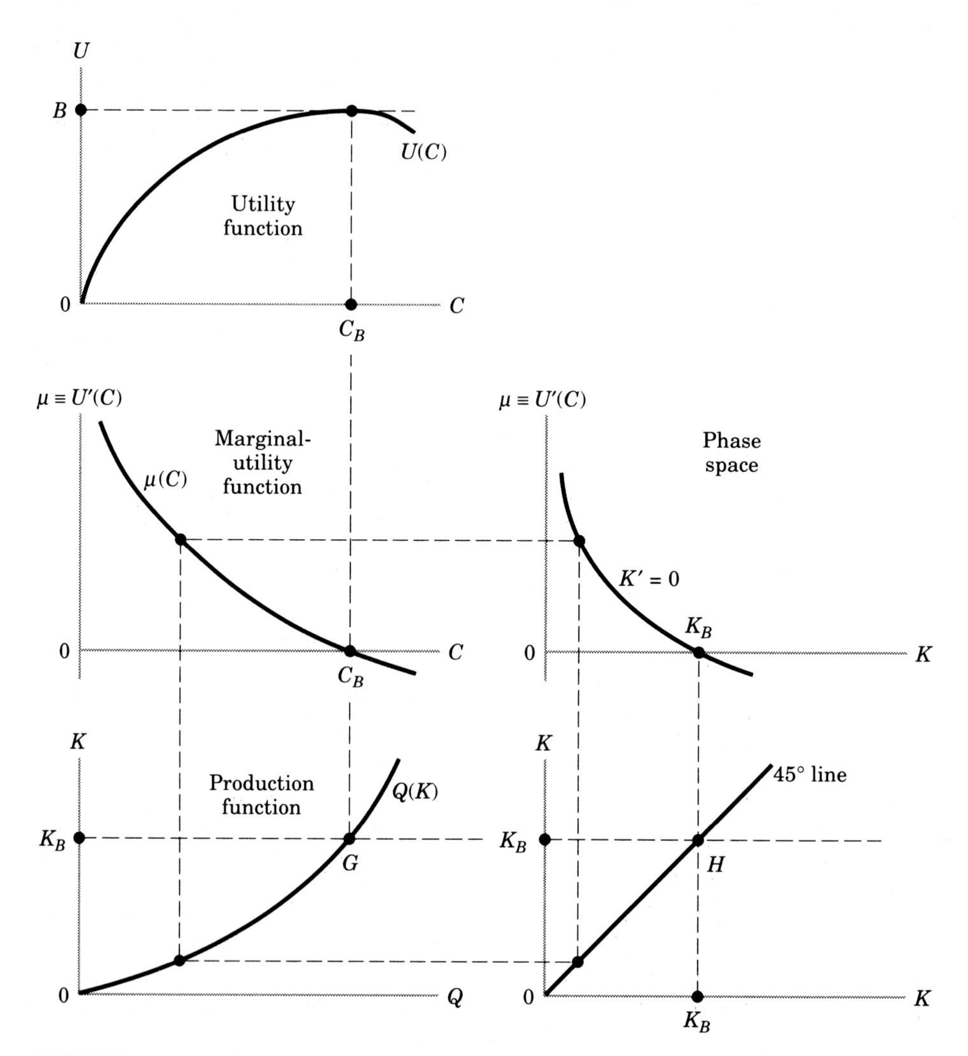

FIGURE 5.5

The Phase Diagram

To construct the phase diagram, we need the $K' = 0$ and $\mu' = 0$ curves. From (5.40), the following two equations readily emerge:

(5.41) $\quad Q(K) = C(\mu) \quad$ [equation for $K' = 0$ curve]

(5.42) $\quad \mu = 0 \quad$ [equation for $\mu' = 0$ curve]

The first of these is self-explanatory. The second results from our assumption of no capital saturation: Since $Q'(K) \neq 0$, we can have $\mu' = 0$ if and only if $\mu = 0$.

To find the proper shape of the $K' = 0$ curve, we equate $Q(K)$ and $C(\mu)$ in accordance with (5.41). It is in order to facilitate this equating process that we have chosen to plot the $Q(K)$ function with the Q variable on the horizontal axis in Fig. 5.5, for then we can equate Q and C graphically by making a vertical movement across the two diagrams. Consider, for instance, the case of $\mu = 0$, that is, the point C_B on the $\mu(C)$ curve. To satisfy (5.41), we simply select point G on the $Q(K)$ curve, the point that lies directly below point C_B. The vertical height of point G may be interpreted as the Bliss level of K; thus we denote it by K_B. The ordered pair $(K = K_B,\ \mu = 0)$ then clearly qualifies as a point on the $K' = 0$ curve in the phase space. Other ordered pairs that qualify as points on the $K' = 0$ curve can be derived similarly.

The easiest way to trace out the entire $K' = 0$ curve is by means of the rectangularly aligned set of our diagrams in the lower part of Fig. 5.5. Aside from the $\mu(C)$ and $Q(K)$ diagrams, we include in this diagram set a 45° line diagram (with K on both axes) and a phase-space diagram which is to serve as the repository of the $K' = 0$ curve being derived. These four diagrams are aligned in a way such that each pair of adjacent diagrams in the same "row" has a common vertical axis, and each pair of adjacent diagrams in the same "column" has a common horizontal axis—except that, in the left column, the C axis and the Q axis are not inherently the same, but rather to be made the same via a deliberate equating process. Now the point $(K = K_B,\ \mu = 0)$ can be traced out via the dashed rectangle, C_BGHK_B, spanned by the $\mu(C)$ curve, the $Q(K)$ curve, and the 45° line. Indeed, starting from any selected point on the $\mu(C)$ curve, if we go straight down to meet the $Q(K)$ curve [thereby satisfying the condition (5.41)], turn right to meet the 45° line, and then go straight up into the phase space (thereby translating the vertical height of point G, or K_B, into an equal horizontal distance in the phase space) to complete the formation of a rectangle, then the fourth corner of this rectangle must be a point on the $K' = 0$ curve. Such a process of rectangle construction yields a $K' = 0$ curve that slopes downward and cuts the horizontal axis of the phase space at $K = K_B$.

The Saddle-Point Equilibrium

We now reproduce the $K' = 0$ curve in Fig. 5.6, and add thereto the $\mu' = 0$ curve, which, according to (5.42), should coincide with the K axis. At the intersection of these two curves is the unique equilibrium point E, which is characterized by $K = K_B$ and $\mu = 0$, and which therefore corresponds to Bliss.

The sketching bars for the $K' = 0$ curve are vertical. Those for the $\mu' = 0$ curve are, however, horizontal, and they lie entirely along that curve. Since the sketching bars coincide with the $\mu' = 0$ curve, the latter will serve not only as a demarcation curve, but also as the locus of some streamlines.

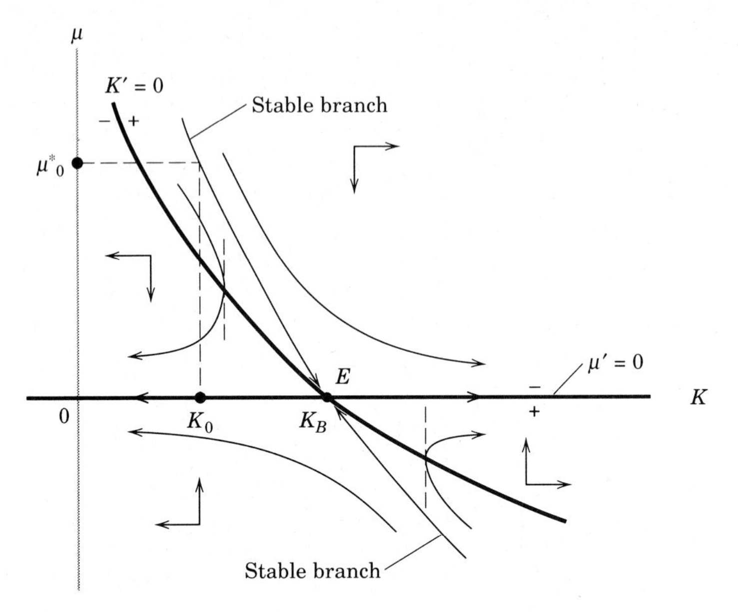

FIGURE 5.6

The sign sequences for K' and μ' can be ascertained from the partial derivatives

$$(5.43) \qquad \frac{\partial K'}{\partial K} = Q'(K) > 0 \quad \text{and} \quad \frac{\partial \mu'}{\partial \mu} = -Q'(K) < 0 \qquad [\text{from } (5.40)]$$

These yield the $(-, 0, +)$ sequence rightward for K', and the $(+, 0, -)$ sequence upward for μ'. Consequently, the K-arrowheads point eastward in the area to the right of the $K' = 0$ curve, and they point westward in the area to its left, whereas the μ-arrowheads point southward in the area above the $\mu' = 0$ curve, and they point northward in the area below it.

The resulting streamlines again produce a saddle point. Given any initial capital K_0, it is thus necessary for us to select an initial marginal utility that will place us on one of the stable branches of the saddle point. A specific illustration, with $K_0 < K_B$, is shown in Fig. 5.6, where there is a unique μ_0 value, μ^*_0, that can land us on the "yellow brick road" leading to E. All the other streamlines will result ultimately in either (1) excessive capital accumulation and failure to attain Bliss, or (2) continual decumulation ("eating up") of capital and, again, failure to reach Bliss. As in the Eisner-Strotz model, we may interpret the strict requirement on the proper selection of an initial μ value as tantamount to the imposition of a transversality condition.

The fact that both the Eisner-Strotz model and the Ramsey model present us with the saddle-point type of equilibrium is not a mere coincidence. The requirement to place ourselves onto a stable branch merely implies that there exists an optimization rule to be followed, much as a profit-maximizing firm must, in the simple static optimization framework, select its output level according to the rule MC = MR. If the equilibrium were a stable node or a stable focus—the "all roads lead to Rome" situation—there would be no specific rule imposed, which is hardly characteristic of an optimization problem. On the other hand, if the equilibrium were an unstable node or focus, there would be no way to arrive at any target level of the state variable at all. This, again, would hardly be a likely case in a meaningful optimization context. In contrast, the saddle-point equilibrium, with a target that is attainable, but attainable only under a specific rule, fits comfortably into the general framework of an optimization problem.

Capital Saturation Versus Consumption Saturation

The preceding discussion is based on the assumption of consumption saturation at C_B, with $U(C_B) = B$ and $\mu(C_B) = 0$. Even though capital saturation does not exist and $Q'(K)$ remains positive throughout, the economy would nevertheless refrain from accumulating capital beyond the level K_B. As a variant of that model, we may also analyze the case in which capital, instead of consumption, is saturated.

Specifically, let $Q(K)$ be a strictly concave curve, with $Q(0) = 0$, and slope $Q'(K) \gtrless 0$ as $K \lessgtr \hat{K}$. Then we have capital saturation at $\hat{K}$. At the same time, let $U(C)$ be an increasing function throughout. Under these new assumptions, it becomes necessary to redraw all the curves in Fig. 5.5 except the 45° line. Consequently, the $K' = 0$ curve in the phase space may be expected to acquire an altogether different shape. However, it is actually also necessary to replace the horizontal $\mu' = 0$ curve, because with consumption saturation assumed away and with $\mu > 0$ throughout, the horizontal axis in the diagram is now strictly off limits. In short, there will be an entirely new phase diagram to analyze.

Instead of presenting the detailed analysis here, we shall leave to the reader the task of deriving the new $K' = 0$ and $\mu' = 0$ curves and pursuing the subsequent steps of determining the $K\mu$-arrowheads and the streamlines. However, some of the major features of the new phase diagram may be mentioned here without spoiling the reader's fun: (1) In the new phase diagram, the $K' = 0$ curve will appear as a U-shaped curve, and the $\mu' = 0$ curve will appear as a vertical line. (2) The new equilibrium point will occur at $K = \hat{K}$, and the marginal utility corresponding to $\hat{K}$ (call it $\hat{\mu}$) will now be positive rather than zero. (3) The equilibrium value of the marginal utility, $\hat{\mu}$, implies a corresponding equilibrium consumption $\hat{C}$ which, in turn, is associated with a corresponding utility level $\hat{U}$. Even though the latter is not the highest conceivable level of utility, it may nevertheless be regarded as "Bliss" in the present context, for, in view of capital saturation, we will never wish to venture beyond $\hat{U}$. The magnitude of this adapted

Bliss is not inherently fixed, but is dependent on the location of the capital-saturation point on the production function. (4) Finally, the equilibrium is again a saddle point.

Yet another variant of the model arises when the $U(C)$ curve is replaced by a concave curve asymptotic to B rather than with a peak at B, as in Fig. 5.3*c*. In this new case, the μ curve will become a convex curve asymptotic to the C axis. What will happen to the $K' = 0$ curve now depends on whether the $Q(K)$ curve is characterized by capital saturation. If it is, the present variant will simply reduce to the preceding variant. If not, and if the $Q(K)$ curve is positively sloped throughout, the $K' = 0$ curve will turn out to be a downward-sloping convex curve asymptotic to the K axis in the phase space. The other curve, $\mu' = 0$, is now impossible to draw because, according to (5.40), μ' cannot attain a zero value unless either $\mu = 0$ or $Q'(K) = 0$, but neither of these two alternatives is available to us. Consequently, we cannot define an intertemporal equilibrium in this case.

Finite Planning Horizon and Turnpike Behavior

The previous variants of the Ramsey model all envisage an infinite planning horizon. But what if the planning period is finite, say, 100 years or 250 years? Can the phase-diagram technique still be used? The answer is yes. In particular, phase-diagram analysis is very useful in explaining the so-called "turnpike" behavior of optimal time paths when *long*, but *finite*, planning horizons are considered.[14]

In the finite-horizon context, we need to have not only a given initial capital K_0 as in the infinite-horizon case, but also a prespecified terminal capital, K_T, to make the solution definite. The magnitude of K_T represents the amount of capital we choose to bequeath to the generations that outlive our planning period. As such, the choice of K_T is purely arbitrary. If we avoid making such a choice and take the problem to be one with a truncated vertical terminal line subject to $K_T \geq 0$, however, the optimal solution path for K will inevitably dictate that $K_T = 0$, because the maximization of utility over our finite planning period, but not beyond, requires that we use up all the capital by time T. More reasonably, we should pick an arbitrary but positive K_T. In Fig. 5.7, we assume that $K_T > K_0$.

The specification of K_0 and K_T is not sufficient to pinpoint a unique optimal solution. In Fig. 5.7*a*, the streamlines labeled $S_1, \ldots, S_4$ (as well as others not drawn), all start with K_0, and are all capable of taking us

[14]See Paul A. Samuelson, "A Catenary Turnpike Theorem Involving Consumption and the Golden Rule," *American Economic Review*, June 1965, pp. 486–496.

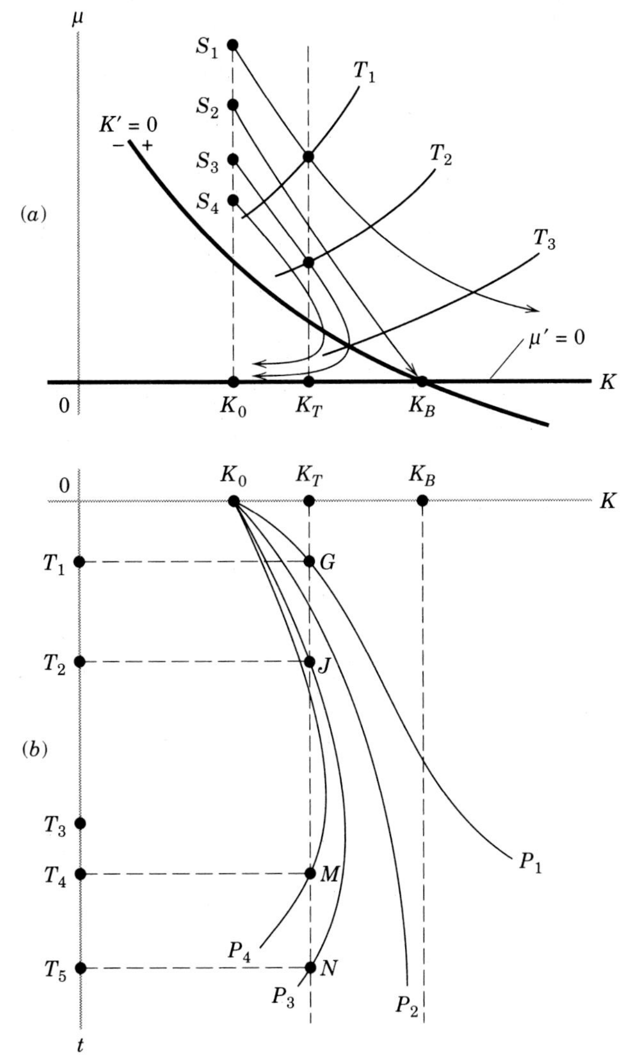

FIGURE 5.7

eventually to the capital level K_T. The missing consideration that can pinpoint the solution is the length of the planning period, T.

Let the T_1 curve be the locus of points that represent positions attained on the various streamlines after exactly T_1 years have elapsed from $t = 0$. Such a curve, referred to as an *isochrone* (equal-time curve), displays a positive slope here for the following reason. Streamline S_1, being the farthest away from the $K' = 0$ curve and hence with the largest K' value among the four, should result in the largest capital accumulation at the end of T_1 years (here, K_T), whereas streamline S_4, at the opposite extreme, should end up at time T_1 with the least amount of capital (here, not much above K_0). If the planning period happens to be T_1 years, streamline S_1 will clearly constitute the best choice, because while all streamlines satisfy the Euler equation, only S_1 satisfies additionally the boundary conditions $K(0) = K_0$ and $K(T_1) = K_T$, with the indicated journey completed in exactly T_1 years.

Streamlines are phase paths, not time paths. Nevertheless, each streamline does imply a specific time path for each of the variables K and μ. The time paths for K corresponding to the four streamlines are illustrated in Fig. 5.7*b*, where, in order to align the K axis with that of Fig. 5.7*a*, we plot the time axis vertically down rather than horizontally to the right. In deriving these time paths, we have also made use of other isochrones, such as T_2 and T_3 (with a larger subscript indicating a longer time). Moving down each streamline in Fig. 5.7*a*, we take note of the K levels attained at various points of time (i.e., at the intersections with various isochrones). When such information is plotted in Fig. 5.7*b*, the time paths $P_1, \dots, P_4$ emerge, with P_i being the counterpart of streamline S_i.

For the problem with finite planning horizon T_1, the optimal time path for K consists only of the segment K_0G on P_1. Time path P_1 is relevant to this problem, because S_1 is already known to be the relevant streamline; we stop at point G on P_1, because that is where we reach capital level K_T at time T_1. However, if the planning horizon is pushed outward to T_2, then we should forsake streamline S_1 in favor of S_3. The corresponding optimal time path for K will then be P_3, or, more correctly, the K_0J segment of it. Other values of T can also be analyzed in an analogous manner. In each case, a different value of T will yield a different optimal time path for K. And, collectively, all the finite-horizon optimal time paths for K are different from the infinite-horizon optimal K path P_2—the Ramsey path—associated with streamline S_2, the stable branch of the saddle point.

With this background, we can discuss the *turnpike behavior* of finite-horizon optimal paths. First, imagine a family traveling by car between two cities. If the distance involved is not too long, it may be the simplest to take some small, but direct, highway for the trip. But if the distance is sufficiently long, then it may be advantageous to take a more out-of-the-way route to get onto an expressway, or turnpike, and stay on it for as long as possible, till an appropriate exit is reached close to the destination city. It turns out that similar behavior characterizes the finite-horizon optimal time paths in Fig. 5.7*b*.

Having specified K_0 as the point of departure and K_T as the destination level of capital, if we successively extend out the planning horizon sufficiently far into the future, then the optimal time path can be made to arch, to any arbitrary extent, toward the Ramsey path P_2 or toward the K_B line—the "turnpike" in that diagram. To see this more clearly, let us compare points M and N. If the planning horizon is T_4, the optimal path is segment K_0M on P_4, which takes us to capital level K_T (point M) at time T_4. If the horizon is extended to T_5, however, we ought to select segment

K_0N on P_3 instead, which takes us to capital level K_T (point N) at time T_5. The fact that this latter path, with a longer planning period, arches more toward the Ramsey path and the K_B line serves to demonstrate the turnpike behavior in the finite-horizon model.

The reader will note that the turnpike analogy is by no means a perfect one. In the case of auto travel, the person actually drives *on* the turnpike. In the finite-horizon problem of capital accumulation, on the other hand, the turnpike is only a standard for comparison, but is not meant to be physically reached. Nevertheless, the graphic nature of the analogy makes it intuitively very appealing.

EXERCISE 5.4

1 In the Eisner-Strotz model, describe the economic consequences of not adhering to the stable branch of the saddle point:

(*a*) With K_0 as given in Fig. 5.4, what will happen if the firm chooses an initial rate of investment greater than I^*_0?

(*b*) What if it chooses an initial investment less than I^*_0?

2 In the Ramsey model with an infinite planning horizon, let there be capital saturation at $\hat{K}$, but no consumption saturation.

(*a*) Would this change of assumption require any modification of the system (5.40)? Why?

(*b*) Should (5.41) and (5.42) be modified? Why? How?

(*c*) Redraw Fig. 5.5 and derive an appropriate new $K' = 0$ curve.

(*d*) In the phase space, also add an appropriate new $\mu' = 0$ curve.

3 On the basis of the new $K' = 0$ and $\mu' = 0$ curves derived in the preceding problem, analyze the phase diagram of the capital-saturation case. In what way(s) is the new equilibrium similar to, and different from, the consumption-saturation equilibrium in Fig. 5.6? [*Hint:* The partial derivatives in (5.43) are not the only ones that can be taken.]

4 Show that when depreciation of capital is considered, the net product can have capital saturation even if the gross product $Q(K)$ does not.

5 In the finite-horizon problem of Fig. 5.7*a*, retain the original K_0, but let capital K_T be located to the right of K_B.

(*a*) Would the streamlines lying below the stable branch remain relevant to the problem? Those lying above? Why?

(*b*) Draw a new phase diagram similar to Fig. 5.7*a*, with four streamlines as follows: S_1 (the stable branch), S_2, S_3, and S_4 (three successive *relevant* streamlines, with S_2 being the closest to S_1). Using isochrones, derive time paths $P_1, \ldots, P_4$ for the K variable, corresponding to the four streamlines.

(*c*) Is turnpike behavior again discernible?

5.5 THE CONCAVITY / CONVEXITY SUFFICIENT CONDITION AGAIN

It was earlier shown in Sec. 4.2 that if the integrand function $F(t, y, y')$ in a fixed-endpoint problem is concave (convex) in the variables (y, y'), then the Euler equation is sufficient for an absolute maximum (minimum) of $V[y]$. Moreover, this sufficiency condition remains applicable when the terminal time is fixed but the terminal state is variable, provided that the supplementary condition

$$\left[F_{y'}(y - y^*)\right]_{t=T} \leq 0$$

is satisfied. For the infinite-horizon case, this supplementary condition becomes[15]

$$\lim_{t \to \infty} \left[F_{y'}(y - y^*)\right] \leq 0 \tag{5.44}$$

In this condition, $F_{y'}$ is to be evaluated along the optimal path, and $(y - y^*)$ represents the deviation of any admissible neighboring path $y(t)$ from the optimal path $y^*(t)$.

Application to the Eisner-Strotz Model

In the Eisner-Strotz model, the integrand function $F(t, K, K')$ yields the following second derivatives, as shown in (5.16):

$$F_{K'K'} = -2ae^{-\rho t} \qquad F_{KK'} = F_{K'K} = 0 \qquad \text{and} \qquad F_{KK} = -2\beta e^{-\rho t}$$

where all the parameters are positive. Applying the sign-definiteness test in (4.9), we thus have

$$|D_1| = F_{K'K'} < 0$$

$$|D_2| = \begin{vmatrix} F_{K'K'} & F_{K'K} \\ F_{KK'} & F_{KK} \end{vmatrix} = 4a\beta e^{-2\rho t} > 0$$

It follows that the quadratic form q associated with these second derivatives is negative definite, and the integrand function F is strictly concave in the variables (K, K').

---

[15]A more formal statement of this condition can be found in G. Hadley and M. C. Kemp, *Variational Methods in Economics*, North-Holland, Amsterdam, 1971, p. 102, Theorem 2.17.5.

For the present model, the supplementary condition (5.44) takes the form

$$\lim_{t \to \infty} \left[F_{K'}(K - K^*) \right] \le 0 \tag{5.45}$$

where the $F_{K'}$ term should specifically be

$$F_{K'} = -(2aK' + b)e^{-\rho t} \qquad \text{[from (5.16)]} \tag{5.46}$$

Substituting the $K^{*\prime}(t)$ expression in (5.23) for the K' term in (5.46), we have

$$F_{K'} = -\left[2ar_2(K_0 - \overline{K})e^{r_2 t} + b\right]e^{-\rho t} \qquad (r_2 < 0) \tag{5.46'}$$

It is clear that $F_{K'}$ tends to zero as t becomes infinite. As for the $(K - K^*)$ component of (5.45), the assumed form of the quadratic profit function depicted in Fig. 5.1a suggests that as t becomes infinite, the difference between the K value on any admissible neighboring path and the K^* value is bounded. Thus, the vanishing of $F_{K'}$ assures us that the supplementary condition (5.45) can be satisfied as an equality. Consequently, the strict concavity of the integrand function would make the Euler equation sufficient for a unique absolute maximum in the total profit $\Pi(K)$.

Application to the Ramsey Model

The integrand function of the (simplified) Ramsey model is

$$F(t, K, K') = B - U(C) = B - U[Q(K) - K']$$

We assume that $U'(C) \ge 0$, $U''(C) < 0$, $Q'(K) > 0$, and $Q''(K) < 0$. From the following first derivatives of F:

$$F_K = -U'(C)Q'(K) \qquad \text{and} \qquad F_{K'} = -U'(C)(-1) = U'(C)$$

the second derivatives are found to be

$$\begin{aligned} F_{K'K'} &= U''(C)(-1) = -U''(C) \\ F_{KK'} &= F_{K'K} = U''(C)Q'(K) \\ F_{KK} &= -U''(C)[Q'(K)]^2 - U'(C)Q''(K) \end{aligned}$$

Again applying the sign-definiteness test in (4.9), we find that

$$|D_1| = F_{K'K'} > 0 \tag{5.47}$$

But since

$$|D_2| = \begin{vmatrix} F_{K'K'} & F_{K'K} \\ F_{KK'} & F_{KK} \end{vmatrix} = U''(C)U'(C)Q''(K) \ge 0 \tag{5.48}$$

is not strictly positive, we cannot claim that the associated quadratic form q is positive definite and that the F function is strictly convex. Nevertheless, it is possible to establish the (nonstrict) convexity of the F function by the sign-semidefiniteness test in (4.12). With reference to the principal minors defined in (4.11), we find that

$$|D^0{}_1| = F_{KK} > 0 \qquad \text{and} \qquad |D^0{}_2| = |D_2| \geq 0$$

These, together with the information in (5.47) and (5.48), imply that

$$|\tilde{D}_1| > 0 \qquad \text{and} \qquad |\tilde{D}_2| \geq 0$$

Thus, by (4.12), we can conclude that the integrand function F is convex.

Turning to the supplementary condition (5.44), we need to confirm that

$$\lim_{t \to \infty} \left[F_{K'}(K - K^*) \right] \leq 0 \tag{5.49}$$

Since $F_{K'} = U'(C)$, it is easy to see that as t becomes infinite (as we move toward Bliss and as the marginal utility steadily declines), $F_{K'}$ tends to zero. As for the $(K - K^*)$ term, it is unfortunately not possible to state generally that the deviation of $K(t)$ from $K^*(t)$ tends to zero, or is bounded, as $t \to \infty$. Since the production function is assumed to have positive marginal product throughout, the $Q(K)$ curve extends upward indefinitely, and there is no bound to the economically meaningful values of K.

In this connection, we see that the assumption of capital saturation would help in the application of the sufficiency condition. If $Q(K)$ contains a saturation point, then $(K - K^*)$ will be bounded as $t \to \infty$, and (5.49) can be satisfied as an equality. In fact, when capital saturation is assumed, condition (5.49) can be satisfied even if there is no consumption saturation. With the capital-saturation point $\hat{K}$ now serving to define "Bliss," all admissible paths must end at $\hat{K}$. Thus the $(K - K^*)$ component of (5.49) must tend to zero as $t \to \infty$. Since $F_{K'} = U'(C)$ is bounded as we approach Bliss, (5.49) is satisfied as an equality. Consequently, given the convexity of the F function, the Euler equation is in this case sufficient for an absolute minimum of the integral $\int_0^\infty [B - U(C)]\, dt$.

CHAPTER

6

CONSTRAINED PROBLEMS

In the previous discussion, constraints have already been encountered several times, even though not mentioned by name. Whenever a problem has a specified terminal curve or a truncated vertical or horizontal terminal line, it imposes a constraint that the solution must satisfy. Such constraints, however, pertain to the endpoint. In the present chapter, we concern ourselves with constraints that more generally prescribe the behavior of the state variables. As a simple example, the input variables K and L may appear in a model as the state variables, but obviously their time paths cannot be chosen independently of each other because K and L are linked to each other by technological considerations via a production function. Thus the model should make room for a production-function constraint. It is also possible to have a constraint that ties one state variable to the time derivative of another. For instance, a model may involve the expected rate of inflation π and the actual rate of inflation p as state variables. The expected rate, π, is subject to revision over time when, in light of experience, it proves to be off the mark either on the up side or on the down side. The revision of the expected rate of inflation may be captured by a constraint such as

$$\frac{d\pi}{dt} = j(p - \pi) \qquad (0 < j \leq 1)$$

as previously encountered in (2.41). While such a constraint can, through substitution, be expunged from a model, it can also be retained and explicitly treated as a constraint. The present chapter explains how this is done.

6.1 FOUR BASIC TYPES OF CONSTRAINTS

Four basic types of constraints will be introduced in this section. As in static optimization problems with constraints, the Lagrange-multiplier method plays an important role in the treatment of constrained dynamic optimization problems.

Equality Constraints

Let the problem be that of maximizing

$$V = \int_0^T F(t, y_1, \ldots, y_n, y_1', \ldots, y_n')\, dt \tag{6.1}$$

subject to a set of m independent but consistent constraints $(m < n)$

$$\begin{aligned} g^1(t, y_1, \ldots, y_n) &= c_1 \\ &\vdots \\ g^m(t, y_1, \ldots, y_n) &= c_m \end{aligned} \qquad (c_1, \ldots, c_m \text{ are constants}) \tag{6.2}$$

and appropriate boundary conditions. By the "independence" of the m constraints, it is meant that there should exist a nonvanishing Jacobian determinant of order m, such as

$$\underset{(m \times m)}{|J|} = \left| \frac{\partial(g^1, \ldots, g^m)}{\partial(y_1, \ldots, y_m)} \right| \neq 0 \tag{6.3}$$

But, of course, any m of the y_j variables can be used in $|J|$, not necessarily the first m of them. Note that, in this problem, the number of constraints, m, is required to be strictly less than the number of state variables, n. Otherwise, with (say) $m = n$, the equation system (6.2) would already uniquely determine the $y_j(t)$ paths, and there would remain no degree of freedom for any optimizing choice. In view of this, this type of constrained dynamic optimization problem ought to contain at least *two* state variables, before a single constraint can meaningfully be accommodated.

In line with our knowledge of Lagrange multipliers in static optimization, we now form a Lagrangian integrand function, $\mathscr{F}$, by augmenting the original integrand F in (6.1) as follows:

$$\begin{aligned} \mathscr{F} &= F + \lambda_1(t)(c_1 - g^1) + \cdots + \lambda_m(t)(c_m - g^m) \\ &= F + \sum_{i=1}^{m} \lambda_i(t)(c_i - g^i) \end{aligned} \tag{6.4}$$

Although the structure of this Lagrangian function appears to be identical with that used in static optimization, there are two fundamental differences. First, whereas in the static problem the Lagrange-multiplier term is added to the original objective function, here the terms with Lagrange multipliers λ_i are added to the integrand function F, not to the objective functional $\int_0^T F\,dt$. Second, in the present framework, the Lagrange multipliers λ_i appear not as constants, but as functions of t. This is because each constraint g^i in (6.2) is supposed to be satisfied at every point of time in the interval $[0, T]$, and to each value of t there may correspond a different value of the Lagrange multiplier λ_i to be attached to the $(c_i - g^i)$ expression. To emphasize the fact that λ_i can vary with t, we write $\lambda_i(t)$. The Lagrangian integrand $\mathscr{F}$ has as its arguments not only the usual t, y_j, and y_j', $(j = 1, \dots, n)$, but also the multipliers λ_i, $(i = 1, \dots, m)$.

The replacement of F by $\mathscr{F}$ in the objective functional gives us the new functional

$$\mathscr{V} = \int_0^T \mathscr{F}\,dt \tag{6.5}$$

which we can maximize as if it is an *un*constrained problem. As long as all of the constraints in (6.2) are satisfied, so that $c_i - g^i = 0$ for all i, then the value of $\mathscr{F}$ will be identical with that of F, and the free extremum of the functional $\mathscr{V}$ in (6.5) will accordingly be identical with the constrained extremum of the original functional V.

It is a relatively simple matter to ensure the requisite satisfaction of all the constraints. Just treat the Lagrange multipliers as additional state variables, each to be subjected to an Euler equation, or the *Euler-Lagrange equation*, as it is sometimes referred to in the present constrained framework. To demonstrate this, let us first note that the Euler-Lagrange equations relating to the y_j variables are simply

(6.6)

$$\mathscr{F}_{y_j} - \frac{d}{dt}\mathscr{F}_{y_j'} = 0 \qquad \text{for all } t \in [0, T] \qquad (j = 1, \dots, n) \qquad [\text{cf. } (2.27)]$$

As applied to the Lagrange multipliers, they are, similarly,

$$\mathscr{F}_{\lambda_i} - \frac{d}{dt}\mathscr{F}_{\lambda_i'} = 0 \qquad \text{for all } t \in [0, T] \qquad (i = 1, \dots, m) \tag{6.7}$$

However, since $\mathscr{F}$ is independent of any λ_i', we have $\mathscr{F}_{\lambda_i'} = 0$ for every i, so that the m equations in (6.7) reduce to

$$\mathscr{F}_{\lambda_i} = 0 \qquad \text{or} \qquad c_i - g^i = 0 \qquad \text{for all } t \in [0, T] \qquad [\text{by } (6.64)] \tag{6.7'}$$

which coincide with the given constraints. Thus, by subjecting all the λ_i

variables to the Euler-Lagrange equation, we can guarantee the satisfaction of all the constraints.

As a matter of practical procedure for solving the present problem, therefore, we can (1) apply the Euler-Lagrange equation to the n state variables y_j only, as in (6.6), (2) take the m constraints exactly as originally given in (6.2), and (3) use these $n + m$ equations to determine the $y_j(t)$ and the $\lambda_i(t)$ paths. In step (1), the resulting n differential equations will contain $2n$ arbitrary constants in their general solutions. These may be definitized by the boundary conditions on the state variables.

EXAMPLE 1 Find the curve with the shortest distance between two given points $A = (0, y_0, z_0)$ and $B = (T, y_T, z_T)$ lying on a surface $\phi(t, y, z) = 0$.

The distance between the given pair of points is measured by the integral $\int_0^T (1 + y'^2 + z'^2)^{1/2}\, dt$, which resembles the integral $\int_0^T (1 + y'^2)^{1/2}\, dt$ for the distance between two points lying in a plane. Thus the problem is to

$$\begin{aligned} &\text{Minimize} && \int_0^T \left(1 + y'^2 + z'^2\right)^{1/2} dt \\ &\text{subject to} && \phi(t, y, z) = 0 \\ &\text{and} && y(0) = y_0,\ y(T) = y_T,\ z(0) = z_0,\ z(T) = z_T \end{aligned}$$

This is a problem with two state variables ($n = 2$) and one constraint ($m = 1$).

As the first step, we form the Lagrangian integrand

$$\begin{aligned} \mathscr{F} &= F + \lambda(t)(0 - \phi) = F - \lambda(t)\phi \\ &= \left(1 + y'^2 + z'^2\right)^{1/2} - \lambda(t)\phi(t, y, z) \end{aligned}$$

Partially differentiating $\mathscr{F}$, we see that

$$\begin{aligned} \mathscr{F}_y &= -\lambda(t)\phi_y & \mathscr{F}_{y'} &= y'\left(1 + y'^2 + z'^2\right)^{-1/2} \\ \mathscr{F}_z &= -\lambda(t)\phi_z & \mathscr{F}_{z'} &= z'\left(1 + y'^2 + z'^2\right)^{-1/2} \end{aligned}$$

These derivatives lead to the two Euler-Lagrange equations

$$-\lambda(t)\phi_y - \frac{d}{dt}\left[y'\left(1 + y'^2 + z'^2\right)^{-1/2}\right] = 0$$

$$-\lambda(t)\phi_z - \frac{d}{dt}\left[z'\left(1 + y'^2 + z'^2\right)^{-1/2}\right] = 0$$

which, together with the constraint

$$\phi(t, y, z) = 0$$

give us three equations to determine the three optimal paths for y, z, and λ.

Differential-Equation Constraints

Now suppose that the problem is to maximize (6.1), subject to a consistent set of m independent constraints $(m < n)$ that are differential equations:[1]

$$\begin{aligned} g^1(t, y_1, \ldots, y_n, y_1', \ldots, y_n') &= c_1 \\ &\vdots \\ g^m(t, y_1, \ldots, y_n, y_1', \ldots, y_n') &= c_m \end{aligned} \tag{6.8}$$

and appropriate boundary conditions. Even though the nature of the constraint equations has been changed, we can still follow essentially the same procedure as before.

The Lagrangian integrand function is still

$$\mathscr{F} = F + \lambda_1(t)(c_1 - g^1) + \cdots + \lambda_m(t)(c_m - g^m)$$

and the Euler-Lagrange equations with respect to the state variables y_j are still in the form of

$$\mathscr{F}_{y_j} - \frac{d}{dt}\mathscr{F}_{y_j'} = 0 \qquad \text{for all } t \in [0, T] \qquad (j = 1, \ldots, n)$$

Moreover, similar equations with respect to the Lagrange multipliers λ_i are again nothing but a restatement of the given constraints. So we have a total of n Euler-Lagrange equations for the state variables, plus the m constraints to determine the $n + m$ paths, $y_j(t)$ and $\lambda_i(t)$, with the arbitrary constants to be definitized by the boundary conditions.

Inequality Constraints

When the constraints are characterized by inequalities, the problem can

---

[1]The independence among the m differential equations in (6.8) means that there should exist a nonvanishing Jacobian determinant of order m, such as

$$\left| \frac{\partial(g^1, \ldots, g^m)}{\partial(y_1', \ldots, y_m')} \right| \neq 0$$

Again, any m of the y_j' can be used in the Jacobian.

generally be stated as follows:

$$
\begin{aligned}
&\text{Maximize} && \int_0^T F(t, y_1, \ldots, y_n, y_1', \ldots, y_n')\, dt \\
&\text{subject to} && g^1(t, y_1, \ldots, y_n, y_1', \ldots, y_n') \le c_1 \\
(6.9)\quad & && \vdots \\
& && g^m(t, y_1, \ldots, y_n, y_1', \ldots, y_n') \le c_m \\
&\text{and} && \text{appropriate boundary conditions}
\end{aligned}
$$

Since inequality constraints are much less stringent then equality constraints, there is no need to stipulate that $m < n$. Even if the number of constraints exceeds the number of state variables, the inequality constraints as a group will not uniquely determine the y_j paths, and hence will not eliminate all degrees of freedom from our choice problem. However, the inequality constraints do have to be consistent with one another, as well as with the other aspects of the problem.

To solve this problem, we may again write the Lagrangian integrand as

$$\mathscr{F} = F + \lambda_1(t)(c_1 - g^1) + \cdots + \lambda_m(t)(c_m - g^m)$$

While the Euler-Lagrange equations for the y_j variables,

$$\mathscr{F}_{y_j} - \frac{d}{dt}\mathscr{F}_{y_j'} = 0 \qquad \text{for all } t \in [0, T] \qquad (j = 1, \ldots, n)$$

are no different from before, the corresponding equations with respect to the Lagrange multipliers must be duly modified to reflect the inequality nature of the constraints. To ensure that all the $\lambda_i(t)(c_i - g^i)$ terms vanish in the solution (so that the optimized values of $\mathscr{F}$ and F are equal), we need a complementary-slackness relationship between the ith multiplier and the ith constraint, for every i (a set of m equations):

$$(6.10) \quad \lambda_i(t)(c_i - g^i) = 0 \qquad \text{for all } t \in [0, T] \qquad (i = 1, \ldots, m)$$

This complementary-slackness relationship guarantees that (1) whenever the ith Lagrange multiplier is nonzero, the ith constraint will be satisfied as a strict equality, and (2) whenever the ith constraint is a strict inequality, the ith Lagrange multiplier will be zero. It is this relationship that serves to maintain the identity between the optimal value of the original integrand F and that of the modified integrand $\mathscr{F}$ in (6.4).

Isoperimetric Problem

The final type of constraint to be considered here is the integral constraint, as exemplified by the equation

$$\int_0^T G(t, y, y')\, dt = k \qquad (k = \text{constant}) \tag{6.11}$$

One of the earliest problems involving such a constraint is that of finding the geometric figure with the largest area that can be enclosed by a curve of some specified length. Since all the figures admissible in the problem must have the same perimeter, the problem is referred to as the *isoperimetric problem*. In later usage, however, this name has been extended to all problems that involve integral constraints, that is, to any problem of the general form

$$\begin{aligned} &\text{Maximize} && \int_0^T F(t, y_1, \ldots, y_n, y_1', \ldots, y_n')\, dt \\ &\text{subject to} && \int_0^T G^1(t, y_1, \ldots, y_n, y_1', \ldots, y_n')\, dt = k_1 \\ &&& \vdots \\ &&& \int_0^T G^m(t, y_1, \ldots, y_n, y_1', \ldots, y_n')\, dt = k_m \\ &\text{and} && \text{appropriate boundary conditions} \end{aligned} \tag{6.12}$$

In the isoperimetric problem, there is again no need to require $m < n$, because even with $m \geq n$, freedom of optimizing choice is not ruled out.

Two characteristics distinguish isoperimetric problems from other constrained problems. First, the constraint in (6.11) is intended, not to restrict the y value at every point of time, but to force the integral of some function G to attain a specific value. In some sense, therefore, the constraint is more indirect. The other characteristic is that the solution values of the Lagrange multipliers $\lambda_i(t)$ are all constants, so that they can simply be written as λ_i.

To verify that the Lagrange multipliers indeed are constants in the solution, let us consider the case of a single state variable y and a single integral constraint ($m = n = 1$). First, let us define a function

$$\Gamma(t) = \int_0^t G(t, y, y')\, dt \tag{6.13}$$

Note that the upper limit of integration here is not a specific value T, but the variable t itself. In other words, this integral is an indefinite integral and hence a function of t. At $t = 0$ and $t = T$, respectively, we have

(6.14)

$$\Gamma(0) = \int_0^0 G\,dt = 0 \qquad \text{and} \qquad \Gamma(T) = \int_0^T G\,dt = k \qquad [\text{by (6.11)}]$$

It is clear that $\Gamma(t)$ measures the accumulation of G from time 0 to time t. For our immediate purpose, we note from (6.13) that the derivative of the $\Gamma(t)$ function is simply the G function. So we can write

$$G(t, y, y') - \Gamma'(t) = 0 \tag{6.15}$$

which is seen to conform to the general structure of the differential constraint $g(t, y, y') = c$, with $g = G - \Gamma'$ and $c = 0$. Thus we have in effect transformed the given integral constraint into a differential-equation constraint. And we can then use the previously discussed approach of solution.

Accordingly, we may write the Lagrangian integrand

$$\begin{aligned} \tilde{F} &= F(t, y, y') + \lambda(t)[0 - G(t, y, y') + \Gamma'(t)] \\ &= F(t, y, y') - \lambda(t)G(t, y, y') + \lambda(t)\Gamma'(t) \end{aligned} \tag{6.16}$$

(We denote this particular Lagrangian integrand by $\tilde{F}$ rather than $\mathscr{F}$, because it is merely an intermediate expression which will later be replaced by a final Lagrangian integrand.) In contrast to the previously encountered $\mathscr{F}$ expressions, we note that $\tilde{F}$ involves an extra variable Γ—although the latter enters into $\tilde{F}$ only in its derivative form $\Gamma'(t)$. This extra variable, whose status is the same as that of a state variable, must also be subjected to an Euler-Lagrange equation. Thus, there arise two parallel conditions in this problem:

$$\tilde{F}_y - \frac{d}{dt}\tilde{F}_{y'} = 0 \qquad \text{for all } t \in [0, T] \tag{6.17}$$

$$\tilde{F}_\Gamma - \frac{d}{dt}\tilde{F}_{\Gamma'} = 0 \qquad \text{for all } t \in [0, T] \tag{6.18}$$

However, inasmuch as $\tilde{F}$ is independent of Γ, and since $\tilde{F}_{\Gamma'} = \lambda(t)$, we see that (6.18) reduces to the condition

$$-\frac{d}{dt}\lambda(t) = 0 \qquad \Rightarrow \qquad \lambda(t) = \text{constant} \tag{6.18'}$$

This then validates our earlier claim about the constancy of λ. In view of this, we may write the Lagrange multiplier of the isoperimetric problem simply as λ.

Turning next to (6.17) and using (6.16), we can obtain a more specific version of that Euler-Lagrange equation:

$$(F_y - \lambda G_y) - \frac{d}{dt}(F_{y'} - \lambda G_{y'}) = 0 \tag{6.19}$$

But a moment's reflection will tell us that this same condition could have been obtained from a modified (abridged) version of $\tilde{F}$ without the $\Gamma'(t)$ term in it, namely,

$$\mathscr{F} = F(t, y, y') - \lambda G(t, y, y') \qquad (\lambda = \text{constant}) \tag{6.20}$$

Thus, in the present one-state-variable problem with a single integral constraint, we can as a practical procedure use the modified Lagrange integrand $\mathscr{F}$ in (6.20) instead of $\tilde{F}$, and then apply the Euler-Lagrange equation to y *alone*, knowing that the Euler-Lagrange equation for λ will merely tell us that λ is a constant, the value of which can be determined from the isoperimetric constraint.

The curious reader may be wondering why the Lagrange-multiplier term λG enters in $\mathscr{F}$ in (6.20) with a minus sign. The answer is that this follows directly from the standard way we write the Lagrangian integrand, where the term $\lambda(c - g) = \lambda c - \lambda g$ has a minus sign attached to λg. When written in this way, the Lagrange multiplier λ would come out with a positive sign in the solution, and can be assigned the economic interpretation of a shadow price.

The foregoing procedure can easily be generalized to the n-state-variable, m-integral-constraint case. For the latter, the modified Lagrangian function is

$$\mathscr{F} = F - (\lambda_1 G^1 + \cdots + \lambda_m G^m) \qquad (\lambda_i \text{ are all constants}) \tag{6.21}$$

EXAMPLE 2 Find a curve AB, passing though two given points $A = (0, y_0)$ and $B = (T, y_T)$ in the ty plane, and having a given length k, that maximizes the area under the curve.

The problem, as illustrated in Fig. 6.1, is to

$$\begin{aligned} &\text{Maximize} && V = \int_0^T y\,dt \\ &\text{subject to} && \int_0^T (1 + y'^2)^{1/2}\,dt = k \\ &\text{and} && y(0) = y_0 \qquad y(T) = y_T \end{aligned}$$

For the problem to be the meaningful, the constant k shall be assumed to be larger then the length of line segment AB, say, L. (If $k < L$, the problem has no solution, for the curve AB in question is impossible to draw. Also, if

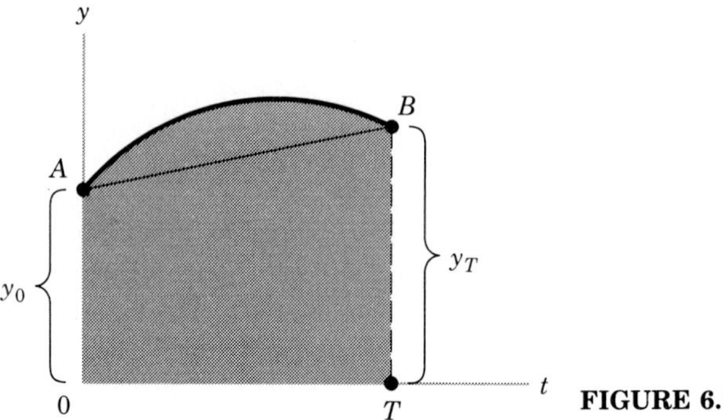

FIGURE 6.1

$k = L$, then the curve AB can only be drawn as a straight line, and no optimizing choice exists.)

To find the necessary condition for the extremal, we first write the Lagrangian integrand à la (6.20):

$$\mathcal{F} = y - \lambda(1 + y'^2)^{1/2} \qquad (\lambda = \text{constant}) \tag{6.22}$$

Since this does not explicitly contain t as an argument, we may take advantage of formula (2.21), and write the Euler-Lagrange equation in a form that has already been partly solved, namely,

$$\mathcal{F} - y'\mathcal{F}_{y'} = c_1 \qquad (c_1 = \text{arbitrary constant}) \tag{6.23}$$

Since $\mathcal{F}_{y'} = -\lambda y'(1 + y'^2)^{-1/2}$, (6.23) can be expressed more specifically as

$$y - \lambda(1 + y'^2)^{1/2} + \lambda y'^2(1 + y'^2)^{-1/2} = c_1 \tag{6.23'}$$

Via a series of elementary operations to be described, this equation can be solved for y'. Transposing terms and simplifying, we can write

$$y - c_1 = \frac{\lambda}{(1 + y'^2)^{1/2}} \qquad \text{or} \qquad (1 + y'^2)^{1/2} = \frac{\lambda}{y - c_1}$$

Squaring both sides and subtracting 1 from both sides, we get

$$y'^2 = \frac{\lambda^2 - (y - c_1)^2}{(y - c_1)^2}$$

The square root of this gives us the following expression of y' in terms of λ, y, and c_1:

$$y' = \frac{\sqrt{\lambda^2 - (y - c_1)^2}}{y - c_1}$$

Since $y' \equiv dy/dt$, however, the last result may alternatively be written as

$$\frac{y - c_1}{\sqrt{\lambda^2 - (y - c_1)^2}}\, dy = dt \tag{6.23''}$$

which the reader will recognize as a nonlinear differential equation with separable variables, solvable by integrating each side in turn.

The integral of dt, the right-hand-side expression in (6.23″), is simply t + constant. And the integral of the left-hand-side expression is $-\sqrt{\lambda^2 - (y - c_1)^2}$ + constant.[2] Thus, by equating the two integrals, consolidating the two constants of integration into a single symbol, $-c_2$, squaring both sides, and rearranging, we finally obtain the general solution

$$(y - c_1)^2 + (t - c_2)^2 = \lambda^2 \tag{6.24}$$

Since this is the equation for a family of circles, with center (c_2, c_1) and radius λ, the desired curve AB must be an arc of a circle. The specific values of the constants c_1, c_2, and λ can be determined from the boundary conditions and the constraint equation.

EXERCISE 6.1

1. Prepare a summary table for this section, with a column for each of the following: (a) type of constraint, (b) mathematical form of the constraint, (c) $m < n$ or $m \gtrless n$? (d) λ_i varying or constant over time?
2. Work out all the steps leading from (6.23) to (6.23″).
3. Find the extremal (general solution) of $V = \int_0^T y'^2\, dt$, subject to $\int_0^T y\, dt = k$.
4. Find the $y(t)$ and $z(t)$ paths (general solutions) that give an extremum of the functional $\int_0^T (y'^2 + z'^2)\, dt$, subject to $y - z' = 0$.
5. With reference to Fig. 6.1, find a curve AB, with a given length and passing through the given points A and B, that produces the largest area between the curve AB and the line segment AB.
6. Find a curve AB with the shortest possible length that passes through two given points A and B, and has the area under the curve equal to a given constant K. [*Note*: This problem serves to illustrate the *principle of reciprocity* in isoperimetric problems, which states that the problem of *maximum area* enclosed by a curve of a *given perimeter* and the problem of *minimum perimeter* for a curve enclosing that area are reciprocal to

---

[2] Let $x \equiv y - c_1$, so that $dx = dy$. The integral of the left-hand-side expression of (6.23″) is then $= \int x/\sqrt{\lambda^2 - x^2}\, dx$. According to standard tables of integrals (e.g., Formula 204 in *CRC Standard Mathematical Tables*, 28th ed., CRC Press, Boca Raton, FL, 1987), this integral is equal to $-\sqrt{\lambda^2 - x^2} = -\sqrt{\lambda^2 - (y - c_1)^2}$, plus a constant of integration.

each other and share the same extremal.] [*Hint*: When working out this problem, it may prove convenient to denote the Lagrange multiplier by μ, and later define $\lambda \equiv 1/\mu$, to facilitate comparison with Example 2 of this section.]

6.2 SOME ECONOMIC APPLICATIONS REFORMULATED

While the Lagrange-multiplier method can play a central role in constrained problems, especially in problems with general functions, it is also possible, and sometimes even simpler, to handle constrained problems by substitution and elimination of variables. Actually, most of the economic models discussed previously contain constraints, and in the solution process we have substituted out certain variables. In the present section, we shall reconsider the Ramsey model and the inflation-unemployment tradeoff model, to illustrate the method of Lagrange multipliers. The results would, of course, be identical regardless of the method used.

The Ramsey Model

In the Ramsey model (Sec. 5.3), the integrand function in the objective functional is

$$F = B - U(C) + D(L) \qquad \text{where } C = Q(K, L) - K'$$

Previously, the F function was considered to contain only two variables, K and L, because we substituted out the variable C. When we took the derivatives F_K and $F_{K'}$, in particular, the chain rule was used to yield the results

$$F_K = -U'(C)Q_K = -\mu Q_K \quad \text{and} \quad F_{K'} = -U'(C)(-1) = \mu \quad [\mu \equiv U'(C)]$$

The use of this rule reduces C to an intermediate variable that disappears from the final scene.

To show how the Lagrange-multiplier method can be applied, let us instead treat C as another variable on the same level as K and L. We then recognize the constraint

$$g(C, L, K, K') = Q(K, L) - K' - C = 0$$

and reformulate the problem as one with an explicit constraint:

$$\begin{aligned} &\text{Minimize} && \int_0^\infty [B - U(C) + D(L)]\, dt \\ (6.25)\qquad &\text{subject to} && Q(K, L) - K' - C = 0 \\ &\text{and} && \text{boundary conditions} \end{aligned}$$

By (6.4), we write the Lagrangian integrand function as

$$(6.26) \qquad \mathscr{F} = B - U(C) + D(L) + \lambda[-Q(K, L) + K' + C]$$

Since we now have three variables (C, L, K) and one Lagrange multiplier λ (not a constant), there should be altogether four Euler-Lagrange equations:

$$(6.27) \qquad \mathscr{F}_C - \frac{d}{dt}\mathscr{F}_{C'} = -U'(C) + \lambda = -\mu + \lambda = 0 \qquad \Rightarrow \qquad \lambda = \mu$$

$$(6.28) \qquad \mathscr{F}_L - \frac{d}{dt}\mathscr{F}_{L'} = D'(L) - \lambda Q_L = D'(L) - \mu Q_L = 0 \qquad [\text{by } (6.27)]$$

$$(6.29) \qquad \mathscr{F}_K - \frac{d}{dt}\mathscr{F}_{K'} = -\lambda Q_K - \frac{d}{dt}\lambda = -\mu Q_K - \frac{d\mu}{dt} = 0 \qquad [\text{by } (6.27)]$$

$$(6.30) \qquad \mathscr{F}_\lambda - \frac{d}{dt}\mathscr{F}_{\lambda'} = -Q(K, L) + K' + C = 0$$

Condition (6.27) tells us that the Lagrange multiplier λ is equal to μ—the symbol we have adopted for the marginal utility of consumption. Condition (6.28) is identical with (5.27). Similarly, (6.29) conveys the same information as (5.28). Lastly, (6.30) merely restates the constraint. Thus the use of the Lagrange-multiplier method produces precisely the same conclusions as before.

The Inflation-Unemployment Tradeoff Model

The inflation-unemployment tradeoff model of Sec. 2.5 will now be reformulated as a problem with two constraints. As might be expected, the presence of two constraints makes the solution process somewhat more involved.

For notational simplicity, we shall use the symbol y to denote $Y_f - Y$ (the deviation of the current national income Y from its full-employment level Y_f). Then the integrand function F—the loss function (2.39)—can be written as

$$\lambda = y^2 + \alpha p^2 \qquad (a > 0)$$

The expectations-augmented Phillips relation (2.40) and the adaptive-expectations equation (2.41), which were previously used to substitute out y and p in favor of the variable π, will now be accepted as two constraints:

$$g^1(t, y, p, \pi, \pi') = \beta y + p - \pi = 0 \qquad [\text{from } (2.40)]$$

$$g^2(t, y, p, \pi, \pi') = j(p - \pi) - \pi' = 0 \qquad [\text{from } (2.41)]$$

The problem thus becomes one where one of the constraints involves a simple equality, and the other constraint involves a differential equation:

$$\text{Minimize} \quad \int_0^T (y^2 + \alpha p^2)e^{-\rho t}\,dt$$

$$\text{(6.31)} \qquad \text{subject to} \quad \beta y + p - \pi = 0$$

$$j(p - \pi) - \pi' = 0$$

$$\text{and} \quad \text{boundary conditions}$$

The Lagrangian integrand function is

(6.32)

$$\mathcal{F} = (y^2 + \alpha p^2)e^{-\rho t} + \lambda_1(-\beta y - p + \pi) + \lambda_2(-jp + j\pi + \pi')$$

with three variables y, p, and π and two (nonconstant) Lagrange multipliers λ_1 and λ_2. The Euler-Lagrange equations are

$$\text{(6.33)} \qquad \mathcal{F}_y - \frac{d}{dt}\mathcal{F}_{y'} = 2ye^{-\rho t} - \beta\lambda_1 = 0$$

$$\text{(6.34)} \qquad \mathcal{F}_p - \frac{d}{dt}\mathcal{F}_{p'} = 2\alpha pe^{-\rho t} - \lambda_1 - j\lambda_2 = 0$$

$$\text{(6.35)} \qquad \mathcal{F}_\pi - \frac{d}{dt}\mathcal{F}_{\pi'} = \lambda_1 + j\lambda_2 - \frac{d}{dt}\lambda_2 = 0$$

$$\text{(6.36)} \qquad \mathcal{F}_{\lambda_1} - \frac{d}{dt}\mathcal{F}_{\lambda_1'} = -\beta y - p + \pi = 0$$

$$\text{(6.37)} \qquad \mathcal{F}_{\lambda_2} - \frac{d}{dt}\mathcal{F}_{\lambda_2'} = -jp + j\pi + \pi' = 0$$

To solve these equations simultaneously requires quite a few steps. One way of doing it is the following. First, solve (6.36) for y, substitute the result into (6.33), and solve for λ_1 to get

$$\text{(6.38)} \qquad \lambda_1 = \frac{2}{\beta^2}(\pi - p)e^{-\rho t}$$

Next, solve (6.37) for the variable p:

$$\text{(6.39)} \qquad p = \pi + \frac{1}{j}\pi'$$

Substitution of (6.39) into (6.38) results in the elimination of π and p in the λ_1 expression:

$$\text{(6.40)} \qquad \lambda_1 = \frac{-2}{\beta^2 j}\pi' e^{-\rho t}$$

Then substituting both (6.39) and (6.40) into (6.34) and solving for λ_2, we obtain (after simplification)

$$\lambda_2 = \left[\frac{2\alpha}{j}\pi + \frac{2(1+\alpha\beta^2)}{\beta^2 j^2}\pi'\right]e^{-\rho t} \tag{6.41}$$

This result implies the total derivative

$$\begin{aligned}\frac{d\lambda_2}{dt} &= -\rho\left[\frac{2\alpha}{j}\pi + \frac{2(1+\alpha\beta^2)}{\beta^2 j^2}\pi'\right]e^{-\rho t} \\ &\quad + \left[\frac{2\alpha}{j}\pi' + \frac{2(1+\alpha\beta^2)}{\beta^2 j^2}\pi''\right]e^{-\rho t} \\ &= 2\left[\frac{-\alpha\rho}{j}\pi + \frac{\alpha\beta^2 j - \rho(1+\alpha\beta^2)}{\beta^2 j^2}\pi' + \frac{(1+\alpha\beta^2)}{\beta^2 j^2}\pi''\right]e^{-\rho t}\end{aligned} \tag{6.42}$$

Finally, we can substitute the λ_1, λ_2, and $d\lambda_2/dt$ expressions of the last three equations into (6.35) to get a single summary statement of the five simultaneous Euler-Lagrange equations. After combining and canceling terms, the complicated expressions reduce to the surprisingly simple result

$$\pi'' - \rho\pi' - \frac{\alpha\beta^2 j(\rho + j)}{1+\alpha\beta^2}\pi = 0 \tag{6.43}$$

This result qualifies as the condensed statement of conditions (6.33) through (6.37) because all five equations have been used in the process and incorporated into (6.43).

Since (6.43) is identical with the earlier result (2.46), we have again verified that the Lagrange-multiplier method leads to the same conclusion. Note, however, that this time the Lagrange-multiplier method entails more involved calculations. This shows that sometimes the more elementary elimination-of-variable approach may work better. It is in situations where the elimination of variables is unfeasible (e.g., with general functions) that the Lagrange-multiplier method shows its powerfulness in the best light.

EXERCISE 6.2

1 The constraint in the Ramsey model (6.25) could have been equivalently written as $C - Q(K, L) + K' = 0$. Write the new Lagrangian integrand function and the new Euler-Lagrange equations. How do the analytical results differ from those in (6.27) through (6.30)?

2 The Eisner-Strotz model of Sec. 5.2 can be reformulated as a constrained problem:

$$\begin{aligned} &\text{Maximize} && \int_0^\infty (\pi - C)e^{-\rho t}\,dt \\ &\text{subject to} && \pi - \alpha K + \beta K^2 = 0 \\ & && C - aK'^2 - bK' = 0 \\ &\text{and} && \text{boundary conditions} \end{aligned}$$

(a) How many variables are there in this problem? How many constraints?
(b) Write the Lagrangian integrand function.
(c) Write the Euler-Lagrange equations and solve them. Compare the result with (5.17).

6.3 THE ECONOMICS OF EXHAUSTIBLE RESOURCES

In the discussion of production functions, there is usually a presumption that all the inputs are inexhaustible, so that as they are being used up, more of them can be obtained for future use. In reality, however, certain resources—such as oil and minerals—are subject to ultimate exhaustion. In the absence of guaranteed success in the discovery of new deposits or substitute resources, the prospect of eventual exhaustion must be taken into account when such a resource is put into use. The optimal production (extraction) of an exhaustible resource over time offers a good illustration of the isoperimetric problem.

The Hotelling Model of Socially Optimal Extraction

In a classic article by Harold Hotelling,[3] the notion of "the social value" of an exhaustible resource is used for judging the desirability of any extraction pattern of the resource. The gross value to society of a marginal unit of output or extraction of the resource is measured by the price society is willing to pay to call forth that particular unit of output, and the net value to society is the gross value less the cost of extracting that unit. If the price of the resource, P, is negatively related to the quantity demanded, as illustrated in Fig. 6.2, then the *gross* social value of an output Q_0 is

---

[3]Harold Hotelling, "The Economics of Exhaustible Resources," *Journal of Political Economy*, April 1931, pp. 137–175.

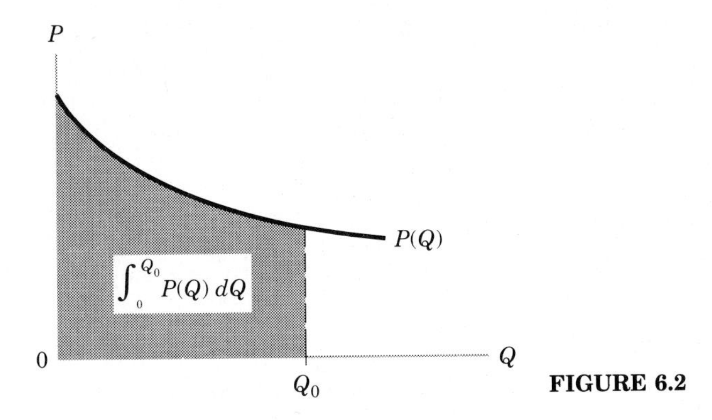

FIGURE 6.2

measured by the shaded area under the curve, or the integral $\int_0^{Q_0} P(Q)\,dQ$. To find the *net* social value, we subtract from the gross social value the total cost of extraction $C(Q)$. Generalizing from a specific output level Q_0 to any output level Q, we may write the net social value of the resource, N, as:[4]

$$N(Q) = \int_0^Q P(Q)\,dQ - C(Q) \tag{6.44}$$

To avoid any confusion caused by using the same symbol Q as the variable of integration as well as the upper limit of integration, we can alternatively write (6.44) as

$$N(Q) = \int_0^Q P(x)\,dx - C(Q) \tag{6.44'}$$

Assuming that the total stock of the exhaustible resource is given at S_0, we have the problem of finding an extraction path $Q(t)$ so as to

$$\begin{aligned} &\text{Maximize} \quad V = \int_0^\infty N(Q)e^{-\rho t}\,dt \\ &\text{subject to} \quad \int_0^\infty Q\,dt = S_0 \end{aligned} \tag{6.45}$$

This, of course, is an isoperimetric problem. Since it is reasonable to expect the net-social-value function to have an upper bound, the improper integral in the objective functional should be convergent.

By (6.20), the Lagrangian integrand is

$$\mathscr{F} = N(Q)e^{-\rho t} - \lambda Q \qquad (\lambda = \text{constant}) \tag{6.46}$$

---

[4]In Hotelling's paper, the extraction cost is subsumed under the symbol P, which is interpreted as the "net price" of the exhaustible resource. Accordingly, the $C(Q)$ term does not appear in his treatment.

Note that the integrand $\mathscr{F}$ does not contain the derivative of Q; hence the extremal to be obtained from the Euler-Lagrange equation may not fit given fixed endpoints. If no rigid endpoints are imposed, however, we can still apply the Euler-Lagrange equation, which, because $\mathscr{F}_{Q'} = 0$, reduces in the present case to the condition $\mathscr{F}_Q = 0$, or

$$N'(Q)e^{-\rho t} - \lambda = 0 \tag{6.47}$$

Upon applying the differentiation formula (2.8) to $N(Q)$ in (6.44'), we have

$$[P(Q) - C'(Q)]e^{-\rho t} - \lambda = 0 \tag{6.47'}$$

The economic interpretation of this condition becomes easier when we recall that, for an isoperimetric problem, the Lagrange multiplier λ is a constant. Along the optimal extraction path, the value of $P(Q) - C'(Q)$ associated with *any* point of time must have a uniform present value, λ. By slightly rearranging the terms, we can reinterpret the condition as requiring that $P(Q) - C'(Q)$ grow at the rate ρ:

$$P(Q) - C'(Q) = \lambda e^{\rho t} \qquad \text{[social optimum]} \tag{6.47''}$$

From the last equation, it is also clear that λ has the connotation of "the initial value of $P(Q) - C'(Q)$." If the $P(Q)$ and the $C(Q)$ functions are specific, we can solve (6.47'') for Q in terms of λ and t, say $Q(\lambda, t)$. The latter, when substituted into the constraint in (6.45), then enables us to solve for λ.

Pure Competition Versus Monopoly

One major conclusion of the Hotelling paper is that pure competition can yield an extraction path identical with the socially optimal one, whereas a monopolistic firm will adopt an extraction path that is more conservationistic, but socially suboptimal. We shall now look into this aspect of the problem.

Assume for simplicity that there are n firms under pure competition. The ith firm extracts at a rate Q_i out of a known total stock of S_i under its control. The firm's problem is to maximize the total discounted profits over time, taking the product price to be exogenously set at P_0:

$$\begin{aligned} &\text{Maximize} && \int_0^\infty [P_0 Q_i - C_i(Q_i)]e^{-\rho t}\,dt \\ &\text{subject to} && \int_0^\infty Q_i\,dt = S_i \end{aligned} \tag{6.48}$$

The Lagrangian integrand now becomes

$$\mathscr{F} = [P_0 Q_i - C_i(Q_i)]e^{-\rho t} - \lambda Q_i$$

and the application of the Euler-Lagrange equation yields the condition

$$[P_0 - C_i'(Q_i)]e^{-\rho t} - \lambda = 0$$

or

$$P_0 - C_i'(Q_i) = \lambda e^{\rho t} \qquad \text{[pure competition]} \tag{6.49}$$

This condition is perfectly consistent with that for social optimum, (6.47″), in that it, too, requires the difference between the price of the exhaustible resource and its marginal cost of extraction to grow exponentially at the rate ρ.

In contrast, the problem of monopolistic profit maximization is

$$\begin{aligned} &\text{Maximize} && \int_0^\infty [R(Q) - C(Q)]e^{-\rho t}\, dt \\ &\text{subject to} && \int_0^\infty Q\, dt = S_0 \end{aligned} \tag{6.50}$$

This time, with the Lagrangian integrand

$$\mathscr{F} = [R(Q) - C(Q)]e^{-\rho t} - \lambda Q$$

the Euler-Lagrange equation leads to the condition

$$R'(Q) - C'(Q) = \lambda e^{\rho t} \qquad \text{[monopoly]} \tag{6.51}$$

which differs from the rule for socially optimal extraction. Here, it is the difference between the marginal revenue (rather than price) and the marginal cost that is to grow at the rate ρ, with λ now representing the initial value of the said difference.

The Monopolist and Conservation

The conclusion of the Hotelling paper is, however, not only that monopolistic production of an exhaustible resource is suboptimal, but also that it is specifically biased toward excessive conservationism. How valid is the latter claim? The answer is: It is under certain circumstances, but not always. This issue has been examined by various people using different specific assumptions. As one might expect, differences in assumptions result in different conclusions. In a paper by Stiglitz,[5] for example, it is shown that if the elasticity of demand increases with time (with the discovery of substitutes), or if the cost of extraction is constant per unit of extraction but

[5]Joseph E. Stiglitz, "Monopoly and the Rate of Extraction of Exhaustible Resources," *American Economic Review*, September 1976, pp. 655–661.

decreases with time (with better technology), then the monopolist tends to be more conservationistic than under the social optimum. But the opposite conclusion is shown to prevail in a paper by Lewis, Matthews, and Burness,[6] under the assumption that the extraction cost does not vary with the rate of extraction (capital costs, leasing fees, etc., are essentially fixed costs), and that the elasticity of demand increases with consumption (sufficiently low prices can attract bulk users at the margin to switch from substitutes).

In one of Stiglitz's submodels, the monopolist faces the (inverse) demand function

$$P = \psi(t)Q^{\alpha-1} \qquad (0 < \alpha < 1) \tag{6.52}$$

with elasticity of demand $1/(1-\alpha)$. The per-unit cost of extraction is constant at any given time, but it can decline over time:

$$C = \phi(t)Q \qquad (\phi' < 0) \tag{6.53}$$

Thus the profit is

$$PQ - C = \psi(t)Q^{\alpha} - \phi(t)Q$$

And the monopolist's dynamic optimization problem is to

$$\begin{aligned} &\text{Maximize} \quad \int_0^\infty [\psi(t)Q^\alpha - \phi(t)Q]e^{-\rho t}\,dt \\ &\text{subject to} \quad \int_0^\infty Q\,dt = S_0 \end{aligned} \tag{6.54}$$

The Lewis-Matthews-Burness model, on the other hand, uses a general (inverse) demand function $P(Q)$ that is stationary, although its elasticity of demand is assumed to increase with consumption. Since the cost of extraction is assumed to be a fixed cost ($= \Phi$), the monopolist's problem is to

$$\begin{aligned} &\text{Maximize} \quad \int_0^\infty [P(Q)Q - \Phi]e^{-\rho t}\,dt \\ &\text{subject to} \quad \int_0^\infty Q\,dt = S_0 \end{aligned} \tag{6.55}$$

The Euler-Lagrange equation is applicable to each of these two problems in a manner similar to the Hotelling model. From that process, we can deduce an optimal rate of growth of Q for the monopolist. Comparing that

[6] Tracy R. Lewis, Steven A. Matthews, and H. Stuart Burness, "Monopoly and the Rate of Extraction of Exhaustible Resources: Note," *American Economic Review*, March 1979, pp. 227–230.

rate to the rate of growth dictated by the social optimum will then reveal whether the monopolist is excessively or insufficiently conservationistic. Instead of analyzing problems (6.54) and (6.55) separately, however, we shall work with a single, more general formulation that consolidates the considerations present in (6.54) and (6.55).

Let the demand function and the cost function be expressed as

$$Q = e^{gt}D(P) \qquad [D'(P) < 0] \tag{6.56}$$

$$C = C(Q, t) \qquad (C_Q \geq 0,\, C_{tQ} \leq 0) \tag{6.57}$$

These functions include those used by Stiglitz and Lewis, Matthews, and Burness as special cases. In the following, we shall use the standard abbreviations for marginal revenue and marginal cost, $\text{MR} \equiv R'(Q)$ and $\text{MC} \equiv C'(Q)$, as well as the notation $r_x \equiv (dx/dt)/x$ for the rate of growth of any variable x.

The condition for social optimum that arises from the Euler-Lagrange equation, (6.47″), requires that $r_{(P-\text{MC})} = \rho$. Using a familiar formula for the rate of growth of a difference,[7] we can rewrite this as

$$\frac{P}{P - \text{MC}} r_P - \frac{\text{MC}}{P - \text{MC}} r_{\text{MC}} = \rho$$

And this implies that

$$r_P = \rho\left(1 - \frac{\text{MC}}{P}\right) + \frac{\text{MC}}{P} r_{\text{MC}} \tag{6.58}$$

We can translate this r_P into a corresponding r_Q by using the fact that r_Q is linked to r_P via the elasticity of demand $\epsilon < 0$, or its negative, $E \equiv |\epsilon|$, as follows:[8]

$$r_Q = g - Er_P \tag{6.59}$$

---

[7]See Alpha C. Chiang, *Fundamental Methods of Mathematical Economics*, 3d ed., McGraw-Hill, New York, 1984, Sec. 10.7.

[8]First take the natural log of both sides of (6.56) to get

$$\ln Q = gt + \ln D(P)$$

Then derive r_Q by differentiating $\ln Q$ with respect to t:

$$r_Q = \frac{d}{dt}\ln Q = g + \frac{d}{dt}\ln D(P) = g + \frac{1}{D(P)}\frac{d}{dt}D(P) = g + \frac{1}{D(P)}D'(P)\frac{dP}{dt}$$

$$= g + \frac{D'(P)P}{D(P)}\frac{dP/dt}{P} = g + \epsilon r_P \equiv g - Er_P \qquad (\text{where } E \equiv |\epsilon|)$$

Substituting (6.58) into (6.59), we then find the socially optimal rate of growth of Q, denoted by r_{Qs} (s for social optimum), to be

$$(6.60) \quad r_{Qs} = g - E\left[\rho\left(1 - \frac{\text{MC}}{P}\right) + \frac{\text{MC}}{P} r_{\text{MC}}\right] \qquad \text{[social optimum]}$$

The counterpart expression for r_{Qm} (m for monopoly) can be derived by a similar procedure. Condition (6.51) requires that $r_{(\text{MR}-\text{MC})} = \rho$, which may be rewritten as

$$\frac{\text{MR}}{\text{MR} - \text{MC}} r_{\text{MR}} - \frac{\text{MC}}{\text{MR} - \text{MC}} r_{\text{MC}} = \rho$$

And this implies that

$$(6.61) \qquad r_{\text{MR}} = \rho\left(1 - \frac{\text{MC}}{\text{MR}}\right) + \frac{\text{MC}}{\text{MR}} r_{\text{MC}}$$

However, since MR and P are related to each other by the equation

$$(6.62) \qquad \text{MR} = P\left(1 - \frac{1}{E}\right)$$

we may derive another expression for r_{MR} from (6.62), namely,

$$(6.63) \quad r_{\text{MR}} \equiv \frac{1}{\text{MR}} \frac{d\text{MR}}{dt} = \frac{1}{P(1 - 1/E)}\left[\left(1 - \frac{1}{E}\right)\frac{dP}{dt} + \frac{P}{E^2}\frac{dE}{dt}\right]$$

$$= \frac{dP/dt}{P} + \frac{1}{(1 - 1/E)E}\frac{dE/dt}{E} = r_P + \frac{1}{E - 1} r_E$$

Equating (6.61) and (6.63), solving for r_P, and then substituting into (6.59), we get the rate of growth of Q under monopoly:

(6.64)

$$r_{Qm} = g - E\left[\rho\left(1 - \frac{\text{MC}}{\text{MR}}\right) + \frac{\text{MC}}{\text{MR}} r_{\text{MC}} - \frac{1}{E - 1} r_E\right] \qquad \text{[monopoly]}$$

A comparison of (6.64) with (6.60) reveals that we can have

$$(6.65) \qquad r_{Qs} = r_{Qm} = g - E\rho \qquad \text{if MC} = r_{\text{MC}} = r_E = 0$$

The situation of MC = 0 occurs when extraction is costless. For $r_{\text{MC}} = 0$, the marginal cost of extraction has to be time invariant [$C_{tQ} = 0$ in (6.57)]. And the condition $r_E = 0$ is met when the demand elasticity is constant over time. This set of conditions—also considered by Stiglitz—is quite stringent, but if the conditions are met, then the monopolist will follow exactly the same extraction path as called for under the social optimum, which is in no

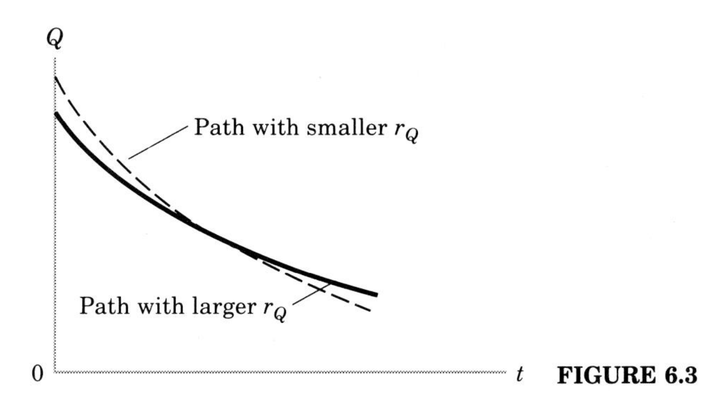

FIGURE 6.3

way overconservationistic. Note that the $-E\rho$ term is negative and implies a declining Q over time, unless offset by a sufficiently large positive value of g. If the demand for the exhaustible resource is stationary ($g = 0$), or shifts downward over time ($g < 0$) as a result of, say, the discovery of substitutes, then r_Q must be negative.

A different outcome emerges, however, when there is a *positive* MC that is output-invariant ($\text{MC} = k > 0$) as well as time-invariant ($r_{\text{MC}} = 0$). Retaining the assumption that $r_E = 0$, we now find

$$r_{Qs} = g - E\left[\rho\left(1 - \frac{k}{P}\right)\right] \quad \text{and} \quad r_{Qm} = g - E\left[\rho\left(1 - \frac{k}{\text{MR}}\right)\right] \tag{6.66}$$

Since $P > \text{MR}$, it follows that k/P is less than k/MR, so that $r_{Qs} < r_{Qm}$ at any level of Q. The social optimizer and the monopolist now follow different $Q(t)$ paths. Recall, however, that both paths must have the same total area under the curve, namely, S_0. To satisfy this constraint, the monopolist's $Q(t)$ path, with the larger r_Q value (e.g., -0.03 as against -0.05), must be characterized by a more gentle negative slope, as illustrated by the solid curve in Fig. 6.3. Since such a curve entails lower initial extraction and higher later extraction than the broken curve, the monopolist indeed emerges in the present case as a conservationist, relative to the social optimizer. We have reached the same conclusion here as Stiglitz, although the assumptions are different.

Nevertheless, it is also possible to envisage the opposite case where the monopolist tends to undertake excessive early extraction. Assume again that $\text{MC} = r_{\text{MC}} = 0$. But let the elasticity of demand E vary with the rate of extraction Q. More specifically, let $E'(Q) > 0$—a case considered by Lewis, Matthews, and Burness. Then, as Q varies over time, E must vary accordingly, even though E is not in itself a function of t. With these assumptions, we now have a nonzero rate of growth of E:

$$r_E \equiv \frac{dE/dt}{E} = \frac{E'(Q)(dQ/dt)}{E} = \frac{E'(Q)Q}{E} r_Q \tag{6.67}$$

Taking the r_Q in this equation to be r_{Qm}, and substituting (6.67) into (6.64), we get

$$r_{Qm} = g - E\rho + \frac{E'(Q)Q}{E-1} r_{Qm}$$

Collecting terms and solving for r_{Qm}, we finally obtain

$$r_{Qm} = \frac{g - E\rho}{1 - E'(Q)Q/(E-1)} \tag{6.68}$$

which should be contrasted against

$$r_{Qs} = g - E\rho \qquad \left[\text{from (6.60) with MC} = r_{\text{MC}} = 0\right]$$

Suppose that $g - E\rho < 0$. Then, if the denominator in (6.68) is a positive fraction, we will end up with $r_{Qm} < r_{Qs}$ (e.g., dividing -0.04 by $\frac{1}{2}$ results in -0.08, a smaller number). And, in view of Fig. 6.3, the monopolist will in such a case turn out to be an anticonservationist. It can be shown[9] that the denominator in (6.68) is a positive fraction when $R''(Q) < 0$, that is, when $d\text{MR}/dQ < 0$. But a declining marginal revenue is a commonly acknowledged characteristic of a monopoly situation. Therefore, we can see from the present analysis that monopoly is by no means synonymous with conservationism.

EXERCISE 6.3

1. In (6.47''), we interpret λ to mean the initial value of $P(Q) - C'(Q)$. Find an alternative economic interpretation of λ from (6.47').
2. If the maximization problem (6.45) is subject instead to the weak inequality $\int_0^\infty Q\,dt \le S_0$, what modification(s) must be made to the result of the analysis?

---

[9] By differentiating (6.62) with respect to Q, and letting the derivative be negative, we have

$$\left(\frac{d\text{MR}}{dQ} = \right) P'(Q)\left(1 - \frac{1}{E}\right) + P\frac{1}{E^2}E'(Q) < 0$$

Multiplying both sides of this inequality by E^2Q/P, and making use of the fact that $P'(Q)Q/P = -1/E$, we can transform the inequality to the form $-(E-1) + E'(Q)Q < 0$, or

$$E'(Q)Q < E - 1$$

In the latter inequality, the left-hand side is positive by virtue of our assumption that $E'(Q) > 0$; the right-hand side is positive because (6.51) requires MR to be positive, and, in view of (6.62), the positivity of MR in turn dictates that $E > 1$. Hence, $E'(Q)Q/(E-1)$ is a positive fraction, and so is the denominator in (6.68).

3 If the transversality conditions (5.5) and (5.7) are applied to the Lagrangian integrands $\mathscr{F}$ for social optimum and monopoly, respectively, what terminal conditions will emerge? Are these conditions economically reasonable?

4 Compare r_{Qs} and r_{Qm} in (6.60) and (6.64), assuming that (1) MC $= r_{\text{MC}} = 0$, (2) the elasticity of demand is output-invariant, and (3) the demand becomes more elastic over time because of the emergence of substitutes.

5 Assume that (1) the demand elasticity is output-invariant and time-invariant, and (2) the MC is positive and output-invariant at any given time, but increases over time. Write out the appropriate expressions for r_{Qs} and r_{Qm}, find the difference $r_{Qm} - r_{Qs}$, study its algebraic sign, and indicate under what conditions the monopolist will be *more* conservationistic and *less* conservationistic than the social optimizer.

6 Let $q \equiv$ the cumulative extraction of an exhaustible resource: $q(t) = \int_0^t Q(t)\,dt$. Verify that the current extraction rate Q is related to q by the equation $Q(t) = q'(t)$.

(*a*) Assume that the cost of extraction C is a function of the cumulative extraction q as well as the current rate of extraction q'. How must the $N(Q)$ expression in (6.44) be modified?

(*b*) Reformulate the social optimizer's problem (6.45), expressing it in terms of the variables q and $q'(t)$.

(*c*) Write the Lagrangian integrand, and find the Euler-Lagrange equation.

7 Using the definition of q in the preceding problem, and retaining the assumption that $C = C(q, q')$, reformulate the competitive firm's problem (6.48) in terms of the variables q and $q'(t)$. Show that the Euler-Lagrange equation for the competitive firm is consistent with that of the social optimizer.

PART 3

OPTIMAL CONTROL THEORY

CHAPTER

7

OPTIMAL CONTROL: THE MAXIMUM PRINCIPLE

The calculus of variations, the classical method for tackling problems of dynamic optimization, like the ordinary calculus, requires for its applicability the differentiability of the functions that enter in the problem. More importantly, only interior solutions can be handled. A more modern development that can deal with nonclassical features such as corner solutions, is found in *optimal control theory*. As its name implies, the optimal-control formulation of a dynamic optimization problem focuses upon one or more control variables that serve as the instrument of optimization. Unlike the calculus of variations, therefore, where our goal is to find the optimal time path for a *state* variable y, optimal control theory has as its foremost aim the determination of the optimal time path for a *control* variable, u. Of course, once the optimal control path, $u^*(t)$, has been found, we can also find the optimal state path, $y^*(t)$, that corresponds to it. In fact, the optimal $u^*(t)$ and $y^*(t)$ paths are usually found in the same process. But the presence of a control variable at center stage does alter the basic orientation of the dynamic optimization problem.

A couple of questions immediately suggest themselves. What makes a variable a "control" variable? And how does it fit into a dynamic optimization problem? To answer these questions, let us consider a simple economic illustration. Suppose there is in an economy a finite stock of an exhaustible resource S (such as coal or oil), as discussed in the Hotelling model, with $S(0) = S_0$. As this resource is being extracted (and used up), the resource stock will be reduced according to the relation

$$\frac{dS(t)}{dt} = -E(t)$$

where $E(t)$ denotes the rate of extraction of the resource at time t. The $E(t)$ variable qualifies as a control variable because it possesses the following two properties. First, it is something that is subject to our discretionary choice. Second, our choice of $E(t)$ impinges upon the variable $S(t)$ which indicates the state of the resource at every point of time. Consequently, the $E(t)$ variable is like a steering mechanism which we can maneuver so as to "drive" the state variable $S(t)$ to various positions at any time t via the differential equation $dS/dt = -E(t)$. By the judicious steering of such a control variable, we can therefore aim to optimize some performance criterion as expressed by the objective functional. For the present example, we may postulate that society wants to maximize the total utility derived from using the exhaustible resource over a given time period $[0, T]$. If the terminal stock is not restricted, the dynamic optimization problem may take the following form:

$$\begin{aligned}
&\text{Maximize} && \int_0^T U(E)e^{-\rho t}\,dt \\
&\text{subject to} && \frac{dS}{dt} = -E(t) \\
&\text{and} && S(0) = S_0 \qquad S(T) \text{ free} \qquad (S_0, T \text{ given})
\end{aligned}$$

In this formulation, it happens that only the control variable E enters into the objective functional. More generally, the objective functional can be expected to depend on the state variable(s) as well as the control variable(s). Similarly, it is fortuitous that in this example the movement of the state variable S depends only on the control variable E. In general, the course of movement of a state variable over time may be affected by the state variable(s) as well as the control variable(s), and indeed even by the t variable itself.

With this background, we now proceed to the discussion of the method of optimal control.

7.1 THE SIMPLEST PROBLEM OF OPTIMAL CONTROL

To keep the introductory framework simple, we first consider a problem with a single state variable y and a single control variable u. As suggested earlier, the control variable is a policy instrument that enables us to

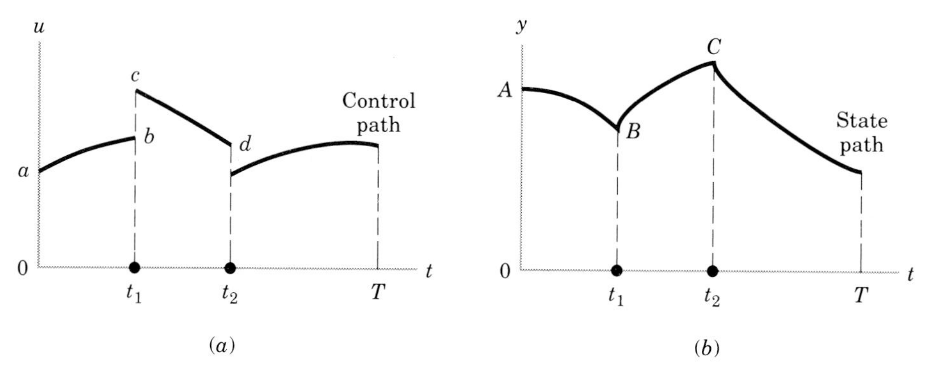

FIGURE 7.1

influence the state variable. Thus any chosen control path $u(t)$ will imply an associated state path $y(t)$. Our task is to choose the optimal admissible control path $u^*(t)$ which, along with the associated optimal admissible state path $y^*(t)$, will optimize the objective functional over a given time interval $[0, T]$.

Special Features of Optimal Control Problems

It is a noteworthy feature of optimal control theory that a control path does not have to be continuous in order to become admissible; it only needs to be *piecewise continuous*. This means that it is allowed to contain jump discontinuities, as illustrated in Fig. 7.1*a*, although we cannot permit any discontinuities that involve an infinite value of u. A good illustration of piecewise-continuous control in daily life is the on-off switch on the computer or lighting fixture. Whenever we turn the switch on ($u = 1$) and off ($u = 0$), the control path experiences a jump.

The state path $y(t)$, on the other hand, does have to be continuous throughout the time period $[0, T]$. But, as illustrated in Fig. 7.1*b*, it is permissible to have a finite number of sharp points, or corners. That is to say, to be admissible, a state path only needs to be *piecewise differentiable*.[1] Note that each sharp point on the state path occurs at the time the control path makes a jump. The reason for this timing coincidence lies in the procedure by which the solution to the problem is obtained. Once we have

[1]Sharp points on a state path can also be accommodated in the calculus of variations via the Weierstrass-Erdmann conditions. We have not discussed the subject in this book, because of the relative infrequency of its application in economics. The interested reader can consult any book on the calculus of variations.

found that the segment of the optimal control path for the time interval $[0, t_1)$ is, say, the curve ab in Fig. 7.1a, we then try to determine the corresponding segment of the optimal state path. This may turn out to be, say, the curve AB in Fig. 7.1b, whose initial point satisfies the given initial condition. For the next time interval, $[t_1, t_2)$, we again determine the optimal state-path segment on the basis of the pre-found optimal control, curve cd, but this time we must take point B as the "initial point" of the optimal state-path segment. Therefore, point B serves at once as the terminal point of the first segment, and the initial point of the second segment, of the optimal state path. For this reason, there can be no discontinuity at point B, although it may very well emerge as a sharp point. Like admissible control paths, admissible state paths must have a finite y value for every t in the time interval $[0, T]$.

Another feature of importance is that optimal control theory is capable of directly handling a constraint on the control variable u, such as the restriction $u(t) \in \mathscr{U}$ for all $t \in [0, T]$, where $\mathscr{U}$ denotes some bounded control set. The control set can in fact be a closed, convex set, such as $u(t) \in [0, 1]$. The fact that $\mathscr{U}$ can be a *closed* set means that corner solutions (boundary solutions) can be admitted, which injects an important nonclassical feature into the framework. When this feature is combined with the possibility of jump discontinuities on the control path, an interesting phenomenon called a *bang-bang solution* may result. Assuming the control set to be $\mathscr{U} = [0, 1]$, if for instance the optimal control path turns out to jump as follows:

$$\begin{aligned} u^*(t) &= 1 \qquad \text{for } t \in [0, t_1) \\ u^*(t) &= 0 \qquad \text{for } t \in [t_1, t_2) \qquad (t_1 < t_2) \\ u^*(t) &= 1 \qquad \text{for } t \in [t_2, T] \qquad (t_2 < T) \end{aligned}$$

then we are "banging" against one boundary of the set $\mathscr{U}$, and then against the other, in succession; hence, the name "bang-bang."

Finally, we point out that the simplest problem in optimal control theory, unlike in the calculus of variations, has a free terminal state (vertical terminal line) rather than a fixed terminal point. The primary reason for this is as follows: In the development of the fundamental first-order condition known as *the maximum principle*, we shall invoke the notion of an *arbitrary* Δu. Any arbitrary Δu must, however, imply an associated Δy. If the problem has a fixed terminal state, we need to pay heed to whether the associated Δy will lead ultimately to the designated terminal state. Hence, the choice of Δu may not be fully and truly arbitrary. If the problem has a free terminal state (vertical terminal line), on the other hand, then we can let the arbitrary Δu lead to wherever it may without having to worry about the final destination of y. And that simplifies the problem.

The Simplest Problem

Based on the preceding discussion, we may state the simplest problem of optimal control as

$$
\begin{aligned}
&\text{Maximize} && V = \int_0^T F(t, y, u)\, dt \\
(7.1)\quad &\text{subject to} && \dot{y} = f(t, y, u) \\
& && y(0) = A \qquad y(T) \text{ free} \qquad (A, T \text{ given}) \\
&\text{and} && u(t) \in \mathscr{U} \qquad \text{for all } t \in [0, T]
\end{aligned}
$$

Here, as in the subsequent discussion, we shall deal exclusively with the *maximization* problem. This way, the necessary conditions for optimization can be stated with more specificity and less confusion. When a minimization problem is encountered, we can always reformulate it as a maximization problem by simply attaching a minus sign to the objective functional. For example, minimizing $\int_0^T F(t, y, u)\, dt$ is equivalent to maximizing $\int_0^T -F(t, y, u)\, dt$.

In (7.1), the objective functional still takes the form of a definite integral, but the integrand function F no longer contains a y' argument as in the calculus of variations. Instead, there is a new argument u. The presence of the control variable u necessitates a linkage between u and y, to tell us how u will specifically affect the course taken by the state variable y. This information is provided by the equation $\dot{y} = f(t, y, u)$, where the dotted symbol $\dot{y}$, denoting the time derivative dy/dt, is an alternative notation to the y' symbol used heretofore.[2] At the initial time, the first two arguments in the f function must take the given value $t = 0$ and $y(0) = A$, so only the third argument is up to us to choose. For some chosen policy at $t = 0$, say, $u_1(0)$, this equation will yield a specific value for $\dot{y}$, say $\dot{y}_1(0)$, which entails a specific direction the y variable is to move. A different policy, $u_2(0)$, will in general give us a different value, $\dot{y}_2(0)$, via the f function. And a similar argument should apply to other points of time. What this equation does, therefore is to provide the mechanism whereby our choice of the control u can be translated into a specific pattern of movement of the state variable y. For this reason, this equation is referred to as the *equation of motion* for the state variable (or the *state equation* for short). Normally, the linkage between u and y can be adequately described by a first-order differential equation $\dot{y} = f(t, y, u)$. However, if it happens that the pattern of change of the state variable y cannot be captured by the first

---

[2] Even though the $\dot{y}$ and y' symbols are interchangeable, we shall exclusively use $\dot{y}$ in the context of optimal control theory, to make a visual distinction from the calculus-of-variations context.

derivative $\dot{y}$ but requires the use of the second derivative $\ddot{y} \equiv d^2y/dt^2$, then the state equation will take the form of a second-order differential equation, which we must transform into a pair of first-order differential equations. The complication is that, in the process, an additional state variable must be introduced into the problem. An example of such a situation can be found in Sec. 8.4.

We shall consistently use the lowercase letter f as the function symbol in the equation of motion, and reserve the capital-letter F for the integrand function in the objective functional. Both the F and f functions are assumed to be continuous in all their arguments, and possess continuous first-order partial derivatives with respect to t and y, but not necessarily with respect to u.

The rest of problem (7.1) consists of the specifications regarding the boundaries and control set. While the vertical-terminal-line case is the simplest, other terminal-point specifications can be accommodated, too. As to the control set, the simplest case is for $\mathscr{U}$ to be the open set $\mathscr{U} = (-\infty, +\infty)$. If so, the choice of u will in effect be unconstrained, in which case we can omit the statement $u(t) \in \mathscr{U}$ from the problem altogether.

A Special Case

As a special case, consider the problem where the choice of u is unconstrained, and where the equation of motion takes the particularly simple form

$$\dot{y} = u$$

Then the optimal control problem becomes

$$\begin{aligned}
&\text{Maximize} && V = \int_0^T F(t, y, u)\, dt \\
(7.2)\qquad &\text{subject to} && \dot{y} = u \\
&\text{and} && y(0) = A \qquad y(T) \text{ free} \qquad (A, T \text{ given})
\end{aligned}$$

By substituting the equation of motion into the integrand function, however, we can eliminate $\dot{y}$, and rewrite the problem as

$$\begin{aligned}
&\text{Maximize} && V = \int_0^T F(t, y, \dot{y})\, dt \\
(7.2')\qquad &\text{subject to} && y(0) = A \qquad y(T) \text{ free} \qquad (A, T \text{ given})
\end{aligned}$$

This is precisely the problem of the calculus of variations with a vertical terminal line. The fundamental link between the calculus of variations and optimal control theory is thus apparent. But the equations of motion encountered in optimal control problems are generally much more complicated than in (7.2).

7.2 THE MAXIMUM PRINCIPLE

The most important result in optimal control theory—a first-order necessary condition—is known as *the maximum principle*. This term was coined by the Russian mathematician L. S. Pontryagin and his associates.[3] As mentioned earlier in Sec. 1.4, however, the same technique was independently discovered by Magnus Hestenes, a mathematician at the University of California, Los Angeles, who later also extended Pontryagin's results. The statement of the maximum principle involves the concepts of the Hamiltonian function and costate variable. We must therefore first explain these concepts.

The Costate Variable and the Hamiltonian Function

Three types of variables were already presented in the problem statement (7.1): t (time), y (state), and u (control). It turns out that in the solution process, yet another type of variable will emerge. It is called the *costate variable* (or *auxiliary variable*), to be denoted by λ. As we shall see, a costate variable is akin to a Lagrange multiplier and, as such, it is in the nature of a valuation variable, measuring the shadow price of an associated state variable. Like y and u, the variable λ can take different values at different points of time. Thus the symbol λ is really a short version of $\lambda(t)$.

The vehicle through which the costate variable gains entry into the optimal control problem is the *Hamiltonian function*, or simply the *Hamiltonian*, which figures very prominently in the solution process. Denoted by H, the Hamiltonian is defined as

$$H(t, y, u, \lambda) \equiv F(t, y, u) + \lambda(t) f(t, y, u) \tag{7.3}$$

Since H consists of the integrand function F plus the product of the costate variable and the function f, it itself should naturally be a function with four arguments: t, y, u as well as λ. Note that, in (7.3), we have assigned a unitary coefficient to F, which is in contrast to the yet undetermined $\lambda(t)$

[3] L. S. Pontryagin, V. G. Boltyanskii, R. V. Gamkrelidze, and E. F. Mishchenko, *The Mathematical Theory of Optimal Processes*, translated from the Russian by K. N. Trirogoff, Interscience, New York, 1962. This book won the 1962 Lenin Prize for Science and Technology.

coefficient for f. Strictly speaking, the Hamiltonian should have been written as

$$H \equiv \lambda_0 F(t, y, u) + \lambda(t) f(t, y, u) \tag{7.4}$$

where λ_0 is a nonnegative constant, also yet undetermined. For the vertical-terminal-line problem (7.1), it turns out that the constant λ_0 is always nonzero (strictly positive); thus, it can be normalized to a unit value, thereby reducing (7.4) to (7.3). The fact that $\lambda_0 \neq 0$ in the simplest problem is due to two stipulations of the maximum principle. First, the multipliers λ_0 and $\lambda(t)$ cannot vanish simultaneously at *any* point of time. Second, the solution to the vertical-terminal-line problem must satisfy the transversality condition $\lambda(T) = 0$, to be explained in the ensuing discussion. The condition $\lambda(T) = 0$ would require a nonzero value for λ_0 at $t = T$. But since λ_0 is a nonnegative constant, we can conclude that λ_0 is a positive constant, which can then be normalized to unity.

For formulations of the optimal control problem other than (7.1), on the other hand, λ_0 may turn out to be zero, thereby invalidating the Hamiltonian in (7.3). The purist would therefore insist on checking in every problem that λ_0 is indeed positive, before using the Hamiltonian (7.3). The checking process would involve a demonstration that $\lambda_0 = 0$ would lead to a contradiction and violate the aforementioned stipulation that λ_0 and $\lambda(t)$ cannot vanish simultaneously.[4] In reality, however, the eventuality of a zero λ_0 occurs only in certain unusual (some say "pathological") situations where the solution of the problem is actually independent of the integrand function F, that is, where the F function does not matter in the solution process.[5] This is, of course, why the coefficient λ_0 should be set equal to zero, so as to drive out the F function from the Hamiltonian. Since most of the problems encountered in economics are those where the F function does matter, the prevalent practice among economists is simply to assume $\lambda_0 > 0$, then normalize it to unity and use the Hamiltonian (7.3), even when the problem is not one with a vertical terminal line. This is the practice we shall follow.

The Maximum Principle

In contrast to the Euler equation which is a single second-order differential equation in the state variable y, the maximum principle involves two

[4] For specific examples of the checking process, see Akira Takayama, *Mathematical Economics*, 2d ed., Cambridge University Press, Cambridge, 1985, pp. 617–618, 674–675, and 679–680.

[5] An example of such a problem can be found in Morton I. Kamien and Nancy L. Schwartz, *Dynamic Optimization: The Calculus of Variations and Optimal Control in Economics and Management, 2d ed*. Elsevier, New York, 1991, p. 149.

first-order differential equations in the state variable y and the costate variable λ. Besides, there is a requirement that the Hamiltonian be maximized with respect to the control variable u at every point of time. For pedagogical effectiveness, we shall first state and discuss the conditions involved, before providing the rationale for the maximum principle.

For the problem in (7.1), and with the Hamiltonian defined in (7.3), the maximum principle conditions are

$$\begin{aligned}
&\max_u H(t, y, u, \lambda) && \text{for all } t \in [0, T] \\
&\dot{y} = \frac{\partial H}{\partial \lambda} && \text{[equation of motion for } y\text{]} \\
&\dot{\lambda} = -\frac{\partial H}{\partial y} && \text{[equation of motion for } \lambda\text{]} \\
&\lambda(T) = 0 && \text{[transversality condition]}
\end{aligned} \tag{7.5}$$

The symbol $\max_u H$ means that the Hamiltonian is to be maximized with respect to u alone as the choice variable. An equivalent way of expressing this condition is

$$H(t, y, u^*, \lambda) \geq H(t, y, u, \lambda) \qquad \text{for all } t \in [0, T] \tag{7.6}$$

where u^* is the optimal control, and u is any other control value. In the following discussion, we shall, for simplicity, sometimes use the shorter notation "Max H" to indicate this requirement without explicitly mentioning u. The reader will note that it is this requirement of maximizing H with respect to u that gives rise to the name "the maximum principle."

It might appear on first thought that the requirement in (7.6) could have been more succinctly embodied in the first-order condition $\partial H/\partial u = 0$ (properly supported by an appropriate second-order condition). The truth, however, is that the requirement $\max_u H$ is a much broader statement of the requirement. In Fig. 7.2, we have drawn three curves, each indicating a possible plot of the Hamiltonian H against the control variable u at a specific point of time, for specific values of y and λ. The control region is assumed to be the closed interval $[a, c]$. For curve 1, which is differentiable with respect to u, the maximum of H occurs at $u = b$, an interior point of the control region $\mathscr{U}$; in this case, the equation $\partial H/\partial u = 0$ could indeed serve to identify the optimal control at that point of time. But if curve 2 is the relevant curve, then the control in $\mathscr{U}$ that maximizes H is $u = c$, a boundary point of $\mathscr{U}$. Thus the condition $\partial H/\partial u = 0$ does not apply, even though the curve is differentiable. And in the case of curve 3, with the Hamiltonian linear in u, the maximum of H occurs at $u = a$, another boundary point, and the condition $\partial H/\partial u = 0$ is again inapplicable because

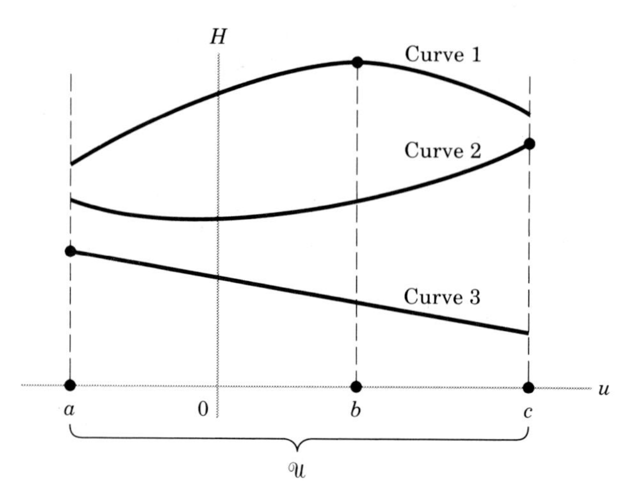

FIGURE 7.2

nowhere is that derivative equal to zero. In short, while the condition $\partial H/\partial u = 0$ may serve the purpose when the Hamiltonian is differentiable with respect to u and yields an interior solution, the fact that the control region may be a closed set, with possible boundary solutions, necessitates the broader statement $\max_u H$. In fact, under the maximum principle the Hamiltonian is not even required to be differentiable with respect to u.

The case where the Hamiltonian is linear in u is of special interest. For one thing, it is an especially simple situation to handle when H plots against u as either a positively sloped or a negatively sloped straight line, since the optimal control is then always to be found at a boundary of $\mathscr{U}$. The only task is to determine which boundary. (If H plots against u as a horizontal straight line, then there is no unique optimal control.) More importantly, this case serves to highlight how a thorny situation in the calculus of variations has now become easily manageable in optimal control theory. In the calculus of variations, whenever the integrand function is linear in y', resulting in $F_{y'y'} = 0$, the Euler equation may not yield a solution satisfying the given boundary conditions. In optimal control theory, in contrast, this case poses no problem at all.

Moving on to the other parts of (7.5), we note that the condition $\dot{y} = \partial H/\partial \lambda$ is nothing but a restatement of the equation of motion for the state variable originally specified in (7.1). To reexpress $\dot{y}$ as a partial derivative of H with respect to the costate variable λ is solely for the sake of showing the symmetry between this equation of motion and that for the costate variable. Note, however, that in the latter equation of motion, $\dot{\lambda}$ is the *negative* of the partial derivative of H with respect to the state variable y. Together, the two equations of motion are referred to collectively as the *Hamiltonian system*, or the *canonical system* (meaning the "standard" system of differential equations) for the given problem. Although we have more than one differential equation to deal with in optimal control theory—one for every state variable and every costate variable—each differential equation is of the first order only. Since the control variable never appears in the derivative form, there is no differential equation for u in the Hamiltonian system. But from the basic solution of (7.5) one can, if desired, derive a differential equation for the control variable. And, in some models, it may turn out to be more convenient to deal with a dynamic system in the variables (y, u) in place of the canonical system in the variables (y, λ).

The last condition in (7.5) is the transversality condition for the free-terminal-state problem—one with a vertical terminal line. As we would expect, such a condition only concerns what should happen at the terminal time T.

EXAMPLE 1 Find the curve with the shortest distance from a given point P to a given straight line L. We have encountered this problem before in the calculus of variations. To reformulate it as an optimal control problem, let point P be $(0, A)$, and assume, without loss of generality, that the line L is a vertical line. (If the given position of line L is not vertical, it can always be made so by an appropriate rotation of the axes.) The previously used F function, $(1 + y'^2)^{1/2}$ can be rewritten as $(1 + u^2)^{1/2}$, provided we let $y' = u$, or $\dot{y} = u$. Also, to convert the distance-minimization problem to one of maximization, we must attach a minus sign to the old integrand. Then our problem is to

$$\begin{aligned} &\text{Maximize} && V = \int_0^T -(1+u^2)^{1/2}\,dt \\ (7.7)\qquad &\text{subject to} && \dot{y} = u \\ &\text{and} && y(0) = A \qquad y(T)\text{ free} \qquad (A, T\text{ given}) \end{aligned}$$

Note that the control variable is not constrained, so the optimal control will be an interior solution.

Step i We begin by writing the Hamiltonian function

$$H = -(1+u^2)^{1/2} + \lambda u \tag{7.8}$$

Observing that H is differentiable and nonlinear, we can apply the first-order condition $\partial H/\partial u = 0$ to get

$$\frac{\partial H}{\partial u} = -\frac{1}{2}(1+u^2)^{-1/2}(2u) + \lambda = 0$$

This yields the solution[6]

$$u(t) = \lambda(1 - \lambda^2)^{-1/2} \tag{7.9}$$

Further differentiation of $\partial H/\partial u$ using the product rule yields

$$\frac{\partial^2 H}{\partial u^2} = -(1 + u^2)^{-3/2} < 0$$

Thus the result in (7.9) does maximize H. Since (7.9) expresses u in terms of λ, however, we must now look for a solution for λ.

Step ii To do that, we resort to the equation of motion for the costate variable $\dot{\lambda} = -\partial H/\partial y$ in (7.5). But since (7.8) shows that H is independent of y, we have

$$\dot{\lambda} = -\frac{\partial H}{\partial y} = 0 \qquad \Rightarrow \qquad \lambda(t) = \text{constant} \tag{7.10}$$

Conveniently, the transversality condition $\lambda(T) = 0$ in (7.5) is sufficient for definitizing the constant. For if λ is a constant, then its value at $t = T$ is also its value for all t. Thus,

$$\lambda^*(t) = 0 \qquad \text{for all } t \in [0, T] \tag{7.10'}$$

Looking back to (7.9), we can now also conclude that

$$u^*(t) = 0 \qquad \text{for all } t \in [0, T] \tag{7.11}$$

Step iii From the equation of motion $\dot{y} = u$, we are now able to write

$$\dot{y} = 0 \qquad \Rightarrow \qquad y(t) = \text{constant} \tag{7.12}$$

Moreover, the initial condition $y(0) = A$ enables us to definitize this constant and write

$$y^*(t) = A \tag{7.12'}$$

---

[6] The $\partial H/\partial u = 0$ equation can be written as

$$u(1 + u^2)^{-1/2} = \lambda$$

Squaring both sides, multiplying through by $(1 + u^2)$, and collecting terms, we get

$$u^2(1 - \lambda^2) = \lambda^2$$

This result implies that $\lambda^2 \neq 1$, for otherwise the equation would become $0 = 1$, which is impossible. Dividing both sides by the nonzero quantity $(1 - \lambda^2)$, and taking the square root, we finally arrive at (7.9).

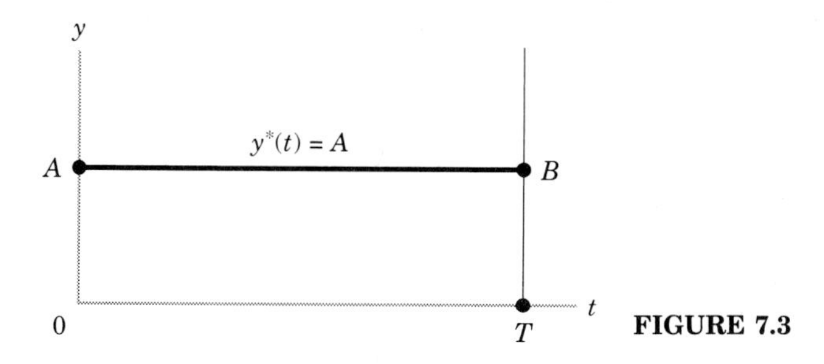

FIGURE 7.3

This y^* path, illustrated in Fig. 7.3, is a horizontal straight line. Alternatively, it may be viewed as a path orthogonal to the vertical terminal line.

EXAMPLE 2 Find the optimal control that will

$$\begin{aligned} &\text{Maximize} && V = \int_0^2 (2y - 3u)\, dt \\ (7.13)\qquad &\text{subject to} && \dot{y} = y + u \\ & && y(0) = 4 \qquad y(2) \text{ free} \\ &\text{and} && u(t) \in \mathscr{U} = [0, 2] \end{aligned}$$

Since this problem is characterized by linearity in u and a closed control set, we can expect boundary solutions to occur.

Step i The Hamiltonian of (7.13), namely,

$$H = 2y - 3u + \lambda(y + u) = (2 + \lambda)y + (\lambda - 3)u$$

is linear in u, with slope $\partial H/\partial u = \lambda - 3$. If at a given point of time, we find $\lambda > 3$, then an upward-sloping curve like curve 1 in Fig. 7.4 will prevail; to maximize H, we have to choose $u^* = 2$. If $\lambda < 3$, on the other hand, then curve 2 will prevail, and we must choose $u^* = 0$ instead. In short,

$$(7.14)\qquad u^*(t) = \begin{Bmatrix} 2 \\ 0 \end{Bmatrix} \qquad \text{if } \lambda(t) \begin{Bmatrix} > \\ < \end{Bmatrix} 3$$

Both $u^* = 2$ and $u^* = 0$ are, of course, boundary solutions. Note that, because H is linear in u, the usual first-order condition $\partial H/\partial u = 0$ is inapplicable in our search for u^*.

Step ii It is our next task to determine $\lambda(t)$, as needed in (7.14). From the equation of motion for λ, we have the differential equation

$$\dot{\lambda} = -\frac{\partial H}{\partial y} = -2 - \lambda \qquad \text{or} \qquad \dot{\lambda} + \lambda = -2$$

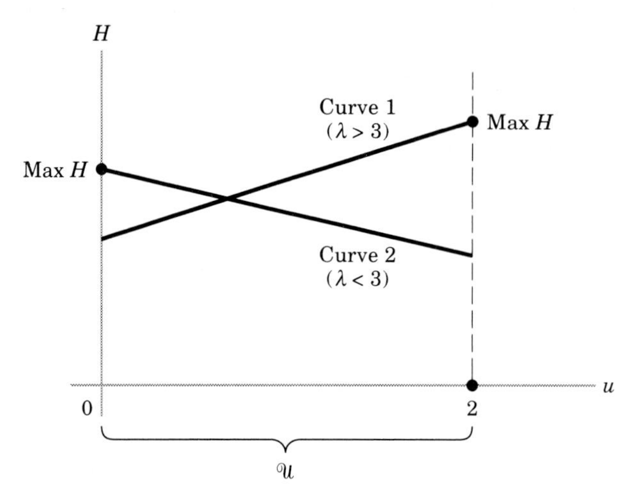

FIGURE 7.4

Its general solution is[7]

$$\lambda(t) = ke^{-t} - 2 \qquad (k \text{ arbitrary})$$

Since the arbitrary constant k can be definitized to $k = 2e^2$ by using the transversality condition $\lambda(T) = \lambda(2) = 0$, we can write the definite solution as

$$\lambda^*(t) = 2e^2e^{-t} - 2 = 2e^{2-t} - 2 \tag{7.15}$$

Note that $\lambda^*(t)$ is a decreasing function, falling steadily from an initial value $\lambda^*(0) = 2e^2 - 2 \simeq 12.778$ to a terminal value $\lambda^*(2) = 2 - 2 = 0$. Thus, λ^* exceeds 3 at first, but eventually falls below 3. The critical point of time, when $\lambda^* = 3$ and when the optimal control should be switched from $u^* = 2$ to $u^* = 0$, can be found by setting $\lambda^*(t) = 3$ in (7.15) and solving for t. Denoting that particular t by the Greek letter τ, we have

$$\tau = 2 - \ln 2.5 \simeq 1.084 \tag{7.16}$$

Consequently, the optimal control in (7.14) can be restated more specifically in two phases:

$$\begin{aligned} &\text{Phase I:} \qquad u^*_{\text{I}} \equiv u^*[0, \tau) = 2 \\ &\text{Phase II:} \qquad u^*_{\text{II}} \equiv u^*[\tau, 2] = 0 \end{aligned} \tag{7.17}$$

[7]First-order linear differential equations of this type are explained in Sec. 14.1 of Alpha C. Chiang, *Fundamental Methods of Mathematical Economics*, 3d ed., McGraw-Hill, New York, 1984.

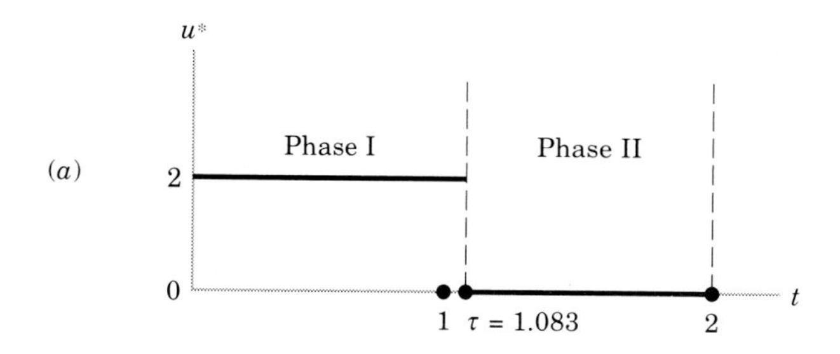

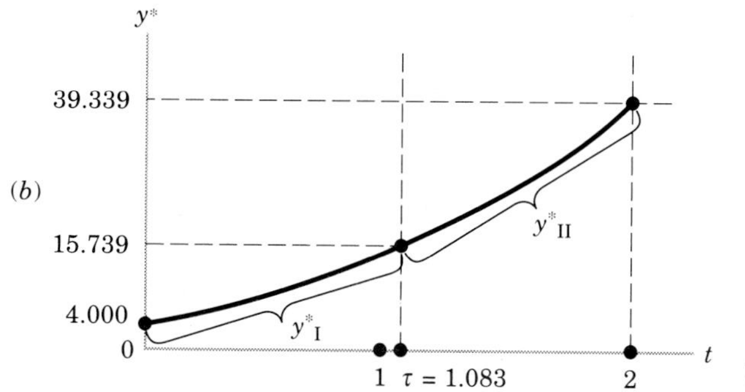

FIGURE 7.5

As graphically depicted in Fig. 7.5*a*, this optimal control exemplifies a simple variety of bang-bang.

Step iii Even though the problem only asks for the optimal control path, we can also find the optimal state path, in two phases. In phase I, the equation of motion for y is $\dot{y} = y + u = y + 2$, or

$$\dot{y} - y = 2 \qquad \text{with initial value } y(0) = 4$$

Its solution is

$$y^*_{\text{I}} \equiv y^*[0, \tau) = 2(3e^t - 1) \tag{7.18}$$

In phase II, the equation of motion for y is $\dot{y} = y + 0$, or

$$\dot{y} - y = 0$$

with general solution

$$y^*_{\text{II}} \equiv y^*[\tau, 2] = ce^t \qquad (c \text{ arbitrary}) \tag{7.19}$$

Note that the constant c cannot be definitized by the initial condition $y(0) = 4$ given in (7.13) because we are already in phase II, beyond $t = 0$. Nor can it be definitized by any terminal condition because the terminal state is free. However, the reader will recall that the optimal y path is required to be continuous, as illustrated in Fig. 7.1*b*. Consequently, the initial value of y^*_{II} must be set equal to the value of y^*_{I} evaluated at τ. Inasmuch as

$$y^*_{I}(\tau) = 2(3e^{\tau} - 1) \qquad [\text{by } (7.18)]$$

and

$$y^*_{II}(\tau) = ce^{\tau} \qquad [\text{by } (7.19)]$$

we find, by equating these two expressions and solving for c, that $c = 2(3 - e^{-\tau})$, so that the optimal y path in phase II is

$$y^*_{II} = 2(3 - e^{-\tau})e^{t} \simeq 5.324e^{t} \tag{7.19'}$$

The value of y^* at the switching time τ is approximately $2(3e^{1.096} - 1) = 15.739$.

By joining the two paths (7.18) and (7.19′), we obtain the complete y^* path for the time interval $[0, 2]$, as shown in Fig. 7.5*b*. In this particular example, the joined path happens to look like one single exponential curve, but the two segments are in fact parts of two separate exponential functions.

EXERCISE 7.2

1 In Example 2, $\lambda(t)$ is a decreasing function, and attains the value 3 at only one point of time, τ. What would happen if it turned out that $\lambda(t) = 3$ for all t?

2 Find the optimal paths of the control, state, and costate variables to

$$\begin{aligned} &\text{Maximize} && \int_0^4 3y\,dt \\ &\text{subject to} && \dot{y} = y + u \\ &&& y(0) = 5 \qquad y(4) \text{ free} \\ &\text{and} && u(t) \in [0, 2] \end{aligned}$$

Be sure to check that the Hamiltonian is maximized rather than minimized.

3 Find the optimal paths of the control, state, and costate variables to

$$\begin{aligned} &\text{Maximize} && \int_0^2 (y - u^2)\,dt \\ &\text{subject to} && \dot{y} = u \\ &\text{and} && y(0) = 0 \qquad y(2) \text{ free} \qquad u(t) \text{ unconstrained} \end{aligned}$$

Make sure that the Hamiltonian is maximized rather than minimized.

4 Find the optimal paths of the control, state, and costate variables to

$$\begin{aligned} &\text{Maximize} && \int_0^1 -\frac{1}{2}(y^2 + u^2)\,dt \\ &\text{subject to} && \dot{y} = u - y \\ &\text{and} && y(0) = 1 \qquad y(1) \text{ free} \qquad u(t) \text{ unconstrained} \end{aligned}$$

Make sure that the Hamiltonian is maximized rather than minimized.

[*Hint*: The two equations of motion should be solved simultaneously. Review the material on simultaneous differential equations in Alpha C. Chiang, *Fundamental Methods of Mathematical Economics*, 3d ed., McGraw-Hill, New York, 1984, Sec. 18.2.]

7.3 THE RATIONALE OF THE MAXIMUM PRINCIPLE

We shall now explain the rationale underlying the maximum principle. What we plan to do is not to give a detailed proof—the complete proof given by Pontryagin and his associates (Chap. 2 of their book) runs for as many as 40 pages—but to present a variational view of the problem, to make the maximum principle plausible. This will later be reinforced by a comparison of the conditions of the maximum principle with the Euler equation and the other conditions of the calculus of variations.

A Variational View of the Control Problem

To make things simple, it is assumed here that the control variable u is unconstrained, so that u^* is an interior solution. Moreover, the Hamiltonian function is assumed to be differentiable with respect to u, and the $\partial H/\partial u = 0$ condition can be invoked in place of the condition "$\underset{u}{\text{Max}}\ H$." As usual, we take the initial point to be fixed, but the terminal point is allowed to vary. This will enable us to derive certain transversality conditions in the process of the discussion. The problem is then to

$$\begin{aligned} &\text{Maximize} && V = \int_0^T F(t, y, u)\,dt \\ &\text{subject to} && \dot{y} = f(t, y, u) \qquad (7.20) \\ &\text{and} && y(0) = y_0 \quad \text{(given)} \end{aligned}$$

Step i As the first step in the development of the maximum principle, let us incorporate the equation of motion into the objective functional, and then reexpress the functional in terms of the Hamiltonian.

The reader will observe that, if the variable y always obeys the equation of motion, then the quantity $[f(t, y, u) - \dot{y}]$ will assuredly take a zero value for all t in the interval $[0, T]$. Thus, using the notion of Lagrange multipliers, we can form an expression $\lambda(t)[f(t, y, u) - \dot{y}]$ for each value of t, and still get a zero value. Although there is an infinite number of values of t in the interval $[0, T]$, summing $\lambda(t)[f(t, y, u) - \dot{y}]$ over t in the period $[0, T]$ would still yield a total value of zero:

$$\int_0^T \lambda(t)[f(t, y, u) - \dot{y}]\, dt = 0 \tag{7.21}$$

For this reason, we can augment the old objective functional by the integral in (7.21) without affecting the solution. That is, we can work with the new objective functional

$$\begin{aligned} \mathscr{V} &\equiv V + \int_0^T \lambda(t)[f(t, y, u) - \dot{y}]\, dt \\ &= \int_0^T \{F(t, y, u) + \lambda(t)[f(t, y, u) - \dot{y}]\}\, dt \end{aligned} \tag{7.22}$$

confident that $\mathscr{V}$ will have the same value as V, as long as the equation of motion in (7.20) is adhered to at all times.

Previously, we have defined the Hamiltonian function as

$$H(t, y, u, \lambda) \equiv F(t, y, u) + \lambda(t) f(t, y, u)$$

The substitution of the H function into (7.22) can simplify the new functional to the form

$$\begin{aligned} \mathscr{V} &= \int_0^T [H(t, y, u, \lambda) - \lambda(t)\dot{y}]\, dt \\ &= \int_0^T H(t, y, u, \lambda)\, dt - \int_0^T \lambda(t)\dot{y}\, dt \end{aligned} \tag{7.22'}$$

It is important to distinguish clearly between the second term in the Hamiltonian, $\lambda(t) f(t, y, u)$, on the one hand, and the Lagrange-multiplier expression, $\lambda(t)[f(t, y, u) - \dot{y}]$, on the other. The latter explicitly contains $\dot{y}$, whereas the former does not. When the last integral in (7.22′) is integrated by parts,[8] we find that

$$-\int_0^T \lambda(t)\dot{y}\,dt = -\lambda(T)y_T + \lambda(0)y_0 + \int_0^T y(t)\dot{\lambda}\,dt$$

Hence, through substitution of this result, the new objective functional can be further rewritten as

$$(7.22'')\quad \mathscr{V} = \underbrace{\int_0^T \left[H(t,y,u,\lambda) + y(t)\dot{\lambda}\right] dt}_{\Omega_1} \underbrace{- \lambda(T)y_T}_{\Omega_2} \underbrace{+ \lambda(0)y_0}_{\Omega_3}$$

The $\mathscr{V}$ expression comprises three additive component terms, Ω_1, Ω_2, and Ω_3. Note that while the Ω_1 term, an integral, spans the entire planning period $[0, T]$, the Ω_2 term is exclusively concerned with the terminal time T, and Ω_3 is concerned only with the initial time.

Step ii The value of $\mathscr{V}$ depends on the time paths chosen for the three variables y, u, and λ, as well as the values chosen for T and y_T. In the present step, we shall focus on λ.

The λ variable, being a Lagrange multiplier, differs fundamentally from u and y, for the choice of the $\lambda(t)$ path will produce no effect on the value of $\mathscr{V}$, so long as the equation of motion $\dot{y} = f(t, y, u)$ is strictly adhered to, that is, so long as

$$(7.23)\qquad \dot{y} = \frac{\partial H}{\partial \lambda} \qquad \text{for all } t \in [0, T]$$

So, to relieve us from further worries about the effect of $\lambda(t)$ on $\mathscr{V}$, we simply impose (7.23) as a necessary condition for the maximization of $\mathscr{V}$. This accounts for one of the three conditions of the maximum principle. This, of course, is hardly an earth-shaking step, since the equation of motion is actually given as a part of the problem itself.

---

[8]The formula for integrating a definite integral by parts has been given in (2.15). Here, we replace the symbol u in (2.15) by w, because u is now used to denote the control variable. Let

$$v = \lambda(t) \qquad (\text{implying that } dv = \dot{\lambda}\,dt)$$
$$w = y(t) \qquad (\text{implying that } dw = \dot{y}\,dt)$$

Then, since $\lambda(t)\dot{y}\,dt = v\,dw$, we have

$$-\int_0^T \lambda(t)\dot{y}\,dt = -[\lambda(t)y(t)]_0^T + \int_0^T y(t)\dot{\lambda}\,dt$$

which leads to the result in the text.

Step iii We now turn to the $u(t)$ path and its effect on the $y(t)$ path. If we have a known $u^*(t)$ path, and if we perturb the $u^*(t)$ path with a perturbing curve $p(t)$, we can generate "neighboring" control paths

$$u(t) = u^*(t) + \epsilon p(t) \tag{7.24}$$

one for each value of ϵ. But, in accordance with the equation of motion $\dot{y} = f(t, y, u)$, there will then occur for each ϵ a corresponding perturbation in the $y^*(t)$ path. The neighboring y paths can be written as

$$y(t) = y^*(t) + \epsilon q(t) \tag{7.25}$$

Furthermore, if T and y_T are variable, we also have

$$T = T^* + \epsilon\,\Delta T \qquad \text{and} \qquad y_T = y_T^* + \epsilon\,\Delta y_T \tag{7.26}$$

$$\left(\text{implying } \frac{dT}{d\epsilon} = \Delta T \text{ and } \frac{dy_T}{d\epsilon} = \Delta y_T\right)$$

In view of the u and y expressions in (7.24) and (7.25), we can express $\mathscr{V}$ in terms of ϵ, so that we can again apply the first-order condition $d\mathscr{V}/d\epsilon = 0$. The new version of $\mathscr{V}$ is

$$\begin{aligned} \mathscr{V} = &\int_0^{T(\epsilon)} \{H[t, y^* + \epsilon q(t), u^* + \epsilon p(t), \lambda] + \dot{\lambda}[y^* + \epsilon q(t)]\}\, dt \\ &- \lambda(T) y_T + \lambda(0) y_0 \end{aligned} \tag{7.27}$$

Step iv We now apply the condition $d\mathscr{V}/d\epsilon = 0$. In the differentiation process, the integral term yields, by formula (2.11), the derivative

$$\int_0^{T(\epsilon)} \left\{\left[\frac{\partial H}{\partial y} q(t) + \frac{\partial H}{\partial u} p(t)\right] + \dot{\lambda} q(t)\right\} dt + \left[H + \dot{\lambda} y\right]_{t=T} \frac{dT}{d\epsilon} \tag{7.28}$$

And the derivative of the second term in (7.27) with respect to ϵ is, from the product rule,

$$-\lambda(T)\frac{dy_T}{d\epsilon} - y_T \frac{d\lambda(T)}{dT}\frac{dT}{d\epsilon} = -\lambda(T)\,\Delta y_T - y_T \dot{\lambda}(T)\,\Delta T \tag{7.29}$$

[by (7.26)]

On the other hand, the $\lambda(0) y_0$ term in (7.27) drops out in differentiation. Thus $d\mathscr{V}/d\epsilon$ is the sum of (7.28) and (7.29). In adding these two expressions, however, we note that one component of (7.28) can be rewritten as follows:

$$\left[\dot{\lambda} y\right]_{t=T} \frac{dT}{d\epsilon} = \dot{\lambda}(T) y_T\,\Delta T \qquad \text{[by (7.26)]}$$

Thus, when the sum of (7.28) and (7.29) is set equal to zero, the first-order condition emerges (after rearrangement) as

$$\frac{d\mathscr{V}}{d\epsilon} = \int_0^T \left[\left(\frac{\partial H}{\partial y} + \dot{\lambda} \right) q(t) + \frac{\partial H}{\partial u} p(t) \right] dt + [H]_{t=T} \Delta T - \lambda(T) \Delta y_T = 0 \tag{7.30}$$

The three components of this derivative relate to different arbitrary things: The integral contains arbitrary perturbing curves $p(t)$ and $q(t)$, whereas the other two involve arbitrary ΔT and Δy_T, respectively. Consequently, each of the three must individually be set equal to zero in order to satisfy (7.30). By setting the integral component equal to zero, we can deduce two conditions:

$$\dot{\lambda} = -\frac{\partial H}{\partial y} \qquad \text{and} \qquad \frac{\partial H}{\partial u} = 0$$

The first gives us the *equation of motion* for the costate variable λ (or the *costate equation* for short). And the second represents a weaker version of the "$\max_u H$" condition—weaker in the sense that it is predicated on the assumption that H is differentiable with respect to u and there is an interior solution. Since the simplest problem has a fixed T and free y_T, the ΔT term in (7.30) is automatically equal to zero, but Δy_T is not. In order to make the $-\lambda(T)\,\Delta y_T$ expression vanish, we must impose the restriction

$$\lambda(T) = 0$$

This explains the transversality condition in (7.5).

Note that although the $\lambda(t)$ path was earlier, in Step ii, brushed aside as having no effect on the value of the objective functional, it has now made an impressive comeback. We see that, in order for the maximum principle to work, the $\lambda(t)$ path is not to be arbitrarily chosen, but is required to follow a prescribed equation of motion, and it must end with a terminal value of zero if the problem has a free terminal state.

7.4 ALTERNATIVE TERMINAL CONDITIONS

What will happen to the maximum principle when the terminal condition specifies something other than a vertical terminal line? The general answer is that the first three conditions in (7.5) will still hold, but the transversality condition will assume some alternative form.

Fixed Terminal Point

The reason why the problem with a fixed terminal point (with both the terminal state and the terminal time fixed) does not qualify as the "simplest" problem in optimal control theory is that the specification of a fixed terminal point entails a complication in the notion of an "arbitrary" perturbing curve $p(t)$ for the control variable u. If the perturbation of the $u^*(t)$ path is supposed to generate through the equation of motion $\dot{y} = f(t, y, u)$ a corresponding perturbation in the $y^*(t)$ path that has to end at a preset terminal state, then the choice of the perturbing curve $p(t)$ is not truly arbitrary. The question then arises as to whether we can still legitimately deduce the condition $\partial H/\partial u = 0$ from (7.30).

Fortunately, the validity of the maximum principle is not affected by this compromise in the arbitrariness of $p(t)$. For simplicity, however, we shall not go into details to demonstrate this point. For our purposes, it suffices to state that, with a fixed terminal point, the transversality is replaced by the condition

$$y(T) = y_T \qquad (T, y_T \text{ given})$$

Horizontal Terminal Line (Fixed-Endpoint Problem)

If the problem has a horizontal terminal line (with a free terminal time but a fixed "endpoint," meaning a fixed terminal state), then y_T is fixed ($\Delta y_T = 0$), but T is not (ΔT is arbitrary). From the second and third component terms in (7.30), it is easy to see that the transversality condition for this case is

$$[H]_{t=T} = 0 \tag{7.31}$$

The Hamiltonian function must attain a zero value at the optimal terminal time. But there is no restriction on the value of λ at time T.

Terminal Curve

In case a terminal curve $y_T = \phi(T)$ governs the selection of the terminal point, then ΔT and Δy_T are not both arbitrary, but are linked to each other by the relation $\Delta y_T = \phi'(T)\,\Delta T$. Using this to eliminate Δy_T, we can combine the last two terms in (7.30) into a single expression involving ΔT only:

$$[H]_{t=T}\,\Delta T - \lambda(T)\phi'(T)\,\Delta T = [H - \lambda\phi']_{t=T}\,\Delta T$$

It follows that, for an arbitrary ΔT, the transversality condition should be

$$[H - \lambda\phi']_{t=T} = 0 \tag{7.32}$$

Truncated Vertical Terminal Line

Now consider the problem in which the terminal time T is fixed, but the terminal state is free to vary, only subject to $y_T \geq y_{\min}$, where $y_{\min}$ denotes a given minimum permissible level of y.

Only two types of outcome are possible in the optimal solution: $y_T^* > y_{\min}$, or $y_T^* = y_{\min}$. In the former outcome, the terminal restriction is automatically satisfied. Thus, the transversality condition for the problem with a regular vertical terminal line would apply:

$$\lambda(T) = 0 \qquad \text{for } y_T^* > y_{\min} \tag{7.33}$$

In the other outcome, $y_T^* = y_{\min}$, since the terminal restriction is binding, the admissible neighboring y paths consist only of those that have terminal states $y_T \geq y_{\min}$. If we evaluate (7.25) at $t = T$ and let $y_T^* = y_{\min}$, we obtain

$$y_T = y_{\min} + \epsilon q(T)$$

Assuming that $q(T) > 0$ on the perturbing curve $q(t)$,[9] the requirement $y_T \geq y_{\min}$ would dictate that $\epsilon \geq 0$. But, by the Kuhn-Tucker conditions, the nonnegativity of ϵ would alter the first-order condition $d\mathscr{V}/d\epsilon = 0$ to $d\mathscr{V}/d\epsilon \leq 0$ for our maximization problem.[10] It follows that (7.30) would now yield an inequality transversality condition

$$-\lambda(T)\,\Delta y_T \leq 0$$

At the same time, we can see from (7.26) that, given $\epsilon \geq 0$, the requirement of $y_T \geq y_{\min}$—which is the same as $y_T \geq y_T^*$ in the present context—implies $\Delta y_T \geq 0$. Thus the preceding inequality transversality condition reduces to

$$\lambda(T) \geq 0 \qquad \text{for } y_T^* = y_{\min} \tag{7.34}$$

Combining (7.33) and (7.34) and omitting the * symbol, we can finally write a single summary statement of the transversality condition as follows:

$$\lambda(T) \geq 0 \qquad y_T \geq y_{\min} \qquad (y_T - y_{\min})\lambda(T) = 0 \tag{7.35}$$

Note that the last part of this statement represents the familiar complementary-slackness condition from the Kuhn-Tucker conditions. As in the similar problem with a truncated vertical terminal line in the calculus of variations, the practical application of (7.35) is not as complicated as the condition may appear. We can always try the $\lambda(T) = 0$ condition first, and check whether the resulting y_T^* value satisfies the terminal restriction $y_T^* \geq y_{\min}$. If it does, the problem is solved. If not, we then set $y_T^* = y_{\min}$ in order to satisfy the complementary-slackness condition, and treat the problem as one with a given terminal point.

---

[9] This assumption does not bias the final result of the deductive process here.

[10] The Kuhn-Tucker conditions are explained in Alpha C. Chiang, *Fundamental Methods of Mathematical Economics*, 3d ed., McGraw-Hill, New York, 1984, Sec. 21.2.

Truncated Horizontal Terminal Line

Let the terminal state be fixed, but the terminal time T be allowed to vary subject to the restriction that $T^* \le T_{max}$, where T_{max} is the maximum permissible value of T—a preset deadline. Then we either have $T^* < T_{max}$ or $T^* = T_{max}$ in the optimal solution.

In the former outcome, the terminal restriction turns out to be nonbinding, and the transversality condition for the problem with a regular horizontal terminal line would still hold:

$$[H]_{t=T} = 0 \qquad \text{for } T^* < T_{max} \tag{7.36}$$

But if $T^* = T_{max}$, then by implication all the admissible neighboring y paths must have terminal time $T \le T_{max}$. By analogous reasoning to that leading to the result (7.34) for the truncated vertical terminal line, it is possible to establish the transversality condition

$$[H]_{t=T} \ge 0 \qquad \text{for } T^* = T_{max} \tag{7.37}$$

By combining (7.36) and (7.37) and omitting the * symbol, we then obtain the following summary statement of the transversality condition:

$$[H]_{t=T} \ge 0 \qquad T \le T_{max} \qquad (T - T_{max})[H]_{t=T} = 0 \tag{7.38}$$

EXAMPLE 1

$$\text{Maximize} \qquad V = \int_0^1 -u^2\,dt$$

$$\text{subject to} \qquad \dot{y} = y + u$$

$$\text{and} \qquad y(0) = 1 \qquad y(1) = 0$$

With fixed endpoints, we need no transversality condition in this problem.

Step i Since the Hamiltonian function is nonlinear:

$$H = -u^2 + \lambda(y + u)$$

and since u is unconstrained, we can apply the first-order condition

$$\frac{\partial H}{\partial u} = -2u + \lambda = 0$$

This yields the solution $u = \lambda/2$ or, more accurately,

$$u(t) = \tfrac{1}{2}\lambda(t) \tag{7.39}$$

Since $\partial^2 H/\partial u^2 = -2$ is negative, this $u(t)$ solution does maximize rather than minimize H. But since this solution is expressed in terms of $\lambda(t)$, we must find the latter path before $u(t)$ becomes determinate.

Step ii From the costate equation of motion

$$\dot{\lambda} = -\frac{\partial H}{\partial y} = -\lambda$$

we get the general solution

$$\lambda(t) = ke^{-t} \qquad (k \text{ arbitrary}) \tag{7.40}$$

To definitize the arbitrary constant, we try to resort to the boundary conditions, but, unfortunately, for the fixed-terminal-point problem these conditions are linked to the variable y instead of λ. Thus, it now becomes necessary first to look into the solution path for y.

Step iii The equation of motion for y is $\dot{y} = y + u$. Using (7.39) and (7.40), however, we can rewrite this equation as $\dot{y} = y + \frac{1}{2}ke^{-t}$, or

$$\dot{y} - y = \tfrac{1}{2}ke^{-t}$$

This is a first-order linear differential equation with a variable coefficient and a variable term, of the type $dy/dt + u(t)\,y = w(t)$—here with $u(t) = -1$ and $w(t) = \frac{1}{2}ke^{-t}$. Via a standard formula, its solution can be found as follows:[11]

$$\begin{aligned} y(t) &= e^{-\int -1\,dt}\left(c + \int \frac{1}{2}ke^{-t}e^{\int -1\,dt}\,dt\right) \\ &= e^{t}\left(c + \int \frac{1}{2}ke^{-t}e^{-t}\,dt\right) \\ &= e^{t}\left(c - \frac{1}{4}ke^{-2t}\right) \\ &= ce^{t} - \frac{1}{4}ke^{-t} \qquad (c \text{ arbitrary}) \end{aligned} \tag{7.41}$$

[11] See Alpha C. Chiang, *Fundamental Methods of Mathematical Economics*, 3d ed., McGraw-Hill, New York, 1984, Sec. 14.3. In performing the integrations involved in the application of the formula, we have omitted the constants of integration whenever they can be subsumed under other constants. Alternatively, we can find the complementary function and the particular integral separately and then combine them. With a variable term in the differential equation, we can obtain the particular integral by the method of undetermined coefficients (*ibid.*, Sec. 15.6)

Step iv Now the boundary conditions $y(0) = 1$ and $y(1) = 0$ are directly applicable, and they give us the following definite values for c and k:

$$c = \frac{1}{1-e^2} \qquad k = \frac{4e^2}{1-e^2}$$

Consequently, substituting these into (7.41), (7.40), and (7.39), we have the following definite solutions for the three optimal paths:

$$y^*(t) = \frac{1}{1-e^2}e^t - \frac{e^2}{1-e^2}e^{-t}$$

$$\lambda^*(t) = \frac{4e^2}{1-e^2}e^{-t} \qquad \text{and} \qquad u^*(t) = \frac{2e^2}{1-e^2}e^{-t}$$

The search for the $u^*(t)$, $y^*(t)$, and $\lambda^*(t)$ paths in the present problem turns out to be an intertwined process. This is because, unlike the simplest problem of optimal control, where the transversality condition $\lambda(T) = 0$ may enable us to get a definite solution of the costate path $\lambda^*(t)$ at an early stage of the game, the fixed-terminal-point problem does not allow the application of the boundary conditions on $y(0)$ and $y(T)$ until the final stage of the solution process.

EXAMPLE 2 Let us reconsider the preceding example, with the terminal condition $y(1) = 0$ replaced by the restriction

$$T = 1 \qquad y(1) \geq 3$$

The problem is then one with a truncated vertical terminal line, and the appropriate transversality condition is (7.35). We shall first try to solve this problem as if its vertical terminal line is *not* truncated. If $y^*(1)$ turns out to be ≥ 3, then the problem is solved; otherwise, we shall then redo the problem by setting $y(1) = 3$.

Step i The Hamiltonian remains the same as in Example 1, and the solution for the control variable is still

$$u(t) = \tfrac{1}{2}\lambda(t) \qquad \text{[from (7.39)]} \tag{7.42}$$

Step ii Although the general solution for λ is still

$$\lambda(t) = ke^{-t} \qquad \text{[from (7.40)]} \tag{7.43}$$

we now can use the transversality condition $\lambda(T) = 0$ or $\lambda(1) = 0$ to definitize the arbitrary constant. The result is $k = 0$, so that

$$\lambda^*(t) = 0 \tag{7.43'}$$

It then follows from (7.42) that

$$u^*(t) = 0 \tag{7.44}$$

Step iii From the equation of motion for y, we find

$$\dot{y} = y + u = y \qquad [\text{by } (7.44)]$$

The general solution of this differential equation is

$$y(t) = ce^t$$

where the constant c can be definitized to $c = 1$ by the initial condition $y(0) = 1$. Thus the optimal state path is

$$y^*(t) = e^t \tag{7.45}$$

Step iv It remains to check (7.45) against the terminal restriction. At the fixed terminal time $T = 1$, (7.45) gives us $y^*(1) = e$. This, unfortunately, violates the terminal restriction $y(1) \geq 3$. Thus, in order to satisfy the transversality condition (7.35), we now have to set $y(1) = 3$ and resolve the problem as one with a fixed terminal point. Note that had the terminal restriction been $T = 1$, $y(1) \geq 2$, then (7.45) would have been an acceptable solution.

EXAMPLE 3

$$\begin{aligned} \text{Maximize} \quad & V = \int_0^T -1\,dt \\ \text{subject to} \quad & \dot{y} = y + u \\ & y(0) = 5 \qquad y(T) = 11 \qquad T \text{ free} \\ \text{and} \quad & u(t) \in [-1, 1] \end{aligned}$$

This example illustrates the problem with a horizontal terminal line. Moreover, it illustrates the type of problem known as a *time-optimal problem*, whose objective is to reach some preset target in the minimum amount of time. The time-optimal nature of the problem is conveyed by the objective functional:

$$\int_0^T -1\,dt = [-t]_0^T = -T$$

Clearly, to *maximize* this integral is to *minimize* T.

Step i To begin with, form the Hamiltonian

$$H = -1 + \lambda(y + u) \tag{7.46}$$

Because this H function is linear in u, the $\partial H/\partial u = 0$ condition is inapplicable. And, with the control variable confined to the closed interval $[-1, 1]$, the optimal value of u at any point of time is expected to be a boundary value, either -1 or 1. Specifically, if $\lambda > 0$ (H is an increasing function of u), then $u^* = 1$ (upper bound); but if $\lambda < 0$, then $u^* = -1$. As a third possibility, if $\lambda = 0$ at some value of t, then the Hamiltonian will plot as a horizontal line against u, and u^* will become indeterminate at that point of time. This relationship between u^* and λ can be succinctly captured in the so-called *signum function*, denoted by the symbol sgn, and defined as follows:

$$(7.47) \qquad y = \operatorname{sgn} x \qquad \Leftrightarrow \qquad y = \begin{Bmatrix} 1 \\ \text{indeterminate} \\ -1 \end{Bmatrix} \quad \text{if } x \begin{Bmatrix} > \\ = \\ < \end{Bmatrix} 0$$

Note that if y is a signum function of x, then y (if determinate) can only take one of two values, and the *value* of y depends on the *sign* (not magnitude) of x.

Applied to the present problem, this function becomes

$$(7.48) \qquad u^* = \operatorname{sgn} \lambda \qquad \text{or} \qquad u^* = \begin{Bmatrix} 1 \\ -1 \end{Bmatrix} \quad \text{if } \lambda \begin{Bmatrix} > \\ < \end{Bmatrix} 0$$

Once more, we find that a knowledge of λ is necessary before u can be determined.

Step ii The equation of motion for the costate variable is, from (7.46),

$$\dot{\lambda} = -\frac{\partial H}{\partial y} = -\lambda$$

which integrates to

$$(7.49) \qquad \lambda(t) = ke^{-t} \qquad (k \text{ arbitrary})$$

In this result, $\lambda(t)$, being exponential, can take only a single algebraic sign—the sign of the constant k. Consequently, barring the eventuality of $k = 0$ so that $\lambda(t) = 0$ for all t (which eventuality in fact does not occur here), u^* must be determinate and adhere to a single sign—nay, a single constant value—in accordance with the signum function. For this reason, even though the linearity of the Hamiltonian in the control variable u results in a boundary solution in the present example, it produces no bang-bang phenomenon.

It turns out that the clue to the sign of k lies in the transversality condition $[H]_{t=T} = 0$. Using the H in (7.46), the λ in (7.49), and the terminal condition $y(T) = 11$, we can write the transversality condition as

$$-1 + ke^{-T}(11 + u^*) = 0$$

Since u^* is either 1 or -1, the quantity $(11 + u^*)$ must be positive, as is e^{-T}. Therefore, k must be positive in order to satisfy this condition. It then follows that $\lambda(t) > 0$ for all t, and that

$$u^*(t) = 1 \tag{7.50}$$

Step iii With $u^* = 1$ for all t, we can express the equation of motion of the state variable, $\dot{y} = y + u$, as

$$\dot{y} - y = 1$$

This fits the format of the first-order linear differential equation with a constant coefficient and a constant term, $dy/dt + ay = b$—here with $a = -1$ and $b = 1$. Its definite solution gives us the optimal y path[12]

$$y^*(t) = \left[y(0) - \frac{b}{a}\right]e^{-at} + \frac{b}{a} \tag{7.51}$$

$$= 6e^t - 1 \qquad [y(0) = 5]$$

Step iv Having obtained the optimal control and state paths $u^*(t)$ and $y^*(t)$, we next look for $\lambda^*(t)$. For this purpose, we first return to the transversality condition $[H]_{t=T} = 0$ to fix the value of the constant k. In view of (7.50) and (7.51), the transversality condition now reduces to

$$-1 + ke^{-T}(6e^T - 1 + 1) = 0 \qquad \text{or} \qquad 6k = 1$$

Thus $k = \frac{1}{6}$. Substituting this result in (7.49) then yields the optimal λ path

$$\lambda^*(t) = \tfrac{1}{6}e^{-t} \tag{7.52}$$

Step v The three optimal paths in (7.50), (7.51), and (7.52) portray the complete solution to the present problem except for the value of T^*. To calculate that, recall that the terminal state value is stipulated to be $y(T) = 11$. This, in conjunction with the $y^*(t)$ path obtained earlier, tells us that $11 = 6e^T - 1$, or $e^T = 2$. Hence,

$$T^* = \ln 2 \; (\simeq 0.6931)$$

The optimal paths for the various variables are easily graphed. We shall leave this to the reader.

---

[12] This solution formula is derived in Alpha C. Chiang, *Fundamental Methods of Mathematical Economics*, 3d ed., McGraw-Hill, New York, 1984, Sec. 14.1.

The Constancy of the Hamiltonian in Autonomous Problems

All the examples discussed previously share the common feature that the problems are "autonomous;" that is, the functions in the integrand and f in the equation of motion do not contain t as an explicit argument. An important consequence of this feature is that the optimal Hamiltonian—the Hamiltonian evaluated along the optimal paths of y, u, and λ—will have a constant value over time.

To see this, let us first examine the time derivative of the Hamiltonian $H(t, y, u, \lambda)$ in the general case:

$$\frac{dH}{dt} = \frac{\partial H}{\partial t} + \frac{\partial H}{\partial y}\dot{y} + \frac{\partial H}{\partial u}\dot{u} + \frac{\partial H}{\partial \lambda}\dot{\lambda}$$

When H is maximized, we have $\partial H/\partial u = 0$ (for an interior solution) or $\dot{u} = 0$ (for a boundary solution). Thus the third term on the right drops out. Moreover, the maximum principle also stipulates that $\dot{y} = \partial H/\partial \lambda$ and $\dot{\lambda} = -\partial H/\partial y$. So the second and fourth terms on the right exactly cancel out. The net result is that H^*, the Hamiltonian evaluated along the optimal paths of all variables, satisfies the equation

$$\frac{dH^*}{dt} = \frac{\partial H^*}{\partial t} \tag{7.53}$$

This result holds generally, for both autonomous and nonautonomous problems.

In the special case of an autonomous problem, since t is absent from the F and f functions as an explicit argument, the Hamiltonian must not contain the t argument either. Consequently, we have $\partial H^*/\partial t = 0$, so that

$$\frac{dH^*}{dt} = 0 \quad \text{or} \quad H^* = \text{constant} \qquad [\text{for autonomous problems}] \tag{7.54}$$

This result is of practical use in an autonomous problem with a horizontal terminal line. The transversality condition $[H]_{t=T} = 0$ is normally expected to hold at the terminal time only. But if the Hamiltonian is a constant in the optimal solution, then it must be zero for all t, and the transversality condition can be applied at any point of time.

In Example 3, for instance, we find that

$$H^* = -1 + \lambda^*(y^* + u^*) = -1 + \tfrac{1}{6}e^{-t}(6e^t - 1 + 1) = 0$$

This zero value of H^* prevails regardless of the value of t, which shows that the transversality condition is indeed satisfied at all t.

EXERCISE 7.4

1 Find the optimal paths of the control, state, and costate variables to

$$\text{Maximize} \quad \int_0^T -(t^2 + u^2)\,dt$$

$$\text{subject to} \quad \dot{y} = u$$

$$\text{and} \quad y(0) = 4 \qquad y(T) = 5 \qquad T \text{ free}$$

2 Find the optimal paths of the control, state, and costate variables to

$$\text{Maximize} \quad \int_0^4 3y\,dt$$

$$\text{subject to} \quad \dot{y} = y + u$$

$$y(0) = 5 \qquad y(4) \geq 300$$

$$\text{and} \quad 0 \leq u(t) \leq 2$$

3 We wish to move from the initial point $(0, 8)$ in the ty plane to achieve the terminal state value $y(T) = 0$ as soon as possible. Formulate and solve the problem, assuming that $dy/dt = 2u$, and that the control set is the closed interval $[-1, 1]$.

4 Find the optimal control path and the corresponding optimal state path that minimize the distance between the point of origin $(0, 0)$ and a terminal curve $y(T) = 10 - T^2$, $T > 0$. Graph the terminal curve and the $y^*(t)$ path.

5 Demonstrate the validity of the transversality condition (7.37) for the problem with a truncated horizontal terminal line.

7.5 THE CALCULUS OF VARIATIONS AND OPTIMAL CONTROL THEORY COMPARED

We have shown earlier in (7.2) and (7.2′) that a simple optimal control problem can be translated into an equivalent problem of the calculus of variations. One may wonder whether, in such a problem, the optimality conditions required by the maximum principle are also equivalent to those of the calculus of variations. The answer is that they indeed are.

For problem (7.2), the Hamiltonian function is

$$H = F(t, y, u) + \lambda u \tag{7.55}$$

Assuming this function to be differentiable with respect to u, we may list the following conditions by the maximum principle:

$$
\begin{aligned}
&\frac{\partial H}{\partial u} = F_u + \lambda = 0 \\
&\dot{y} = \frac{\partial H}{\partial \lambda} = u \\
&\dot{\lambda} = -\frac{\partial H}{\partial y} = -F_y \\
&\lambda(T) = 0
\end{aligned}
\tag{7.56}
$$

The first equation in (7.56) can be rewritten as $\lambda = -F_u$. But in view of the second equation therein, it can be further rewritten as

$$\lambda = -F_{\dot{y}} \tag{7.57}$$

Differentiation of (7.57) with respect to t yields

$$\dot{\lambda} = -\frac{d}{dt}F_{\dot{y}}$$

However, the third equation in (7.56) gives another expression for $\dot{\lambda}$. By equating the two expressions, we end up with the single condition

$$F_y - \frac{d}{dt}F_{\dot{y}} = 0$$

which is identical with the Euler equation (2.18).

When the Hamiltonian is maximized with respect to u, the condition $\partial H/\partial u$ should be accompanied by the second-order necessary condition $\partial^2 H/\partial u^2 \le 0$. Further differentiation of the $\partial H/\partial u$ expression in (7.56) yields

$$\frac{\partial^2 H}{\partial u^2} = F_{uu} = F_{\dot{y}\dot{y}} \le 0$$

This, of course, is the Legendre necessary condition. Thus the maximum principle is perfectly consistent with the conditions of the calculus of variations.

Now, let us take a look at the transversality conditions. For a control problem with a vertical terminal line, the transversality condition is $\lambda(T) = 0$. By (7.57), however, this may be written as $[-F_{\dot{y}}]_{t=T} = 0$, or, equivalently,

$$[F_{\dot{y}}]_{t=T} = 0$$

Again, this is precisely the transversality condition in the calculus of variations presented in (3.10).

For the problem with a horizontal terminal line, the optimal-control transversality condition is $[H]_{t=T} = 0$. In view of (7.55), this means $[F + \lambda u]_{t=T} = 0$. Using (7.56) again and after substituting $\dot{y}$ for u, however, we can transform this condition into

$$\left[F - F_{\dot{y}}\dot{y}\right]_{t=T} = 0$$

Except for the slight difference in symbology, this is precisely the transversality condition under the calculus of variations given in (3.11).

It can also be shown that the transversality condition for the problem with a terminal curve $y_T = \phi(T)$ under optimal control theory can be translated into the corresponding condition under the calculus of variations, and vice versa. The details of the demonstration are, however, left to the reader.

7.6 THE POLITICAL BUSINESS CYCLE

Applications of the maximum principle to problems of economics mushroomed between 1965 and 1975, and the technique has become fairly common. Its applications range from the more standard areas in macro- and microeconomics all the way to such topics as fishery, city planning, and pollution control. In the present section, we introduce an interesting model of William Nordhaus,[13] which shows that, in a democracy, attempts by an incumbent political party to prevent its rival party (or parties) from ousting it from office may encourage the pursuit of economic policies that result in a particular time profile for the rate of unemployment and the rate of inflation within each electoral period. Over successive electoral periods, the repetition of that pattern will then manifest itself as a series of business cycles rooted solely in the play of politics.

The Vote Function and the Phillips Tradeoff

The incumbent party, in control of the national government, is obliged in a democracy to pursue policies that appeal to a majority of the voters in order to keep on winning at the polls. In the present model, attention is focused on *economic* policies only, and in fact on only *two* economic variables: U (the rate of unemployment) and p (the rate of inflation). Since the ill effects of unemployment and inflation seem to have been the primary economic concerns of the electorate, this choice of focus is certainly reasonable. The reaction of the voters to any realized values of U and p is assumed to be

[13] William D. Nordhaus, "The Political Business Cycle," *Review of Economic Studies*, April 1975, pp. 169–190.

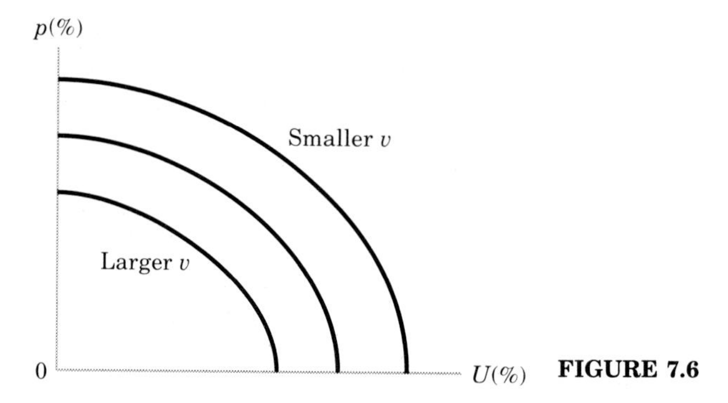

FIGURE 7.6

embodied in an (aggregate) *vote function*

$$v = v(U, p) \qquad (v_U < 0,\ v_p < 0) \tag{7.58}$$

where v is a measure of the vote-getting power of the incumbent party. The partial derivatives of v with respect to each argument are negative, because high values of U and p are both conducive to vote loss. This fact is reflected in Fig. 7.6, where, out of the three *isovote* curves illustrated, the highest one is associated with the lowest v. The notion of the isovote curve underscores the fact that, on the political side, there is a tradeoff between the two variables U and p. If the incumbent party displeases the voters by producing higher inflation, it can hope to recoup the vote loss via a sufficient reduction in the rate of unemployment.

Aside from the political tradeoff, the two variables under consideration are also linked to each other in an economic tradeoff via the expectations-augmented Phillips relation

$$p = \phi(U) + a\pi \qquad (\phi' < 0,\ 0 < a \le 1) \tag{7.59}$$

where π denotes the expected rate of inflation. Expectations are assumed to be formed adaptively, according to the differential equation

$$\dot{\pi} = b(p - \pi) \qquad (b > 0) \tag{7.60}$$

All in all, we now have three variables, U, p, and π. But which of these should be considered as state variables and which as control variables? For a variable to qualify as a state variable, it must come with a given equation of motion in the problem. Since (7.60) constitutes an equation of motion for π, we can take π as a state variable. The variable U, on the other hand, does not come with an equation of motion. But since U can affect p via (7.59) and then dynamically drive π via (7.60), we can use it as a control variable. To use U as a control variable, however, requires the implicit assumption that the government in power does have the ability to implement any target rate of unemployment it chooses at any point of time. As to the remaining variable, p, (7.59) shows that its value at any time t will become determinate once the values of the state and control variables are known. We may thus view it as neither a state variable nor a control variable, but, like v, just a function of the other two variables.

The Optimal Control Problem

Suppose that a party has just won the election at $t = 0$, and the next election is to be held T years later at $t = T$. The incumbent party then has a total of T years in which to impress the voters with its accomplishments (or outward appearances thereof) in order to win their votes. At any time in the period $[0, T]$, the pair of realized values of U and p will determine a specific value of v. Such values of v for different points of time must all enter into the objective functional of the incumbent party. But the various values may have to be weighted differently, depending on the time of occurrence. If the voters have a short collective memory and are influenced more by the events occurring near election time, then the v values of the later part of the period $[0, T]$ should be assigned heavier weights than those that come earlier. We may then formulate the optimal control problem of the incumbent party as follows:

$$
\begin{aligned}
&\text{Maximize} && \int_0^T v(U, p) e^{rt}\, dt \\
(7.61)\qquad &\text{subject to} && p = \phi(U) + a\pi \\
& && \dot{\pi} = b(p - \pi) \\
&\text{and} && \pi(0) = \pi_0 \qquad \pi(T) \text{ free} \qquad (\pi_0, T \text{ given})
\end{aligned}
$$

A few comments on (7.61) may be in order. First, the weighting system for the v values pertaining to different points of time has been given the specific form of the exponential function e^{rt}, where $r > 0$ denotes the rate of decay of memory. This function shows that the v values at later points of time are weighted more heavily. Note that, in contrast to the expression $e^{-\rho t}$, what we have here is not a discount factor, but its reverse. Second, we have retained the expectations-augmented Phillips relation in the problem statement. At the moment, however, we are not yet equipped to deal with such a constraint. Fortunately, the variable p can easily be eliminated by substituting that equation into the vote function and the equation of motion. Then the p equation will disappear as a separate constraint. Third, as the boundary conditions indicate, the incumbent party faces a vertical terminal line, since T (election time) is predetermined. Fourth, although the rate of unemployment is perforce nonnegative, no nonnegativity restriction has actually been placed on the control variable U. The plan—and this is an oft-used strategy—is to impose no restriction and just let the solution fall out however it may. If $U^*(t)$ turns out to have economically acceptable values for all t, then there is no need to worry; if not, and only if not, we shall have to amend the problem formulation.

As stated in (7.61), the problem contains general functions, and thus cannot yield a quantitative solution. To solve the problem quantitatively, Nordhaus assumes the following specific function forms:

$$v(U,p) = -U^2 - hp \qquad (h > 0) \tag{7.62}$$

$$p = (j - kU) + a\pi \qquad (j, k > 0, 0 < a \le 1) \tag{7.63}$$

From (7.62), it can be seen that the partial derivatives of v are indeed negative. In (7.63), we find that the Phillips relation $\phi(U)$ has been linearized. Using these specific functions, and after substituting (7.63) into (7.62), we now have the specific problem:

$$\begin{aligned} &\text{Maximize} && \int_0^T (-U^2 - hj + hkU - ha\pi)e^{rt}\,dt \\ &\text{subject to} && \dot{\pi} = b[j - kU - (1-a)\pi] \\ &\text{and} && \pi(0) = \pi_0 \qquad \pi(T) \text{ free} \qquad (\pi_0, T \text{ given}) \end{aligned} \tag{7.64}$$

Maximizing the Hamiltonian

The Hamiltonian is

$$H = (-U^2 - hj + hkU - ha\pi)e^{rt} + \lambda b[j - kU - (1-a)\pi] \tag{7.65}$$

Maximizing H with respect to the control variable U, we have the equation

$$\frac{\partial H}{\partial U} = (-2U + hk)e^{rt} - \lambda bk = 0$$

This implies the control path

$$U(t) = \tfrac{1}{2}k(h - \lambda b e^{-rt}) \tag{7.66}$$

Since $\partial^2 H/\partial U^2 = -2e^{rt} < 0$, the control path in (7.66) indeed maximizes the Hamiltonian at every point of time, as the maximum principle requires.

The presence of λ in the $U(t)$ solution now necessitates a search for the $\lambda(t)$ path.

The Optimal Costate Path

The search for the costate path begins with the equation of motion

$$\dot{\lambda} = -\frac{\partial H}{\partial \pi} = hae^{rt} + \lambda b(1-a)$$

When rewritten into the form

$$\dot{\lambda} - b(1 - a)\lambda = hae^{rt}$$

the equation is readily recognized as a first-order linear differential equation with a constant coefficient but a variable term. Employing the standard methods of solution,[14] we can find the complementary function λ_c and the particular integral $\bar{\lambda}$ to be, respectively,

$$\lambda_c = Ae^{b(1-a)t} \qquad (A \text{ arbitrary})$$

$$\bar{\lambda} = \frac{ha}{B}e^{rt} \qquad (B \equiv r - b + ab)$$

It follows that the general solution for λ is

$$\lambda(t) = \lambda_c + \bar{\lambda} = Ae^{b(1-a)t} + \frac{ha}{B}e^{rt} \tag{7.67}$$

Note that the two constants A and B are fundamentally different in nature; B is merely a shorthand symbol we have chosen in order to simplify the notation, but A is an arbitrary constant to be definitized.

To definitize A, we can make use of the transversality condition for the vertical-terminal-line problem, $\lambda(T) = 0$. Letting $t = T$ in (7.67), applying the transversality condition, and solving for A, we find that $A = (-ha/B)e^{BT}$. It follows that the definite solution—the optimal costate path—is

$$\lambda^*(t) = \frac{ha}{B}[e^{rt} - e^{BT + b(1-a)t}] \tag{7.67'}$$

The Optimal Control Path

Now that we have found $\lambda^*(t)$, all it takes is to substitute (7.67′) into (7.66) to derive the optimal control path. The result is, upon simplification,

$$U^*(t) = \frac{kh}{2B}[(r - b) + bae^{B(T-t)}] \tag{7.68}$$

It is this control path that the incumbent party should follow in the interest of its reelection in year T.

---

[14] See Alpha C. Chiang, *Fundamental Methods of Mathematical Economics*, 3d ed., McGraw-Hill, New York, 1984, Sec. 14.1 (for the complementary function) and Sec. 15.6 (for the particular integral).

What are the economic implications of this path? First, we note that U^* is a decreasing function of t. Specifically, we have

$$\frac{dU^*}{dt} = -\frac{1}{2}khbae^{B(T-t)} < 0 \tag{7.69}$$

because k, h, b, and a are all positive, as is the exponential expression. The vote-maximizing economic policy is, accordingly, to set a high unemployment level immediately upon winning the election at $t = 0$, and then let the rate of unemployment fall steadily throughout the electoral period $[0, T]$. In fact, the optimal levels of unemployment at time 0 and time T can be exactly determined. They are

$$U^*(0) = \frac{kh}{2B}\left[(r-b) + bae^{BT}\right]$$

$$U^*(T) = \frac{kh}{2B}\left[(r-b) + ba\right] = \frac{kh}{2}$$

Note that the terminal unemployment level, $kh/2$, is a positive quantity. And since $U^*(T)$ represents the lowest point on the entire $U^*(t)$ path, the U^* values at all values of t in $[0, T]$ must uniformly be positive. This means that the strategy of deliberately not imposing any restriction on the variable U does not cause any trouble regarding the sign of U in the present case. However, to be economically meaningful, $U^*(0)$ must be less than unity or, more realistically, less than some maximum tolerable unemployment rate $U_{max} < 1$. Unless the parameter values are such that $U^*(0) \leq U_{max}$, the model needs to be reformulated by including the constraint $U(t) \in [0, U_{max}]$.

The typical optimal unemployment path, $U^*(t)$, is illustrated in Fig. 7.7, where we also show the repetition of similar $U^*(t)$ patterns over successive electoral periods generates political business cycles. However, the curvature of the $U^*(t)$ path does not always have to be concave as in Fig. 7.7. For, by differentiating (7.69) with respect to t, we can see that

$$\frac{d^2U^*}{dt^2} = \frac{1}{2}Bkhbae^{B(T-t)} \gtreqless 0 \quad \text{as} \quad B \gtreqless 0 \tag{7.70}$$

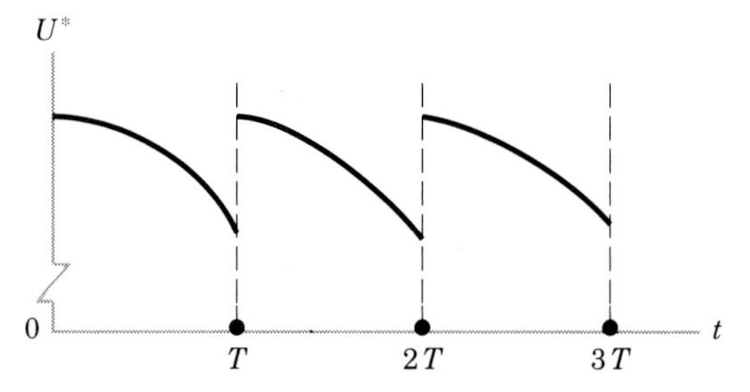

FIGURE 7.7

If, for illustration, $r = 0.03$, $b = 0.30$, and $a = 0.70$, then $B = r - b + ab = -0.06$, and the $U^*(t)$ path will be concave. But a positive value of B will turn the curvature the other way. And, with parameter changes in different electoral periods, the position as well as the curvature of the $U^*(t)$ paths in succeeding electoral periods may very well change. Nevertheless, the political business cycles will tend to persist.

The Optimal State Path

The politically inspired cyclical tendency in the control variable U must also spill over to the state variable π, and hence also to the actual rate of inflation p. The general pattern would be for the optimal rate of inflation to be relatively low at the beginning of each electoral period, but undergo a steady climb. In other words, the time profile of p^* tends to be the opposite of that of U^*. But we shall not go into the actual derivation of the optimal inflation rate path here.

The reader is reminded that the conclusions of the present model—like those of any other model—are intimately tied to the assumptions adopted. In particular, the specific forms chosen for the vote function in (7.62) and the expectations-augmented Phillips relation (7.63) undoubtedly exert an important influence upon the final solutions. Alternative assumptions—such as changing the linear term $-hp$ in (7.62) to $-hp^2$—are likely to modify significantly the $U^*(t)$ solution as well as the $p^*(t)$ solution. But the problem formulation is also likely to become much more complicated.

EXERCISE 7.6

1 (*a*) What would happen in the Nordhaus model if the optimal control path were characterized by $dU^*/dt = 0$ for all t?

(*b*) What values of the various parameters would cause dU^*/dt to become zero?

(*c*) Interpret economically the parameter values you have indicated in part (*b*).

2 What parameter values would, aside from causing $dU^*/dt = 0$, also cause $U^*(t) = 0$ for all t? Explain the economic implications and rationale for such an outcome.

3 How does a change in the value of parameter r (the rate of decay of voter memory) affect the slope of the $U^*(t)$ path? Discuss the economic implications of your result. [*Note:* $B = r - b + ab$.]

4 Eliminate the e^{rt} term in the objective function in (7.64) and write out the new problem.

(*a*) Solve the new problem by carrying out the same steps as those illustrated in the text for the original problem.

(*b*) Check your results by setting $r = 0$ in the results of the original model, especially (7.68) and (7.69).

7.7 ENERGY USE AND ENVIRONMENTAL QUALITY

When an economy is faced with an essential resource that is exhaustible, say, fossil fuel, it certainly behooves its citizenry to be concerned about the question of how the limited supply of the resource is best to be allocated for use over time. We have discussed some of the issues involved in Sec. 6.3 with the method of the calculus of variations. But the citizens of the present-day world are also intensely concerned about the quality of the environment in which they live. If the use of the exhaustible fuel generates pollution as a by-product, then what is the optimal time path for energy use? We shall illustrate how such a question can be tackled by optimal control theory with a model of Bruce A. Forster.[15]

The Social Utility Function

Let $S(t)$ denote the stock of the fuel and $E(t)$ the rate of extraction of fuel (and energy use) at any time t. Then we have

$$\dot{S} = -E \tag{7.71}$$

Energy use, E, makes possible the production of goods and services for consumption, C, which creates utility, but also generates a flow of pollution, P, which creates disutility. Instead of writing a simple utility function $U(E)$ as we did in the introductory section of this chapter, therefore, our new utility function should contain two arguments, $C(E)$ and $P(E)$. Forster specifies the consumption function and the pollution function as follows:

$$C = C(E) \qquad (C' > 0, C'' < 0) \tag{7.72}$$

$$P = P(E) \qquad (P' > 0, P'' > 0) \tag{7.73}$$

While energy use raises consumption at a decreasing rate, it generates pollution at an increasing rate. In this particular model, pollution is assumed for simplicity to be nonaccumulating; that is, it is a flow that

---

[15] Bruce A. Forster, "Optimal Energy Use in a Polluted Environment," *Journal of Environmental Economics and Management*, 1980, pp. 321–333. While this paper presents three different models, here we shall confine our attention exclusively to the first one, which assumes a single energy source producing a nonaccumulating pollutant. Another model, treating pollution as a stock variable and involving two state variables, will be discussed in Sec. 8.5.

dissipates and does not build up into a stock. This is exemplified by the auto-emission type of pollution.

The social utility function depends on consumption and pollution, with derivatives as follows:

$$U = U(C, P) \qquad (U_C > 0,\ U_P < 0,\ U_{CC} < 0,\ U_{PP} < 0,\ U_{CP} = 0) \tag{7.74}$$

The specification of $U_C > 0$ and $U_{CC} < 0$ shows that the marginal utility of consumption is positive but diminishing. In contrast, the specification of $U_P < 0$ and $U_{PP} < 0$ reveals that the marginal utility of pollution is negative and diminishing (given a particular increase in P, U_P may decrease from, say, -2 to -3). In terms of the marginal *dis*utility of $P(\equiv -U_P)$, therefore, $U_{PP} < 0$ signifies *increasing marginal disutility.*

Since both C and P in turn depend on E, the social utility hinges, in the final analysis, on energy use alone—positively via consumption and negatively via pollution. This means that C and P can both be substituted out, leaving E as the prime candidate for the control variable. The only other variable in the model, S, appears in (7.71) in the derivative form. Since it is a variable dynamically driven by the control variable E, it is clear that S plays the role of a state variable here.

The Optimal Control Problem

If an Energy Board is appointed to plan and chart the optimal time path of the energy-use variable E over a specified time period $[0, T]$, the dynamic optimization problem it must solve may take the form

$$\begin{aligned} &\text{Maximize} && \int_0^T U[C(E), P(E)]\, dt \\ &\text{subject to} && \dot{S} = -E \\ &\text{and} && S(0) = S_0 \qquad S(T) \ge 0 \qquad (S_0, T \text{ given}) \end{aligned} \tag{7.75}$$

This particular formulation allows no discount factor in the integrand, a practice in the Ramsey tradition. And it grants the Energy Board the discretion of selecting the terminal stock $S(T)$, subject only to the nature-imposed restriction that it be nonnegative. Since the terminal time is fixed, a truncated vertical terminal line characterizes the problem. With a single control variable E and a single state variable S, the problem can be solved with relative ease.

Maximizing the Hamiltonian

The Hamiltonian function

$$H = U[C(E), P(E)] - \lambda E \tag{7.76}$$

involves nonlinear differentiable functions U, C, and P. Thus we may maximize H with respect to the control variable simply by setting its first derivative equal to zero:

$$\frac{\partial H}{\partial E} = U_C C'(E) + U_P P'(E) - \lambda = 0 \tag{7.77}$$

When solved, this equation expresses E in terms of λ.

To make sure that (7.77) maximizes rather than minimizes the Hamiltonian, we check the sign of $\partial^2 H/\partial E^2$. Since U_C and U_P are, like U, dependent on E, the second derivative is

$$\frac{\partial^2 H}{\partial E^2} = U_{CC}C'^2 + U_C C'' + U_{PP}P'^2 + U_P P'' < 0$$

[by (7.72), (7.73), and (7.74)]

Its negative sign guarantees that (7.77) does maximize H.

The Optimal Costate and Control Paths

To elicit more information about E from (7.77), however, we need to look into the time path of λ. The maximum principle tells us that the equation of motion for λ is

$$\dot{\lambda} = -\frac{\partial H}{\partial S} = 0 \qquad \text{implying } \lambda(t) = c \text{ (constant)} \tag{7.78}$$

To definitize the constant c, we can resort to the transversality condition. For the problem at hand, with a truncated vertical terminal line, the condition takes the form

$$\lambda(T) \geq 0 \qquad S(T) \geq 0 \qquad \lambda(T)S(T) = 0 \qquad \text{[by (7.35)]} \tag{7.79}$$

In practical applications of this type of condition, the standard initial step is to set $\lambda(T) = 0$ (as if the terminal line is not truncated) to see whether the solution will work. Since $\lambda(t)$ is a constant by (7.78), to set $\lambda(T) = 0$ is in effect to set $\lambda(t) = 0$ for all t.

With $\lambda(t) = 0$, (7.77) reduces to an equation in the single variable E,

$$U_C C'(E) + U_P P'(E) = 0 \tag{7.80}$$

which, in principle, can be solved for the optimal control path. Since this equation is independent of the variable t, its solution is constant over time:

$$E^*(t) = E^* \quad \text{(a specific constant)} \qquad [\text{if } \lambda^*(t) = 0] \tag{7.81}$$

Whether this solution is acceptable from the standpoint of the $S(T) \geq 0$ restriction is, however, still a matter to be settled.

Meanwhile, it is useful to examine the economic meaning of (7.80). The first term, $U_C C'(E)$, measures the effect of a change in E on U via C. That is, it represents the marginal utility of energy use through its contribution to consumption. Similarly, the $U_P P'(E)$ term expresses the marginal disutility of energy use through its pollution effect. What (7.80) does is, therefore, to direct the Energy Board to select a E^* value that balances the marginal utility and disutility of energy use, much as the familiar MC = MR rule requires a firm to balance the cost and revenue effects of production.

The Optimal State Path

It remains to check whether the E^* solution in (7.81) can satisfy the $S(T) \geq 0$ restriction. For this purpose, we must find the state path $S(t)$.

With constant energy use, the equation of motion $\dot{S} = -E$ can be readily integrated to yield

$$S(t) = -Et + k \qquad (k \text{ arbitrary})$$

Moreover, by setting $t = 0$ in this result, it is easily seen that k represents the initial fuel stock S_0. Thus the optimal state path can be written as

$$S^*(t) = S_0 - E^*t \tag{7.82}$$

The value of $S^*(t)$ at any time clearly hinges on the magnitude of E^*. Since the functions we have been working with—$U(C, P)$, $C(E)$, and $P(E)$—are all general functions, E^* cannot be assigned a specific numerical value or parametric expression. Nonetheless, we can examine the $S(T) \geq 0$ restriction qualitatively.

Consider the three illustrative values of E^* in Fig. 7.8, where ${E^*}_1 < {E^*}_2 < {E^*}_3$. When the low rate of energy use ${E^*}_1$ is in effect, the optimal stock $S^*(t)$ appears as a gently downward-sloping straight line, such that $S^*(T)$ is positive. With the higher rate of energy use ${E^*}_2$, on the other hand, the fuel stock is brought down to zero at time T. Even so, the Energy Board would still be within the bounds of its given authority. But the other case, ${E^*}_3$, entailing the premature exhaustion of the fuel endowment, would patently violate the $S(T) \geq 0$ stipulation. Thus, if our solution E^* in (7.81)

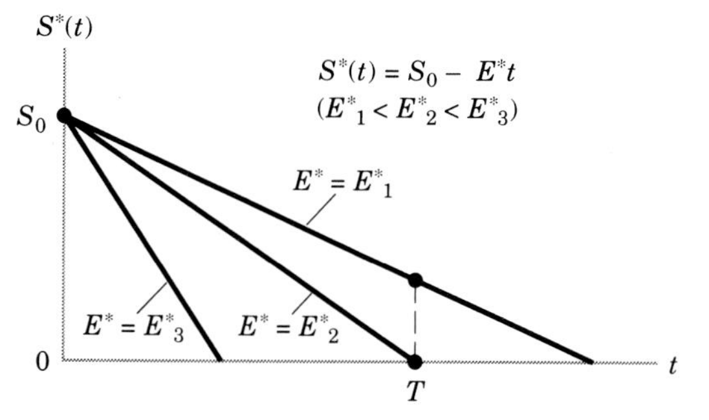

FIGURE 7.8

turns out to be like E^*_1 or E^*_2, then the transversality condition (7.79) is met, and the problem solved. But if it is like E^*_3, then we must set $S(T) = 0$, and solve the problem as one with a given terminal point. In that event, the E^* value can be directly found from (7.82) by setting $t = T$ and $S(T) = 0$:

$$E^* = \frac{S_0}{T} \qquad [\text{if (7.81) violates } S^*(T) \geq 0] \tag{7.83}$$

This new E^* can be illustrated by E^*_2 in Fig. 7.8.

It is a notable feature of this model that E^*, the optimal rate of energy use, is constant over time. This constancy of E^* prevails whether the terminal-stock restriction, $S(T) \geq 0$, is binding [as in (7.83)] or nonbinding [as in (7.81)]. What assumption in the model is responsible for this particular result? The answer lies in the absence of a discount factor. If a discount factor is introduced [see Prob. 3 in Exercise 7.7], the E^* path then will turn out to be decreasing over time, provided that $\lambda^*(t) > 0$. However, in the other case where $\lambda^*(t) = 0$, E^* will still be constant.

EXERCISE 7.7

1 Suppose that the solution of (7.80) turns out to be E^*_3, which fails to satisfy the $S(T) \geq 0$ restriction, and consequently the Energy Board is forced to select the lower rate of energy use, E^*_2, instead.

(*a*) Does E^*_3 satisfy the "marginal utility = marginal disutility" rule?

(*b*) Does E^*_2 satisfy that rule? If not, is E^*_2 characterized by "marginal utility < marginal disutility" or "marginal utility > marginal disutility"? Explain.

2 Let the terminal condition in the Forster model be changed from $S(T) \geq 0$ to $S(T) \geq S_{min} > 0$. How should Fig. 7.8 be modified to show that E^*_1 results in $S(T) > S_{min}$, E^*_2 results in $S(T) = S_{min}$, and E^*_3 results in $S(T) < S_{min}$?

3 Suppose that the Energy Board decides to incorporate a discount factor $e^{-\rho t}$ into the objective functional.

(*a*) Write the new Hamiltonian, and find the condition that will maximize the new Hamiltonian.

(*b*) Examine the optimal costate path. Do you still obtain a constant λ path as in (7.78)?

(*c*) If the transversality condition $\lambda(T) = 0$ applies, what will become of the H-maximizing condition in part (*a*)? Can the condition be simplified to (7.80)? What can you conclude about E^* for this case?

(*d*) If the transversality condition is $\lambda(T) > 0$ and $S(T) = 0$ instead, what will become of the H-maximizing condition in part (*a*)? Find the derivative dE/dt and deduce the time profile of the $E^*(t)$ path for this case.

CHAPTER 8

MORE ON OPTIMAL CONTROL

For a better grasp of mathematical methods and results, it is always helpful to understand the intuitive economic meanings behind them. For this reason, we shall now seek to buttress our mathematical understanding of the maximum principle by an economic interpretation of the various conditions it imposes. After that is accomplished, we shall then move on to some other aspects of optimal control theory such as the current-value Hamiltonian, concavity sufficient conditions, and problems with multiple state and control variables.

8.1 AN ECONOMIC INTERPRETATION OF THE MAXIMUM PRINCIPLE

In a remarkable article, Robert Dorfman shows that each part of the maximum principle can be assigned an intuitively appealing economic meaning, and that each condition therein can be made plausible from a commonsense point of view.[1] We shall draw heavily from that article in this section.

---

[1]Robert Dorfman, "An Economic Interpretation of Optimal Control Theory," *American Economic Review*, December 1969, pp. 817–831.

Consider a firm that seeks to maximize its profits over the time interval $[0, T]$. There is a single state variable, capital stock K. And there is a single control variable u, representing some business decision the firm has to make at each moment of time (such as its advertising budget or inventory policy). The firm starts out at time 0 with capital K_0, but the terminal capital stock is left open. At any moment of time, the profit of the firm depends on the amount of capital it currently holds as well as on the policy u it currently selects. Thus, its profit function is $\pi(t, K, u)$. But the policy selection u also bears upon the rate at which capital K changes over time; that is, $\dot{K}$ is affected by u. It follows that the optimal control problem is to

$$\begin{aligned}
&\text{Maximize} && \Pi = \int_0^T \pi(t, K, u)\, dt \\
(8.1)\qquad &\text{subject to} && \dot{K} = f(t, K, u) \\
&\text{and} && K(0) = K_0 \qquad K(T) \text{ free} \qquad (K_0, T \text{ given})
\end{aligned}$$

The Costate Variable as a Shadow Price

The maximum principle places conditions on three types of variables: control, state, and costate. The control variable u and the state variable K have already been assigned their economic meanings. What about the costate variable λ?

As intimated in an earlier section, λ is in the nature of a Lagrange multiplier and, as such, it should have the connotation of a shadow price. To confirm this, let us adapt (7.22″) to the present context, and plug in the optimal paths or values for all variables, to get

$$\Pi^* = \int_0^T \left[H(t, K^*, u^*, \lambda^*) + K^*(t)\dot{\lambda}^* \right] dt - \lambda^*(T)K^*(T) + \lambda^*(0)K_0$$

Partial differentiation of Π^* with respect to the (given) initial capital and the (optimal) terminal capital yields

$$(8.2)\qquad \frac{\partial \Pi^*}{\partial K_0} = \lambda^*(0) \qquad \text{and} \qquad \frac{\partial \Pi^*}{\partial K^*(T)} = -\lambda^*(T)$$

Thus, $\lambda^*(0)$, the optimally determined initial costate value, is a measure of the sensitivity of the optimal total profit Π^* to the given initial capital. If we had one more (infinitesimal) unit of capital initially, Π^* would be larger by the amount $\lambda^*(0)$. Therefore, the latter expression can be taken as the imputed value or shadow price of a unit of initial capital. In the other partial derivative in (8.2), the terminal value of the optimal costate path,

$\lambda^*(T)$, is seen to be the *negative* of the rate of change of Π^* with respect to the optimal terminal capital stock. If we wished to preserve one more unit (use up one less unit) of capital stock at the end of the planning period, then we would have to sacrifice our total profit by the amount $\lambda^*(T)$. So, again, the λ^* value at time T measures the shadow price of a unit of the terminal capital stock.

In general, then, $\lambda^*(t)$ for any t is the shadow price of capital at that particular point of time. For this interpretation to be valid, however, we have to write the Lagrange-multiplier expression as $\lambda(t)[f(t, K, u) - \dot{K}]$ rather than $\lambda(t)[\dot{K} - f(t, K, u)]$. Otherwise, λ^* would become instead the *negative* of the shadow price of K.

The Hamiltonian and the Profit Prospect

The Hamiltonian of problem (8.1) is

$$H = \pi(t, K, u) + \lambda(t) f(t, K, u) \tag{8.3}$$

The first component on the right is simply the profit function at time t, based on the current capital and the current policy decision taken at that time. We may think of it as representing the "current profit corresponding to policy u." In the second component of (8.3), the $f(t, K, u)$ function indicates the rate of change of (physical) capital, $\dot{K}$, corresponding to policy u, but when the f function is multiplied by the shadow price $\lambda(t)$, it is converted to a monetary value. Hence, the second component of the Hamiltonian represents the "rate of change of capital value corresponding to policy u." Unlike the first term, which relates to the *current-profit effect* of u, the second term can be viewed as the *future-profit effect* of u, since the objective of capital accumulation is to pave the way for the production of profits for the firm in the future. These two effects are in general competing in nature: If a particular policy decision u is favorable to the current profit, then it will normally involve a sacrifice in the future profit. In sum, then, the Hamiltonian represents the *overall profit prospect* of the various policy decisions, with both the immediate and the future effects taken into account.[2]

The maximum principle requires the maximization of the Hamiltonian with respect to u. What this means is that the firm must try at each point

[2]Note that if the Hamiltonian is written out more completely as $H = \lambda_0 \pi(t, K, u) + \lambda(t) f(t, K, u)$ as in (7.4), and if it turns out that $\lambda_0 = 0$, then this would mean economically that the current-profit effect of u somehow does not matter to the firm, and does not have to be weighed against the future-profit effect. Such an outcome is intuitively not appealing. In a meaningful economic model, we would expect λ_0 to be nonzero, in which case λ_0 can be normalized to one.

of time to maximize the overall profit prospect by the proper choice of u. Specifically, this would require the balancing of prospective gains in the current profit against prospective losses in the future profit. To see this more clearly, examine the weak version of the "Max H" condition:

$$\frac{\partial H}{\partial u} = \frac{\partial \pi}{\partial u} + \lambda(t)\frac{\partial f}{\partial u} = 0$$

When it is rewritten into the form

$$\frac{\partial \pi}{\partial u} = -\lambda(t)\frac{\partial f}{\partial u} \tag{8.4}$$

this condition shows that the optimal choice u^* must balance any marginal *increase* in the current profit made possible by the policy [the left-hand-side expression in (8.4)] against the marginal *decrease* in the future profit that the policy will induce via the change in the capital stock [the right-hand-side expression in (8.4)].

The Equations of Motion

The maximum principle involves two equations of motion. The one for the state variable K, included as part of the problem statement (8.1) itself, merely specifies the way the policy decision of the firm will affect the rate of change of capital. The equation of motion for the costate variable is

$$\dot{\lambda} = -\frac{\partial H}{\partial K} = -\frac{\partial \pi}{\partial K} - \lambda(t)\frac{\partial f}{\partial K}$$

or, after multiplying through by -1,

$$-\dot{\lambda} = \frac{\partial \pi}{\partial K} + \lambda(t)\frac{\partial f}{\partial K} \tag{8.5}$$

The left-hand-side of (8.5) denotes the rate of decrease of the shadow price over time, or the rate of depreciation of the shadow price. The equation of motion requires this rate to be equal in magnitude to the sum of the two terms on the right-hand side of (8.5). The first of these, $\partial \pi/\partial K$, represents the marginal contribution of capital to current profit, and the second, $\lambda(\partial f/\partial K)$, represents the marginal contribution of capital to the enhancement of capital value. What the maximum principle requires is that the shadow price of capital depreciate at the rate at which capital is contributing to the current and future profits of the firm.

Transversality Conditions

What about transversality conditions? With a free terminal state $K(T)$ at a fixed terminal time T (vertical terminal line), that condition is

$$\lambda(T) = 0$$

This means that the shadow price of capital should be driven to zero at the terminal time. The reason for this is that the valuableness of capital to the firm emanates solely from its potential for producing profits. Given the rigid planning horizon T, the tacit understanding is that only the profits made within the period $[0, T]$ would matter, and that whatever capital stock that still exists at time T, being too late to be put to use, would have no economic value to the firm. Hence, it is only natural that the shadow price of capital at time T should be set equal to zero. In view of this, we would not expect the firm to engage in capital accumulation in earnest toward the end of the planning period. Rather, it should be trying to use up most of its capital by time T. The situation is not unlike that of a pure egotist—a sort of Mr. Scrooge—who places no value on any material possessions that he himself cannot enjoy and must leave behind upon his demise.

For a firm that intends to continue its existence beyond the planning period $[0, T]$, it may be reasonable to stipulate some minimum acceptable level for the terminal capital, say, $K_{\min}$. In that case, of course, we would have a truncated vertical terminal line instead. The transversality condition now stipulates that

$$\lambda(T) \geq 0 \qquad \text{and} \qquad [K^*(T) - K_{\min}]\lambda(T) = 0 \qquad [\text{from } (7.35)]$$

If $K^*(T)$ turns out to exceed $K_{\min}$, then the restriction placed upon the terminal capital stock proves to be nonbinding. The outcome is the same as if there is no restriction, and the old condition $\lambda(T) = 0$ will still apply. But if the terminal shadow price $\lambda(T)$ is optimally nonzero (positive), then the restriction $K_{\min}$ indeed is binding, in the sense that it is preventing the firm from using up as much of its capital toward the end of the period as it would otherwise do. The amount of terminal capital actually left by the firm will therefore be exactly at the minimum required level, $K_{\min}$.

Finally, consider the case of a horizontal terminal line. In that case, the firm has a prespecified terminal capital level, say K_{T0}, but is free to choose the time to reach the target. The transversality condition

$$[H]_{t=T} = 0$$

simply means that, at the (chosen) terminal time, the sum of the current and future profits pertaining to that point of time must be zero. In other words, the firm should not attain K_{T0} at a time when the sum of immediate and future profits (the value of H) is still positive; rather, it should—after having taken advantage of all possible profit opportunities—attain K_{T0} at a time when the sum of immediate and future profits has been squeezed down to zero.

8.2 THE CURRENT-VALUE HAMILTONIAN

In economic applications of optimal control theory, the integrand function F often contains a discount factor $e^{-\rho t}$. Such an F function can in general be expressed as

$$F(t, y, u) = G(t, y, u)e^{-\rho t} \tag{8.6}$$

so that the optimal control problem is to

$$\begin{aligned} &\text{Maximize} && V = \int_0^T G(t, y, u)e^{-\rho t}\, dt \\ &\text{subject to} && \dot{y} = f(t, y, u) \\ &\text{and} && \text{boundary conditions} \end{aligned} \tag{8.7}$$

By the standard definition, the Hamiltonian function takes the form

$$H = G(t, y, u)e^{-\rho t} + \lambda f(t, y, u) \tag{8.8}$$

But since the maximum principle calls for the differentiation of H with respect to u and y, and since the presence of the discount factor adds complexity to the derivatives, it may be desirable to define a new Hamiltonian that is free of the discount factor. Such a Hamiltonian is called the *current-value Hamiltonian*, where the term "current-value" (as against "present-value") serves to convey the "undiscounted" nature of the new Hamiltonian.

The concept of the current-value Hamiltonian necessitates the companion concept of the current-value Lagrange multiplier. Let us therefore first define a new (current-value) Lagrange multiplier m:

$$m = \lambda e^{\rho t} \qquad \text{(implying } \lambda = me^{-\rho t}\text{)} \tag{8.9}$$

Then the current-value Hamiltonian, denoted by H_c, can be written as

(8.10)

$$H_c \equiv He^{\rho t} = G(t, y, u) + mf(t, y, u) \qquad \text{[by (8.8) and (8.9)]}$$

As intended, H_c is now free of the discount factor. Note that (8.10) implies

$$H \equiv H_c e^{-\rho t} \tag{8.10'}$$

The Maximum Principle Revised

If we choose to work with H_c instead of H, then all the conditions of the maximum principle should be reexamined to see whether any revision is needed.

The first condition in the maximum principle is to maximize H with respect to u at every point of time. When we switch to the current-value Hamiltonian $H_c = He^{\rho t}$, the condition is essentially unchanged except for the substitution of H_c for H. This is because the exponential term $e^{\rho t}$ is a constant for any given t. The particular u that maximizes H will therefore also maximize H_c. Thus the revised condition is simply

$$\max_u H_c \qquad \text{for all } t \in [0, T] \tag{8.11}$$

The equation of motion for the state variable originally appears in the canonical system as $\dot{y} = \partial H/\partial \lambda$. Since $\partial H/\partial \lambda = f(t, y, u) = \partial H_c/\partial m$ [by (8.8) and (8.10)], this equation should now be revised to

$$\dot{y} = \frac{\partial H_c}{\partial m} \qquad \text{[equation of motion for } y\text{]} \tag{8.12}$$

To revise the equation of motion for the costate variable, $\dot{\lambda} = -\partial H/\partial y$, we shall first transform each side of this equation into an expression involving the new Lagrange multiplier m, and then equate the two results. For the left-hand side, we have, by differentiating (8.9),

$$\dot{\lambda} = \dot{m}e^{-\rho t} - \rho m e^{-\rho t}$$

Using the definition of H in (8.10′), we can rewrite the right-hand side as

$$-\frac{\partial H}{\partial y} = -\frac{\partial H_c}{\partial y}e^{-\rho t}$$

Upon equating these two results and canceling the common factor $e^{-\rho t}$, we arrive at the following revised equation of motion:

$$\dot{m} = -\frac{\partial H_c}{\partial y} + \rho m \qquad \text{[equation of motion for } m\text{]} \tag{8.13}$$

Note that, compared with the original equation of motion for λ, the new one for m involves an extra term, ρm.

It remains to examine the transversality conditions. We shall do this for the vertical terminal line and the horizontal terminal line only. For the former, we can deduce that

$$\begin{aligned} (8.14) \qquad \lambda(T) = 0 \quad &\Rightarrow \quad [me^{-\rho t}]_{t=T} = 0 \qquad \text{[by (8.9)]} \\ &\Rightarrow \quad m(T)e^{-\rho T} = 0 \end{aligned}$$

For the latter, similar reasoning shows that

$$(8.15)\qquad [H]_{t=T}=0 \quad \Rightarrow \quad \left[H_c e^{-\rho t}\right]_{t=T} = 0 \qquad [\text{by } (8.10')]$$
$$\Rightarrow \quad [H_c]_{t=T} e^{-\rho T} = 0$$

Autonomous Problems

As a special case of problem (8.7), both the G and f functions may contain no t argument; that is, they may take the forms

$$G = G(y,u) \qquad \text{and} \qquad f = f(y,u)$$

Then the problem becomes

$$(8.16)\qquad \begin{aligned} &\text{Maximize} && V = \int_0^T G(y,u)e^{-\rho t}\,dt \\ &\text{subject to} && \dot{y} = f(y,u) \\ &\text{and} && \text{boundary conditions} \end{aligned}$$

Since the integrand $G(y,u)e^{-\rho t}$ still explicitly contains t, the problem is, strictly speaking, nonautonomous. However, by using the current-value Hamiltonian, we can in effect take the discount factor $e^{-\rho t}$ out of consideration. It is for this reason that economists tend to view problem (8.16) as an autonomous problem—in the special sense that the t argument explicitly enters in the problem via the discount factor only.

The current-value Hamiltonian is, of course, as applicable to problem (8.16) as it is to (8.7). And all the revised maximum-principle conditions (8.11) to (8.15) still hold. But the current-value Hamiltonian of the autonomous problem (8.16) has an additional property not available in problem (8.7). Since H_c now specializes to the form

$$H_c = G(y,u) + mf(y,u)$$

which is free of the t argument, its value evaluated along the optimal paths of all variables must be constant over time. That is,

$$(8.17)\qquad \frac{dH_c^*}{dt} = 0 \quad \text{or} \quad H_c^* = \text{constant} \qquad [\text{autonomous problem}]$$

This result is, of course, nothing but a replay of (7.54).

Another View of the Eisner-Strotz Model

Consider the following autonomous control problem adapted from the Eisner-Strotz model, originally studied as a calculus-of-variations problem with an infinite horizon in Sec. 5.2:

$$\begin{aligned}
&\text{Maximize} && \int_0^T [\pi(K) - C(I)]e^{-\rho t}\,dt \\
(8.18)\quad &\text{subject to} && \dot{K} = I \\
&\text{and} && \text{boundary conditions}
\end{aligned}$$

The meanings of the symbols are: π = profit, K = capital stock, C = adjustment cost, and I = net investment. The only state variable is K, and the only control variable is I. The π and C functions have derivatives

$$\pi''(K) < 0 \qquad C'(I) > 0 \qquad \text{and} \qquad C''(I) > 0 \qquad [\text{see Fig. 5.1}]$$

From the Hamiltonian function

$$H = [\pi(K) - C(I)]e^{-\rho t} + \lambda I$$

there emerge the maximum-principle conditions (not including the transversality condition)

$$\frac{\partial H}{\partial I} = -C'(I)e^{-\rho t} + \lambda = 0$$

$$\dot{K} = \frac{\partial H}{\partial \lambda} = I$$

$$\dot{\lambda} = -\frac{\partial H}{\partial K} = -\pi'(K)e^{-\rho t}$$

If we decide to work with the current-value Hamiltonian, however, we have

$$(8.19) \qquad H_c = \pi(K) - C(I) + mI \qquad [\text{by (8.10)}]$$

with equivalent maximum-principle conditions

$$(8.20) \qquad \frac{\partial H_c}{\partial I} = -C'(I) + m = 0 \qquad [\text{by (8.11)}]$$

$$(8.21) \qquad \dot{K} = \frac{\partial H_c}{\partial m} = I \qquad [\text{by (8.12)}]$$

$$(8.22) \qquad \dot{m} = -\frac{\partial H_c}{\partial K} + \rho m = -\pi'(K) + \rho m \qquad [\text{by (8.13)}]$$

The latter version is simpler because it is free of the discount factor $e^{-\rho t}$.

From (8.20), we see that

$$m = C'(I) > 0 \qquad \Rightarrow \qquad \frac{dm}{dI} = C''(I) > 0$$

that is, m is a monotonically increasing function of I. Accordingly, we should be able to write an inverse function $\psi = C'^{-1}$:

(8.23) $$I = \psi(m) \qquad (\psi' > 0)$$

Substituting (8.23) into (8.21), we can express $\dot{K}$ as

(8.24) $$\dot{K} = \psi(m)$$

Paired together, (8.24) and (8.22) constitute a system of two simultaneous equations in the variables K and m. After solving these for the $K^*(t)$ and $m^*(t)$ paths, and definitizing the arbitrary constants by the boundary and transversality conditions, we can find the optimal control path $I^*(t)$ via (8.23).

EXERCISE 8.2

Find the revised transversality conditions stated in terms of the current-value Hamiltonian for the following:

1 A problem with terminal curve $y_T = \phi(T)$.

2 A problem with a truncated vertical terminal line.

3 A problem with a truncated horizontal terminal line.

8.3 SUFFICIENT CONDITIONS

The maximum principle furnishes a set of necessary conditions for optimal control. In general, these conditions are not sufficient. However, when certain concavity conditions are satisfied, then the conditions stipulated by the maximum principle are sufficient for maximization. We shall present here only two such sufficiency theorems, those of Mangasarian and Arrow.

The Mangasarian Sufficiency Theorem

A basic sufficiency theorem due to O. L. Mangasarian[3] states that for the optimal control problem

(8.25)
$$\begin{aligned} &\text{Maximize} && V = \int_0^T F(t, y, u)\, dt \\ &\text{subject to} && \dot{y} = f(t, y, u) \\ &\text{and} && y(0) = y_0 \qquad (y_0, T \text{ given}) \end{aligned}$$

---

[3] O. L. Mangasarian, "Sufficient Conditions for the Optimal Control of Nonlinear Systems," *SIAM Journal on Control*, Vol. 4, February 1966, pp. 139–152.

the necessary conditions of the maximum principle are also sufficient for the global maximization of V, if (1) both the F and f functions are differentiable and concave in the variables (y, u) jointly, and (2) in the optimal solution it is true that

$$\lambda(t) \geq 0 \qquad \text{for all } t \in [0, T] \qquad \text{if } f \text{ is nonlinear in } y \text{ or in } u \tag{8.26}$$

[If f is linear in y and in u, then $\lambda(t)$ needs no sign restriction.]

As a preliminary to demonstrating the validity of this theorem, let us first remind ourselves that with the Hamiltonian

$$H = F(t, y, u) + \lambda f(t, y, u)$$

the optimal control path $u^*(t)$—along with the associated $y^*(t)$ and $\lambda^*(t)$ paths—must satisfy the maximum principle, so that

$$\left.\frac{\partial H}{\partial u}\right|_{u^*} = F_u(t, y^*, u^*) + \lambda^* f_u(t, y^*, u^*) = 0$$

This implies that

$$F_u(t, y^*, u^*) = -\lambda^* f_u(t, y^*, u^*) \tag{8.27}$$

Moreover, from the costate equation of motion, $\dot{\lambda} = -\partial H/\partial y$, we should have

$$\dot{\lambda}^* = -F_y(t, y^*, u^*) - \lambda^* f_y(t, y^*, u^*)$$

which implies that

$$F_y(t, y^*, u^*) = -\dot{\lambda}^* - \lambda^* f_y(t, y^*, u^*) \tag{8.28}$$

Finally, assuming for the time being that the problem has a vertical terminal line, the initial condition and the transversality condition should give us

$$y_0^* = y_0 \text{ (given)} \qquad \text{and} \qquad \lambda^*(T) = 0 \tag{8.29}$$

These relations will prove instrumental in the following development.

Now let both the F and f functions be concave in (y, u). Then, for two distinct points (t, y^*, u^*) and (t, y, u) in the domain, we have

$$\begin{aligned} F(t, y, u) - F(t, y^*, u^*) \leq\; & F_y(t, y^*, u^*)(y - y^*) \\ & + F_u(t, y^*, u^*)(u - u^*) \end{aligned} \tag{8.30}$$

$$\begin{aligned} f(t, y, u) - f(t, y^*, u^*) \leq\; & f_y(t, y^*, u^*)(y - y^*) \\ & + f_u(t, y^*, u^*)(u - u^*) \qquad \text{[cf. (4.4)]} \end{aligned} \tag{8.30'}$$

Upon integrating both sides of (8.30) over $[0, T]$, that inequality becomes

(8.31)

$$V - V^* \leq \int_0^T \left[F_y(t, y^*, u^*)(y - y^*) + F_u(t, y^*, u^*)(u - u^*) \right] dt$$

$$= \int_0^T \left[-\dot{\lambda}^*(y - y^*) - \lambda^* f_y(t, y^*, u^*)(y - y^*) - \lambda^* f_u(t, y^*, u^*)(u - u^*) \right] dt \qquad \text{[by (8.28) and (8.27)]}$$

The first component of the last integral, relating to the expression $-\dot{\lambda}^*(y - y^*)$, can be integrated by parts to yield[4]

$$\int_0^T -\dot{\lambda}^*(y - y^*)\, dt = \int_0^T \lambda^* [f(t, y, u) - f(t, y^*, u^*)]\, dt$$

This result enables us, upon substitution into (8.31), to write

$$(8.31') \quad V - V^* \leq \int_0^T \lambda^* \left[f(t, y, u) - f(t, y^*, u^*) - f_y(t, y^*, u^*)(y - y^*) - f_u(t, y^*, u^*)(u - u^*) \right] dt$$

$$\leq 0$$

The last inequality follows from the assumption of $\lambda^* \geq 0$ in (8.26), and the fact that the bracketed expression in the integrand is ≤ 0 by (8.30′). Consequently, the final result is

$$(8.31'') \qquad V \leq V^*$$

which establishes V^* to be a (global) maximum, as claimed in the theorem.

Note that if the f function is linear in (y, u), then (8.30′) becomes a strict equality. In that case, the bracketed expression in the integrand in

---

[4]Let $u = -\lambda^*$ and $v = y - y^*$. Then $du = -\dot{\lambda}^*\, dt$ and $dv = (\dot{y} - \dot{y}^*)\, dt$. So,

$$\int_0^T -\dot{\lambda}^*(y - y^*)\, dt \left(= \int_0^T v\, du \right)$$

$$= [-\lambda^*(y - y^*)]_0^T - \int_0^T -\lambda^*(\dot{y} - \dot{y}^*)\, dt$$

$$= -\lambda^*(T)(y_T - y_T^*) + \lambda^*(0)(y_0 - y_0^*) + \int_0^T \lambda^*(\dot{y} - \dot{y}^*)\, dt$$

$$= \int_0^T \lambda^*(\dot{y} - \dot{y}^*)\, dt \qquad \text{[by (8.29)]}$$

$$= \int_0^T \lambda^*[f(t, y, u) - f(t, y^*, u^*)]\, dt \qquad \text{[by (8.25)]}$$

(8.31′) is zero, and we can have the desired result $V - V^* \leq 0$ regardless of the sign of λ^*, so the restriction in (8.26) can be omitted.

The above theorem is based on the F and f functions being concave. If those functions are *strictly* concave, the weak inequalities in (8.30) and (8.30′) will become strict inequalities, as will the inequalities in (8.31), (8.31′), and (8.31″). The maximum principle will then be sufficient for a *unique* global maximum of V.

Although the proof of the theorem has proceeded on the assumption of a vertical terminal line, the theorem is also valid for other problems with a fixed T (fixed terminal point or truncated vertical terminal line). To see this, recall that in the proof of this theorem the transversality condition $\lambda^*(T) = 0$ in (8.29) is applied in the integration-by-parts process (see the associated footnote) to make the expression $-\lambda^*(T)(y_T - y_T^*)$ vanish. But the latter expression will also vanish if the problem has a fixed terminal state y_{T0}, for then the said expression will become $-\lambda^*(T)(y_{T0} - Y_T^*)$, and it must be zero because y_T^* has to be equal to y_{T0}. Moreover, if the problem has a truncated vertical terminal line, then either the transversality condition $\lambda^*(T) = 0$ is satisfied (if the truncation point is nonbinding) or we must treat the problem as one with a fixed terminal state at the truncation point. In either event, the said expression will vanish. Thus the Mangasarian theorem is applicable as long as T is fixed.

In the application of this theorem, it is possible to combine Mangasarian's conditions (1) and (2) into a single concavity condition on the Hamiltonian. If the F and f functions are both concave in (y, u), and if λ is nonnegative, then the Hamiltonian, $H = F + \lambda f$, being the sum of two concave functions, must be concave in (y, u), too. Hence, the theorem can be restated in terms of the concavity of H.

The Arrow Sufficiency Theorem

Another sufficiency theorem, due to Kenneth J. Arrow,[5] uses a weaker condition than Mangasarian's theorem, and can be considered as a generalization of the latter. Here, we shall describe its essence without reproducing the proof.

At any point of time, given the values of the state and costate variables y and λ, the Hamiltonian function is maximized by a particular u, u^*,

[5]This theorem appears without proof as Proposition 5 in Kenneth J. Arrow, "Applications of Control Theory to Economic Growth," in George B. Dantzig and Arthur F. Veinott, Jr., eds., *Mathematics of the Decision Sciences*, Part 2, American Mathematical Society, Providence, RI, 1968, p. 92. It also appears with proof as Proposition 6 in Kenneth J. Arrow and Mordecai Kurz, *Public Investment, The Rate of Return, and Optimal Fiscal Policy*, published for Resources for the Future, Inc., by the Johns Hopkins Press, Baltimore, MD, 1970, p. 45. A proof of the theorem is also provided in Morton I. Kamien and Nancy L. Schwartz, "Sufficient Conditions in Optimal Control Theory," *Journal of Economic Theory*, Vol. 3, 1971, pp. 207–214.

which depends on t, y, and λ:

$$u^* = u^*(t, y, \lambda) \tag{8.32}$$

When (8.32) is substituted into the Hamiltonian, we obtain what is referred to as the *maximized Hamiltonian function*

$$H^0(t, y, \lambda) = F(t, y, u^*) + \lambda f(t, y, u^*) \tag{8.33}$$

Note that the concept of H^0 is different from that of the optimal Hamiltonian H^* encountered in (7.53) and (7.54). Since H^* denotes the Hamiltonian evaluated along all the optimal paths, that is, evaluated at $y^*(t)$, $u^*(t)$, and $\lambda^*(t)$ for every point of time, the y, u, and λ arguments can all be substituted out, leaving H^* as a function of t alone: $H^* = H^*(t)$. In contrast, H^0 is evaluated along $u^*(t)$ only; thus, while the u argument is substituted out, the other arguments remain, so that $H^0 = H^0(t, y, \lambda)$ is still a function with three arguments.

The Arrow theorem states that, in the optimal control problem (8.25), the conditions of the maximum principle are sufficient for the global maximization of V, if the maximized Hamiltonian function H^0 defined in (8.33) is concave in the variable y for all t in the time interval $[0, T]$, for given λ.

The reason why the Arrow theorem can be considered as a generalization of the Mangasarian theorem—or the latter as a special case of the former—is as follows: If both the F and f functions are concave in (y, u) and $\lambda \geq 0$, as stipulated by Mangasarian, then $H \equiv F + \lambda f$ is also concave in (y, u), and from this it follows that H^0 is concave in y, as stipulated by Arrow. But H^0 can be concave in y even if F and f are not concave in (y, u), which makes the Arrow condition a weaker requirement.

Like the Mangasarian theorem, the validity of the Arrow theorem actually extends to problems with other types of terminal conditions as long as T is fixed. Also, although the theorem has been couched in terms of the regular Hamiltonian H and its "maximized" version H^0, it can also be rephrased using the current-value Hamiltonian H_c and its "maximized" version H_c^0, which differ from H and H^0, respectively, only by a discount factor $e^{-\rho t}$.

EXAMPLE 1 In Sec. 7.2, Example 1, we discussed the shortest-distance problem

$$\text{Maximize} \quad 2V = \int_0^T -(1 + u^2)^{1/2}\, dt$$

$$\text{subject to} \quad \dot{y} = u$$

$$\text{and} \quad y(0) = A \qquad y(T) \text{ free} \qquad (A, T \text{ given}) \qquad [\text{from } (7.7)]$$

Let us apply both sufficiency theorems.

For the Mangasarian theorem, we note that neither the F function nor the f function depends on y, so the concavity condition relates to u alone. From the F function, we obtain

$$F_u = -u(1+u^2)^{-1/2} \qquad \text{and} \qquad F_{uu} = -(1+u^2)^{-3/2} < 0$$

[by the product rule]

Thus F is concave in u. As to the f function, $f = u$, since it is linear in u, it is automatically concave in u. Besides, the fact that f is linear makes condition (8.26) irrelevant. Consequently, the conditions of Mangasarian are satisfied, and the optimal solution found earlier does maximize V (and minimize the distance) globally.

Once the Mangasarian conditions are satisfied, it is no longer necessary to check the Arrow condition. But if we do wish to apply the Arrow theorem, we can proceed to check whether the maximized Hamiltonian H^0 is concave in y. In the present example, the Hamiltonian is

$$H = -(1+u^2)^{1/2} + \lambda u$$

When the optimal control

$$u(t) = \lambda(1-\lambda^2)^{-1/2} \qquad [\text{from } (7.9)]$$

is substituted into H to eliminate u, the resulting H^0 expression contains λ alone, with no y. Thus H^0 is linear and hence concave in y for given λ, and it satisfies the Arrow sufficient condition.

EXAMPLE 2 In the problem of Example 2 in Sec. 7.2:

$$\begin{aligned} &\text{Maximize} && V = \int_0^2 (2y - 3u)\,dt \\ &\text{subject to} && \dot{y} = y + u \\ &&& y(0) = 4 \qquad y(2) \text{ free} \\ &\text{and} && u(t) \in [0, 2] \qquad [\text{from } (7.13)] \end{aligned}$$

both the F and f functions are linear in (y, u). As a result, all the second-order partial derivatives of F and f are zero. With reference to the test for sign semidefiniteness in (4.12), we have here $|\tilde{D}_1| = |\tilde{D}_2| = 0$, which establishes the negative semidefiniteness of the quadratic form involved. Hence, both F and f are concave in (y, u). Moreover, as in Example 1, the stipulation in (8.26) is irrelevant. The conditions of the Mangasarian theorem are therefore satisfied.

As to the Arrow condition, we recall from (7.14) that the optimal control depends on λ, and that u^* takes the boundary values 2 and 0 in the two phases of the solution as follows:

$$u^*_{\text{I}} = 2 \qquad \text{and} \qquad u^*_{\text{II}} = 0 \qquad \text{[from (7.17)]}$$

Therefore, we should have two maximized Hamiltonian functions, one for each phase. From the Hamiltonian

$$H = 2y - 3u + \lambda(y + u)$$

we obtain, upon eliminating u,

$$H^0_{\text{I}} = 2y - 6 + \lambda(y + 2) = (2 + \lambda)y - 6 + 2\lambda$$
$$H^0_{\text{II}} = 2y + \lambda y = (2 + \lambda)y$$

In either phase, H^0 is linear in y for given λ, so the Arrow condition is satisfied.

EXAMPLE 3 Let us now check whether the maximum principle is sufficient for the control problem adapted from the Eisner-Strotz model (Sec. 8.2, Example 1):

$$\text{Maximize} \qquad \int_0^T [\pi(K) - C(I)]e^{-\rho t}\, dt$$
$$\text{subject to} \qquad \dot{K} = I$$
$$\text{and} \qquad \text{boundary conditions} \qquad \text{[from (8.18)]}$$

where $\pi''(K) < 0$, $C'(I) > 0$, and $C''(I) > 0$.

For the Mangasarian conditions, we can immediately see that the f function, $f = I$, is linear and concave in I. To check the concavity of the F function in the variables (K, I), we need the second-order partial derivatives of $F = [\pi(K) - C(I)]e^{-\rho t}$. These are found to be

$$F_{KK} = \pi''(K)e^{-\rho t} < 0$$
$$F_{KI} = F_{IK} = 0$$
$$F_{II} = -C''(I)e^{-\rho t} < 0$$

Thus, with reference to the test for sign definiteness in (4.8) and (4.9), we find here $|D_1| < 0$ and $|D_2| > 0$, indicating that F is strictly concave in (K, I). Since condition (8.26) is again irrelevant, the Mangasarian conditions are fully met.

To check the Arrow sufficient condition, we recall from (8.20) and (8.23) that, with the current-value Hamiltonian

$$H_c = \pi(K) - C(I) + mI \qquad \text{[from (8.19)]}$$

we can, by setting $\partial H_c/\partial I = 0$, solving for m in terms of I, and then writing I as an inverse function of m, express the optimal control in the form

$$I = \psi(m) \qquad [\text{from } (8.23)]$$

Substitution of this optimal control into H_c yields the maximized current-value Hamiltonian

$$H_c^0 = \pi(K) - C[\psi(m)] + m\psi(m)$$

Since $\partial H_c^{\ 0}/\partial K = \pi'(K)$ and $\partial^2 H_c^{\ 0}/\partial K^2 = \pi''(K) < 0$, the maximized current-value Hamiltonian is strictly concave in the state variable K. The Arrow sufficient condition is thus satisfied.

The checking of concavity in the present example is slightly more laborious than in the preceding two examples, because even though the model is simple, it involves general functions. For more elaborate general-function models, checking concavity—especially for the maximized Hamiltonian—can be a tedious process.[6] Frequently, therefore, models are constructed with the appropriate concavity/convexity properties incorporated or assumed at the outset, thereby obviating the need to check sufficient conditions.

EXERCISE 8.3

Check the Mangasarian and Arrow sufficient conditions for the following:

1 Example 1 in Sec. 7.4.

2 Example 3 in Sec. 7.4.

3 Problem 1 in Exercise 7.4.

4 Problem 3 in Exercise 7.4.

5 The Nordhaus model in Sec. 7.6.

6 The Forster model in Sec. 7.7.

8.4 PROBLEMS WITH SEVERAL STATE AND CONTROL VARIABLES

For simplicity, we have so far confined our attention to problems with one state variable and one control variable. The generalization to problems with multiple state variables and control variables is, in principle, very straightforward. But the solution procedure naturally becomes more complicated.

[6] For more examples of checking the Arrow sufficient condition, see Morton I. Kamien and Nancy L. Schwartz, *Dynamic Optimization: The Calculus of Variations and Optimal Control in Economics and Management*, 2d ed., Elsevier, New York, 1991, pp. 222–225; Atle Seierstad and Knut Sydsæter, *Optimal Control Theory with Economic Applications*, Elsevier, New York, 1987, pp. 112–129.

The Problem

Let there be n state variables, $y_1, \ldots, y_n$, and m control variables, $u_1, \ldots, u_m$. The two numbers n and m are not restricted in relative magnitudes; we can have either $n < m$, $n = m$, or $n > m$. Each state variable y_j must come with an equation of motion, describing how $\dot{y}_j$ specifically depends on t, on the values of y_j and the other state variables at time t, as well as on the values of each control variable u_i chosen at time t. Thus, there should be in the problem n differential equations in the form

$$\begin{aligned} \dot{y}_1 &= f^1(t, y_1, \ldots, y_n, u_1, \ldots, u_m) \\ &\cdots\cdots\cdots\cdots\cdots\cdots \\ \dot{y}_n &= f^n(t, y_1, \ldots, y_n, u_1, \ldots, u_m) \end{aligned}$$

There should be n initial conditions on the y variables. Assuming the problem to be one with fixed terminal points, we should also have n given terminal conditions on the state variables. In contrast, the control variables are not accompanied by equations of motion, but each control variable u_i may be subject to the restriction of a control region $\mathscr{U}_i$. Assuming fixed initial and terminal points for the time being, the optimal control problem can be expressed as

$$\begin{aligned} &\text{Maximize} && V = \int_0^T F(t, y_1, \ldots, y_n, u_1, \ldots, u_m)\, dt \\ &\text{subject to} && \dot{y}_1 = f^1(t, y_1, \ldots, y_n, u_1, \ldots, u_m) \\ &&& \cdots\cdots\cdots\cdots\cdots\cdots \\ (8.34)\quad &&& \dot{y}_n = f^n(t, y_1, \ldots, y_n, u_1, \ldots, u_m) \\ &&& y_1(0) = y_{10}, \ldots, y_n(0) = y_{n0} \\ &&& y_1(T) = y_{1T}, \ldots, y_n(T) = y_{nT} \\ &\text{and} && u_1(t) \in \mathscr{U}_1, \ldots, u_m(t) \in \mathscr{U}_m \end{aligned}$$

This rather lengthy statement of the problem can be shortened somewhat by the use of index subscripts. Using the index j for the state variables and the index i for the control variables, we may restate the problem as

$$\begin{aligned} &\text{Maximize} && V = \int_0^T F(t, y_1, \ldots, y_n, u_1, \ldots, u_m)\, dt \\ (8.34')\quad &\text{subject to} && \dot{y}_j = f^j(t, y_1, \ldots, y_n, u_1, \ldots, u_m) \\ &&& y_j(0) = y_{j0} \quad y_j(T) = y_{jT} \\ &\text{and} && u_i(t) \in \mathscr{U}_i \quad (i = 1, \ldots, m,\ j = 1, \ldots, n) \end{aligned}$$

To have an even more compact statement of the problem, we may employ vector notation. Define a y vector and a u vector

$$y \equiv \begin{bmatrix} y_1 \\ \vdots \\ y_n \end{bmatrix} \qquad u \equiv \begin{bmatrix} u_1 \\ \vdots \\ u_m \end{bmatrix}$$

Then the integrand function F in (8.34′) can be abbreviated to $F(t, y, u)$. By the same token, the arguments in the f^j functions can be simplified to (t, y, u). If we define two more vectors

$$\dot{y} \equiv \begin{bmatrix} \dot{y}_1 \\ \vdots \\ \dot{y}_n \end{bmatrix} \qquad f(t, y, u) \equiv \begin{bmatrix} f^1(t, y, u) \\ \vdots \\ f^n(t, y, u) \end{bmatrix}$$

then the equations of motion can be collectively written in a single vector equation: $\dot{y} = f(t, y, u)$. Obvious extensions of the idea to the boundary conditions would enable us to write the vector equations $y(0) = y_0$ and $y(T) = y_T$, where $y(0)$, y_0, $y(T)$, and y_T are all $n \times 1$ vectors. And, finally, we can express the control-region restrictions by the vector statement $u(t) \in \mathscr{U}$, where $u(t)$ and $\mathscr{U}$ are both $m \times 1$ vectors.

In vector notation, therefore, the statement of the optimal control problem is simply

$$\begin{aligned} &\text{Maximize} && V = \int_0^T F(t, y, u)\, dt \\ (8.34'')\quad &\text{subject to} && \dot{y} = f(t, y, u) \\ &&& y(0) = y_0 \qquad y(T) = y_T \\ &\text{and} && u(t) \in \mathscr{U} \end{aligned}$$

which, in appearance, is not at all different from the problem with one state variable and one control variable. The only difference is that, in (8.34″), several symbols represent vectors, and therefore involve much more than what is visible on the surface. Note, however, that even in the vector statement of the problem, not all the symbols represent vectors. The symbol V is clearly a scalar, and so are the symbols t and T.

The Maximum Principle

The extension of the maximum principle to the multivariable case is comparably straightforward. First of all, to form the Hamiltonian, we introduce a

Lagrange multiplier for every f^j function in the problem. Thus, we have

$$H \equiv F(t, y, u) + \sum_{j=1}^{n} \lambda_j f^j(t, y, u) \tag{8.35}$$

where y and u are the state and costate vectors defined earlier. The summation term on the right can, if desired, also be written in vector notation. Once we define another $n \times 1$ vector

$$\lambda \equiv \begin{bmatrix} \lambda_1(t) \\ \vdots \\ \lambda_n(t) \end{bmatrix} \qquad \text{with transpose } \lambda' \equiv \begin{bmatrix} \lambda_1(t) & \cdots & \lambda_n(t) \end{bmatrix}$$

the Hamiltonian can be rewritten as

$$H(t, y, u, \lambda) = F(t, y, u) + \lambda' f(t, y, u) \tag{8.35'}$$

which looks strikingly similar to the Hamiltonian in a one-variable case, except that, here, the last term is a scalar product—the product of the row vector λ' and the column vector $f(t, y, u)$. Note that the Hamiltonian itself is also a scalar.

The requirement that the Hamiltonian be maximized at every point of time with respect to the control variables stands as before. We can thus still write the condition

$$\max_u H \tag{8.36}$$

But u is now an m-vector, so we must at each point of time make a choice of m control values, $u_1^*, \ldots, u_m^*$.

As to the equations of motion, those for the state variables always come with the problem; they are just the n differential equations appearing in (8.34). Moreover, these n equations are still expressible in terms of the derivatives of the Hamiltonian:

$$\dot{y}_j = \frac{\partial H}{\partial \lambda_j} \qquad (j = 1, \ldots, n) \tag{8.37}$$

Similarly, the equations of motion for the costate variables are simply

$$\dot{\lambda}_j = -\frac{\partial H}{\partial y_j} \qquad (j = 1, \ldots, n) \tag{8.38}$$

Together, (8.37) and (8.38)—representing a total of $2n$ differential equations—constitute the canonical system of the present problem. The $2n$ arbitrary constants that are expected to arise from these differential equations can be definitized by using the $2n$ boundary conditions.

If desired, the canonical system can also be translated into vector notation. The equivalent statement for (8.37) is

$$\dot{y} = \frac{\partial H}{\partial \lambda'} \qquad [\lambda' = \text{transpose of } \lambda] \tag{8.37'}$$

The noteworthy thing about this is that H is differentiated with respect to λ' (a row vector), not λ (a column vector). A look at (8.35′) would make it clear why this should be the case. More to the point, however, is the fact that (8.37′) is in line with the mathematical rule that the derivative of a *scalar* with respect to a *row* (*column*) vector is a *column* (*row*) vector. Accordingly, the vector-notation translation of (8.38) should be

$$\dot{\lambda}' = -\frac{\partial H}{\partial y} \qquad \left[\dot{\lambda}' \equiv \text{transpose of } \dot{\lambda}\right] \tag{8.38'}$$

Transversality Conditions

Problem (8.34) assumes fixed endpoints. If the terminal point is variable, we again need appropriate transversality conditions.

Referring back to (7.30), we see that, for the one-state-variable problem, the transversality conditions are derived from the last two terms of the $dV/d\epsilon$ expression set equal to zero: $[H]_{t=T}\,\Delta T - \lambda(T)\,\Delta y_T = 0$. When there are n state variables in the problem, those two terms will expand to an expression with $(n+1)$ terms:

$$[H]_{t=T}\,\Delta T - \lambda_1(T)\,\Delta y_{1T} - \lambda_2(T)\,\Delta y_{2T} - \cdots - \lambda_n(T)\,\Delta y_{nT} = 0 \tag{8.39}$$

From this, the following two basic transversality conditions arise:

$$[H]_{t=T} = 0 \qquad [\text{if } T \text{ is free}] \tag{8.40}$$

$$\lambda_j(T) = 0 \qquad \left[\text{if } y_{jT} \text{ is free}\right] \tag{8.41}$$

Clearly, these conditions are essentially no different from the transversality conditions for the one-state-variable case. The only differences are that (1) the Hamiltonian H in (8.40) for the present n-variable problem contains more terms than its one-variable counterpart, and (2) there will be as many conditions of the $\lambda_j(T) = 0$ type in (8.41) as there are state variables with free terminal states.

The transversality conditions for the terminal-curve and truncated-terminal-line problems are also essentially similar to those for the one-state-variable case. Suppose that the terminal time is free, and two of the state variables, y_1 and y_2, are required to have their terminal values tied to the terminal time by the relations

$$y_{1T} = \phi_1(T) \qquad \text{and} \qquad y_{2T} = \phi_2(T) \tag{8.42}$$

Then, for a small ΔT, we may expect the following to hold:

$$\Delta y_{1T} = \phi_1'(T)\,\Delta T \qquad \text{and} \qquad \Delta y_{2T} = \phi_2'(T)\,\Delta T$$

Using these to eliminate Δy_{1T} and Δy_{2T} in (8.39), we can rewrite the latter as

$$(8.43) \quad [H - \lambda_1\phi_1' - \lambda_2\phi_2']_{t=T}\,\Delta T - \lambda_3(T)\,\Delta y_{3T} - \cdots - \lambda_n(T)\,\Delta y_{nT} = 0$$

The transversality condition that emerges for this terminal-curve problem is

$$(8.44) \qquad [H - \lambda_1\phi_1' - \lambda_2\phi_2']_{t=T} = 0 \qquad [\text{cf. } (7.32)]$$

Condition (8.44)—which should replace (8.40)—together with the two equations in (8.42) provide three relations to determine the three unknowns T, y_{1T}, and y_{2T}. The other state variables, $(y_3, \ldots, y_n)$, assumed here to have free terminal values, are still subject to the type of transversality condition in (8.41).

If the terminal time is fixed, and all the state variables have truncated vertical terminal lines, then the transversality condition is

$$(8.45) \qquad \lambda_j(T) \geq 0 \qquad y_{jT} \geq y_{j,\min} \qquad (y_{jT} - y_{j,\min})\lambda_j(T) = 0$$
$$(j = 1, \ldots, n) \qquad [\text{cf. } (7.35)]$$

This condition serves as a replacement for (8.41).

Finally, for the problem with a maximum permissible terminal time, $T_{\max}$, the transversality condition is

$$(8.46) \qquad [H]_{t=T} \geq 0 \qquad T \leq T_{\max} \qquad (T - T_{\max})[H]_{t=T} = 0$$

This condition is another type of replacement for (8.40).

An Example from Pontryagin

While the conceptual extension of the maximum principle to the multivariable problem is straightforward, the actual solution process can become much more involved. For this reason, economic applications with more than one state variable are not nearly as common as those with one state variable. A taste of the increasing complexity can be had from a purely mathematical example adapted from the work of Pontryagin et al.,[7] with two state variables and one control variable.

This example actually starts out with a single state variable y, which we wish to move from an initial state $y(0) = y_0$ to the terminal state

[7] L. S. Pontryagin et al., *The Mathematical Theory of Optimal Processes*, Interscience, New York, 1962, pp. 23–27.

$y(T) = 0$ in the shortest possible time. In other words, we have a time-optimal problem. Instead of the usual equation of motion in the form of $\dot{y} = f(t, y, u)$, however, we have this time a second-order differential equation, $\ddot{y} = u$, where $\ddot{y} \equiv d^2y/dt^2$, and where u is assumed to be constrained to the control region $[-1, 1]$. Since the second-order derivative $\ddot{y}$ is not admissible in the standard format of the optimal control problem, we have to translate this second-order differential equation into an equivalent pair of first-order differential equations. It is this translation process that causes the problem to become one with two state variables. Thus, this example also serves to illustrate how to handle a problem that contains a higher-order differential equation in the state equation of motion.

To effect the translation, we introduce a second state variable z:

$$z \equiv \dot{y} \qquad \text{implying} \qquad \dot{z} \equiv \ddot{y}$$

Then the equation $\ddot{y} = u$ can be rewritten as $\dot{z} = u$. And we now have two first-order equations of motion

$$\dot{y} = z \qquad \text{and} \qquad \dot{z} = u$$

one for each state variable. The problem is then to

$$\begin{aligned} &\text{Maximize} && \int_0^T -1\,dt \\ &\text{subject to} && \dot{y} = z \\ &(8.47) && \dot{z} = u \\ & && y(0) = y_0 \text{ (given)} \qquad y(T) = 0 \qquad T \text{ free} \\ & && z(0) = z_0 \\ &\text{and} && u(t) \in [-1, 1] \end{aligned}$$

The boundary condition on the new variable z is equivalent to the specification of an initial value for $\dot{y}$. The specification of both $y(0)$ and $\dot{y}(0)$ is a normal ingredient of a problem involving a second-order differential equation in y.

Step i The Hamiltonian function is

$$H = -1 + \lambda_1 z + \lambda_2 u \tag{8.48}$$

Being linear in u, this function leads to corner solutions for u. In view of the given control region, the optimal control should be

$$u^* = \operatorname{sgn} \lambda_2 \qquad [\text{cf. } (7.48)] \tag{8.49}$$

Step ii Since u^* depends on λ_2, we must next turn to the equations of motion for the costate variables:

$$\dot{\lambda}_1 = -\frac{\partial H}{\partial y} = 0 \qquad \text{implying } \lambda_1(t) = c_1 \text{ (constant)}$$

$$\dot{\lambda}_2 = -\frac{\partial H}{\partial z} = -\lambda_1 = -c_1$$

Thus, the time path for λ_2 is linear:

$$\lambda_2(t) = -c_1 t + c_2 \qquad (c_1, c_2 \text{ arbitrary}) \tag{8.50}$$

Excepting the special case of $c_1 = 0$, this path plots as a straight line that intersects the t axis. At that intersection, λ_2 switches its algebraic sign and, according to (8.49), the optimal control u^* must make a concomitant switch from 1 to -1, or the other way around, thereby yielding a bang-bang solution. However, since the linear function $\lambda_2(t)$ can cross the t axis at most once, there cannot be more than one switch in the value of u^*.

Step iii Next, we examine the state variables. The variable z has the equation of motion $\dot{z} = u$. Inasmuch as u can optimally take only two possible values, $u^* = \pm 1$, there are only two possible forms that equation can take: $\dot{z} = 1$ and $\dot{z} = -1$.

Possibility 1 $u^* = 1$ and $\dot{z} = 1$. In this case, the time path for z is

$$z(t) = t + c_3 \tag{8.51}$$

Then, since $\dot{y} = z = t + c_3$, straight integration yields the following quadratic time path for y:

$$y(t) = \tfrac{1}{2}t^2 + c_3 t + c_4 \qquad (c_3, c_4 \text{ arbitrary}) \tag{8.52}$$

The initial condition $y(0) = y_0$ tells us that $c_4 = y_0$. But c_3 is not as easy to definitize from the terminal condition $y(T) = 0$, because T is not yet known. Glancing at (8.51), however, we note that c_3 represents the value of z at $t = 0$. If we denote $z(0)$ by z_0, then $c_3 = z_0$. Thus, (8.51) and (8.52) may be definitized to

$$z(t) = t + z_0 \tag{8.53}$$

$$y(t) = \tfrac{1}{2}t^2 + z_0 t + y_0 \tag{8.54}$$

The way these last two equations are written, y and z appear as two separate functions of time. But it is possible to condense them into one single equation by eliminating t and expressing y in terms of z. This will enable us to depict the movement of y and z in a phase diagram in the yz

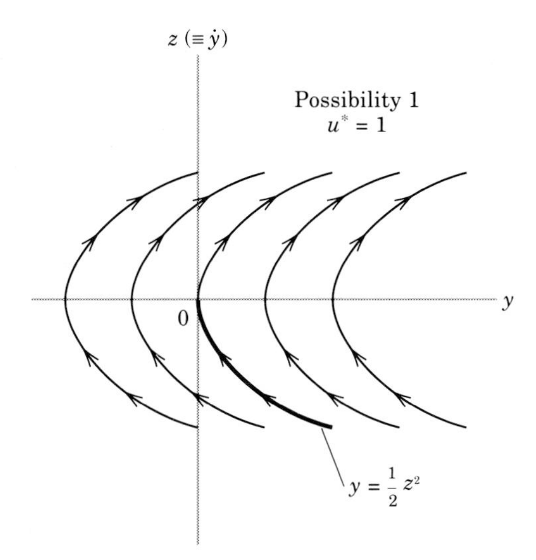

FIGURE 8.1

plane. Since

$$z^2 = (t + z_0)^2 = t^2 + 2z_0 t + z_0^2$$

we can solve this for t^2 and rewrite (8.54) as

$$y = \tfrac{1}{2}z^2 + \left(y_0 - \tfrac{1}{2}z_0^2\right) = \tfrac{1}{2}z^2 + k \qquad \left(k \equiv y_0 - \tfrac{1}{2}z_0^2\right) \tag{8.54'}$$

In the yz plane, this equation plots as a family of parabolas as illustrated in Fig. 8.1. Each parabola is associated with a specific value of k which is based on the values of y_0 and z_0. In particular, when $k = 0$, we get the parabola that passes through the point of origin. Of course, once y_0 and z_0 are numerically specified, we can pinpoint not only the relevant parabola, but also the exact starting point on that parabola.

Figure 8.1 is a one-variable phase diagram, with $z \equiv dy/dt$ on the vertical axis and y on the horizontal axis. Above the horizontal axis, with $\dot{y} > 0$, the directional arrowheads attached to the curves should uniformly point eastward. The opposite is true below the horizontal axis. Note that, according to these arrowheads, the designated terminal state $y(T) = 0$ is attainable only if we start at some point on the heavy arc of the parabola that passes through the point of origin. Note also that the terminal value of z is $z(T) = 0$.

Possibility 2 $u^* = -1$ and $\dot{z} = -1$. Under the second possibility, the equation of motion for z integrates to

$$z(t) = -t + c_5 = -t + z_0 \qquad \text{[cf. (8.53)]} \tag{8.55}$$

Since this result also represents the $\dot{y}$ path, further integration and defini-

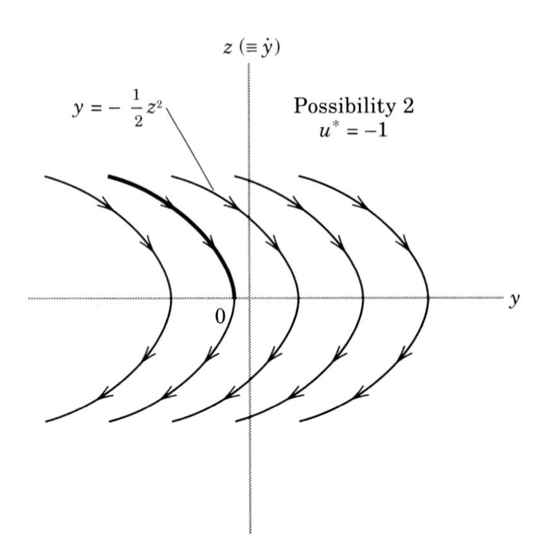

FIGURE 8.2

tization of the constant gives us

$$y(t) = -\tfrac{1}{2}t^2 + z_0 t + y_0 \qquad \text{[cf. (8.54)]} \tag{8.56}$$

Combining the last two equations, we obtain the single result

$$y = -\tfrac{1}{2}z^2 + \left(y_0 + \tfrac{1}{2}z_0{}^2\right) = -\tfrac{1}{2}z^2 + h \qquad \left(h \equiv y_0 + \tfrac{1}{2}z_0{}^2\right) \tag{8.56'}$$

which is comparable to (8.54′) in nature. Note that the two shorthand symbols k and h are not the same. Like (8.54′), this new condensed equation plots as a family of parabolas in the yz plane. But because of the negative coefficient of the z^2 term, the curves should all bend the other way, as shown in Fig. 8.2. Each parabola is uniquely associated with a value of h; for example, the one passing through the origin corresponds to $h = 0$. As before, the directional arrowheads uniformly point eastward above the horizontal axis, but westward below. Here, again, only the points located on the heavy arc can lead to the designated terminal state $y(T) = 0$. The terminal value of z is, as may be expected, still $z(T) = 0$.

Although we have analyzed the two cases of $u^* = 1$ and $u^* = -1$ as separate possibilities, both possibilities will, in a bang-bang solution, become realities, one sequentially after the other. It remains to establish the link between the two and to find out how and when the switch in the optimal control is carried out, if needed.

Step iv This link can best be developed diagrammatically. Since the heavy arcs in Figs. 8.1 and 8.2 are the only routes that can lead ultimately to the desired terminal state, let us place them together in Fig. 8.3. Even though the two arcs appear to have been merged into a single curve, they should be

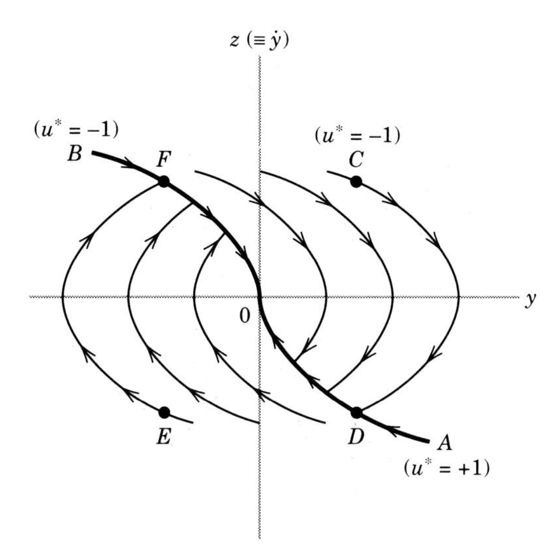

FIGURE 8.3

kept distinct in our thinking; the arrowheads go in opposite directions on the two arcs, and they call for different optimal controls—$u^* = +1$ for arc $A0$ and $u^* = -1$ for arc $B0$.

Should the actual initial point be located anywhere on arc $A0$ or arc $B0$, the selection of the proper control would start us off on a direct journey toward the origin. No switch in the control is needed. For an initial point located off the two heavy arcs, however, the journey to the origin would consist of two legs. The first leg is to take us from the initial point to arc $A0$ or $B0$, as the case may be, and the second leg is to take us to the origin along the latter arc. For this reason, there is no need to reproduce in Fig. 8.3 every part of every parabola in Figs. 8.1 and 8.2. All we need are the portions of the parabolas flowing toward and culminating in arc $A0$ and arc $B0$, respectively. If, for instance, point C is the initial point, then the first leg is to follow a parabolic trajectory to point D. Since the parabola involved is taken from Fig. 8.2 under possibility 2, the control appropriate to the circumstance is $u^* = -1$. Once point D is reached, however, we should detour onto arc $A0$. Since the latter comes from Fig. 8.1 under possibility 1, the control called for is $u^* = 1$. By observing these u^* values, and only by so doing, we are able to proceed on course along the optimal path $CD0$.

Other initial points off the heavy arcs (such as point E) can be similarly analyzed. For each such initial point, there is a unique optimal path leading to the designated terminal state, and that path must involve one (and only one) switch in the optimal control. Thus, in such cases, the optimal movement of y over time is never monotonic; an initial rise must be followed by a subsequent fall, and vice versa. The optimal movement of the control variable over time, on the other hand, traces out a step function composed of two linear segments.

More about the Switching

Having demonstrated the general pattern of the bang-bang solution, Pontryagin et al. dropped the example at this point. However, we can proceed a step further with the discussion of the $CD0$ path, to find the coordinates of the switching point D and also to determine the exact time, τ, at which the switch in the control should be made.

Point D is the intersection of two curves. One of these, arc $A0$, being part of a parabola taken from Fig. 8.1, falls under possibility 1, and is described by (8.54′) with $k = 0$. Specifically, we have

$$y = \tfrac{1}{2}z^2 \qquad (z < 0) \tag{8.57}$$

The other is a parabola taken from Fig. 8.2 under possibility 2, and is described instead by (8.56′). Since the intercept on the y axis is positive, h is positive. Hence, we have

$$y = -\tfrac{1}{2}z^2 + h \qquad (h > 0) \tag{8.58}$$

At point D, both (8.57) and (8.58) are satisfied. Solving these two equations simultaneously yields $y = \frac{1}{2}h$ and $z = -\sqrt{h}$ (the positive root is inadmissible). Thus,

$$\text{Point } D = \left(\tfrac{1}{2}h, -\sqrt{h}\right) \tag{8.59}$$

And the switch in u^* should be made precisely when $y = \frac{1}{2}h$ and $z(= \dot{y}) = -\sqrt{h}$.

The calculation of the switching time τ can be undertaken from the knowledge that, at point D, and hence at time τ, z attains the value $-\sqrt{h}$. From the initial point C, the movement of z follows the pattern given in (8.55): $z(t) = -t + z_0$. At time τ, its value should be

$$z(\tau) = -\tau + z_0$$

Setting $z(\tau)$ equal to $-\sqrt{h}$, we then find that

$$\tau = z_0 + \sqrt{h} \tag{8.60}$$

This result shows that the larger the values of z_0 and h, the longer it would take to reach switching time. A look at Fig. 8.3 will confirm the reasonableness of this conclusion: A larger z_0 and/or a larger h would mean an initial point farther away from arc $A0$, and hence a longer journey to reach arc $A0$, where the switching point is to be found. It should be realized, of course, that (8.60) is appropriate only for switching points located on arc $A0$, because the assumptions of a negative z and a positive h used in its derivation are applicable only to arc $A0$. For switching points located on arc $B0$, the expression for τ is different, although the procedure for finding it is the same.

It is also not difficult to calculate the amount of time required to travel from point D to the final destination at the origin. On that second leg of the journey, the value of z goes from $z(\tau) = -\sqrt{h}$ to zero. The relevant equation for that movement is (8.53), where, for the present purpose, we should consider τ as the initial time and $z(\tau) = -\sqrt{h}$ as the initial value z_0. In other words, the equation should be modified to: $z(t) = t - \sqrt{h}$. At the terminal time—which we shall denote for the time being as T'—the left-hand-side expression becomes $z(T') = 0$, and the right-hand-side one becomes $T' - \sqrt{h}$. Thus, we obtain

$$T' = \sqrt{h} \tag{8.61}$$

The reason we use the symbol T' instead of T is that the symbol T is supposed to represent the total time of travel from the initial point C to the origin, whereas the quantity shown in (8.61) tells only the travel time from point D to the origin. To get the optimal value of T, (8.60) and (8.61) should be added together. That is,

$$T^* = \tau + T' = z_0 + 2\sqrt{h} \tag{8.62}$$

The preceding discussion covers most aspects of the time-optimal problem under consideration. About the only thing not yet resolved is the matter of the arbitrary constants c_1 and c_2 in (8.50). The purpose of finding the values of c_1 and c_2 is only to enable us to definitize the $\lambda_2(t)$ path, so as to know when λ_2 will change sign and thereby trigger the switch in the optimal control. Since we have already found τ in (8.60) via another route, there is little point in worrying about c_1 and c_2. For the interested reader, however, we mention here that c_1 and c_2 can be definitized by means of the following two relations: First, when $t = \tau$, λ_2 must be zero (at the juncture of a sign change); hence, from (8.50), we have $-c_1\tau + c_2 = 0$. Second, the transversality condition

$$[H]_{t=T} = [-1 + \lambda_1 z + \lambda_2 u]_{t=T} = 0$$

reduces to $-1 + \lambda_2(T) = 0$, since $z(T) = 0$ and $u(T) = 1$. Substituting (8.50) into this simplified transversality condition yields another equation relating c_1 to c_2. Together, these two relations enable us to definitize c_1 and c_2.

EXERCISE 8.4

1 On the basis of the optimal path $CD0$ in Fig. 8.3, plot the time path $u^*(t)$.

2 On the basis of the optimal path $CD0$ in Fig. 8.3, describe the way y^* changes its values over time, bearing in mind that the value of z represents the rate of change of y. From your description, plot the time path $y^*(t)$.

3 For each of the following pairs of y_0 and z_0 values, indicate whether the resulting optimal path in Fig. 8.3 would contain a switch in the control, and explain why:

(*a*) $(y_0, z_0) = (2, 1)$
(*b*) $(y_0, z_0) = (2, -2)$
(*c*) $(y_0, z_0) = (-\frac{1}{2}, 1)$

4 Let point E in Fig. 8.3 be the initial point.

(*a*) What is the optimal path in the figure?
(*b*) Describe the time path $u^*(t)$.
(*c*) Describe the time path $y^*(t)$, taking into account the fact that $z \equiv \dot{y}$. Plot the $y^*(t)$ path.

5 Let point E in Fig. 8.3 be the initial point.

(*a*) Find the coordinates of the switching point on arc $B0$.
(*b*) Find the switching time τ.
(*c*) Find the optimal total time for traveling from point E to the destination at the origin.

8.5 ANTIPOLLUTION POLICY

In the Forster model on energy use and environmental quality in Sec. 7.7, pollution is taken to be a *flow variable*.[8] This is exemplified by auto emission, which, while currently detrimental to the environment, dissipates quickly and does not accumulate into a long-lasting stock. But in other types of pollution, such as radioactive waste and oil spills, the pollutants do emerge as a stock and produce lasting effects. A formulation with pollution as a *stock variable* is considered by Forster in the same paper cited previously. This formulation contains two state variables and two control variables.

The Pollution Stock

As before, we use the symbol E to represent the extraction of fuel and energy use. But the symbol P will now denote the stock (rather than flow) of pollution, with $\dot{P}$ as its flow. The use of energy generates a flow of pollution. If the amount of pollution flow is directly proportional to the amount of energy used, then we can write $\dot{P} = \alpha E$, $(\alpha > 0)$. Let A stand for the level of antipollution activities, and assume that A can reduce the pollution stock in a proportionate manner. Then, from this consideration, we have $\dot{P} = -\beta A$, $(\beta > 0)$. Further, if the pollution stock is subject to

[8] Bruce A. Forster, "Optimal Energy Use in a Polluted Environment," *Journal of Environmental Economics and Management*, 1980, pp. 321–333.

exponential decay at the rate $\delta > 0$, then we have $\dot{P}/P = -\delta$, so that $\dot{P} = -\delta P$. Combining these factors that affect P, we can write

$$\dot{P} = \alpha E - \beta A - \delta P \qquad (\alpha, \beta > 0,\ 0 < \delta < 1) \tag{8.63}$$

The Stock of Energy Resource

The implementation of antipollution activities A in itself requires the use of energy. That is, A causes a reduction in S, the stock of the energy resource. Assuming a proportional relationship between A and $\dot{S}$, we can, by the proper choice of unit A, simply write $\dot{S} = -A$. But since S is also reduced by the use of energy in other economic activities, we should also have $\dot{S} = -E$ [same as (7.71)]. Combining these considerations, we have

$$\dot{S} = -A - E \tag{8.64}$$

The Control Problem

Since the relations in (8.63) and (8.64) delineate the dynamic changes of P and S, respectively, these equations can evidently serve as the equations of motion in this model. This then points to P (pollution stock) and S (fuel stock) as the state variables. A further look at (8.63) and (8.64) also reveals that E (energy use) and A (antipollution activities) should play the role of control variables in the present analysis.

If we adhere to the same utility function used before:

$$U = U[C(E), P] \qquad (U_C > 0,\ U_P < 0,\ U_{CC} < 0,\ U_{PP} < 0,\ C' > 0,\ C'' < 0)$$
$$[\text{from } (7.74) \text{ and } (7.72)]$$

then the dynamic optimization problem may be stated as

$$\begin{aligned}
&\text{Maximize} && \int_0^T U[C(E), P]\, dt \\
&\text{subject to} && \dot{P} = \alpha E - \beta A - \delta P \\
&(8.65) && \dot{S} = -A - E \\
& && P(0) = P_0 > 0 \qquad P(T) \geq 0 \text{ free} \qquad (T \text{ given}) \\
& && S(0) = S_0 > 0 \qquad S(T) \geq 0 \text{ free} \\
&\text{and} && E \geq 0 \qquad 0 \leq A \leq \hat{A}
\end{aligned}$$

Two aspects of this problem are comment-worthy. First, while the initial values of P and S are fixed, as is the terminal time T, the terminal values of both the pollution stock P and the energy resource stock S are left free, subject only to nonnegativity. This means that there is a truncated vertical terminal line for P, and another one for S. Second, both control variables E and A are confined to their respective control regions. For E, the control region is $[0, \infty)$. And for A, the control region is $[0, \hat{A}]$, where $\hat{A}$ denotes the maximum feasible level of antipollution activities. Since budget considerations and other factors are likely to preclude unlimited pursuit of environmental purification, the assumption of an upper bound $\hat{A}$ does not seem unreasonable.

Maximizing the Hamiltonian

As usual, we start the solution process by writing the Hamiltonian function:

$$H = U[C(E), P] + \lambda_P(\alpha E - \beta A - \delta P) - \lambda_S(A + E) \tag{8.66}$$

where the subscript of each costate variable λ indicates the state variable associated with it. To maximize H with respect to the control variable E, where $E \geq 0$, the Kuhn–Tucker condition is $\partial H/\partial E \leq 0$, with the complementary-slackness proviso that $E(\partial H/\partial E) = 0$. But inasmuch as we can rule out the extreme case of $E = 0$ (which would totally halt consumption production), we must postulate $E > 0$. It then follows from complementary slackness that we must satisfy the condition

$$\frac{\partial H}{\partial E} = U_C C'(E) + \alpha\lambda_P - \lambda_S = 0 \tag{8.67}$$

Note that $\partial^2 H/\partial E^2 = U_{CC}C'^2 + U_C C'' < 0$; so H indeed is maximized rather than minimized.

In addition, H should be maximized with respect to A. As (8.66) shows, H is linear in the variable A, with

$$\frac{\partial H}{\partial A} = -\beta\lambda_P - \lambda_S$$

Besides, E is restricted to the closed control set $[0, \hat{A}]$. Thus, to maximize H, the left-hand-side boundary solution $A^* = 0$ should be chosen if $\partial H/\partial A$ is negative, and the right-hand-side boundary $A^* = \hat{A}$ should be selected if $\partial H/\partial A$ is positive. That is,

$$\beta\lambda_P + \lambda_S \begin{Bmatrix} > \\ < \end{Bmatrix} 0 \quad \Rightarrow \quad A^* = \begin{Bmatrix} 0 \\ \hat{A} \end{Bmatrix} \tag{8.68}$$

From (8.67), however, we see that

$$\lambda_S = U_C C'(E) + \alpha\lambda_P$$

Substitution of this into (8.68) results in the alternative condition

$$(8.68') \qquad U_C C'(E)\left\{\begin{matrix} > \\ < \end{matrix}\right\} - (\alpha + \beta)\lambda_P \qquad \Rightarrow \qquad A^* = \left\{\begin{matrix} 0 \\ \hat{A} \end{matrix}\right\}$$

The optimal choice of A thus depends critically on λ_P.

The Policy Choices

Categorically, the optimal choice of antipollution activities A is either an *interior* solution or a *boundary* solution. Forster shows that it is not possible to have an interior solution in the present model.

To see this, consider the equations of motion of the costate variables:

$$(8.69) \qquad \dot{\lambda}_P = -\frac{\partial H}{\partial P} = -U_P + \delta\lambda_P$$

$$(8.70) \qquad \dot{\lambda}_S = -\frac{\partial H}{\partial S} = 0 \qquad \Rightarrow \qquad \lambda_S = \text{constant}$$

If A^* is an interior solution, then

$$\beta\lambda_P + \lambda_S = 0 \qquad [\text{by (8.68)}]$$

Since λ_S is a constant by (8.70), this last equation shows that λ_P must also be a constant, which in turn implies that

$$(8.71) \qquad \dot{\lambda}_P = 0 \qquad \Rightarrow \qquad \delta\lambda_P = U_P \qquad [\text{by (8.69)}]$$

But the constancy of λ_P requires U_P to be constant, too. Since U is monotonic in P, there can only be one value of P that would make U_P take any particular constant value. Thus P, too, must be a constant if A^* is an interior solution.

Given the initial pollution stock $P_0 > 0$, to have a constant P is to fix the terminal pollution stock at $P(T) = P_0 > 0$. For a problem with a truncated terminal line, the transversality condition includes the stipulation that

$$(8.72) \qquad P(T)\lambda_P(T) = 0$$

With a positive $P(T)$, it is required that $\lambda_P(T) = 0$, which, since λ_P is a constant by (8.71), means that

$$\lambda_P(t) = 0 \qquad \text{for all } t \in [0, T]$$

But the zero value for λ_P would imply, by (8.68′), that $U_C C'(E) = 0$, which contradicts the assumptions that U_C and C' are both positive. Consequently, an interior solution for A^* must be ruled out in the present model.

Boundary Solutions for Control Variable A

The only viable policies are therefore the boundary solutions $A^* = 0$ (no attempt to fight pollution) or $A^* = \hat{A}$ (abatement of pollution at the maximal feasible rate). These two policies share the common feature that

$$U_C C'(E) = \lambda_S - \alpha\lambda_P \qquad \text{[from (8.67)]} \tag{8.73}$$

Interpreted economically, the effect of energy use on utility through consumption, $U_C C'(E)$, should under both policies be equated to the shadow value of the depletion of the energy resource, measured by λ_S, adjusted for the shadow value of pollution via the $-\alpha\lambda_P$ term. But the two policies differ in that

$$A^* = \begin{Bmatrix} 0 \\ \hat{A} \end{Bmatrix} \quad \text{is associated with} \quad \lambda_S \begin{Bmatrix} > \\ < \end{Bmatrix} -\beta\lambda_P \qquad \text{[by (8.68)]} \tag{8.74}$$

The economic meaning of the first line in (8.74) is that a hands-off policy on pollution is suited to the situation where the shadow price of energy resource, λ_S, exceeds that of pollution abatement, measured by $-\beta\lambda_P$. In this situation, it is not worthwhile to expend the resource on antipollution activities because the resource cost is greater than the benefit. But, as shown in the second line, fighting pollution at the maximal rate is justified in the opposite situation. In distinguishing between these two situations, the parameter β, which measures the efficacy of antipollution activities, plays an important role.

Let us look further into the implications of the $A^* = 0$ case. Since the pollution stock is left alone in this case, the terminal stock of pollution is certainly positive. With $P(T) > 0$, the transversality stipulation $P(T)\lambda_P(T) = 0$ in (8.72) mandates that $\lambda_P(T) = 0$. This suggests that λ_P, the shadow price of pollution, which is initially negative, should increase over time to a terminal value of zero to satisfy the complementary-slackness condition. That is,

$$\dot{\lambda}_P > 0$$

Now if we differentiate (8.73) totally with respect to t, we find

$$\left(U_{CC}C'^2 + U_C C''\right)\dot{E} = -\alpha\dot{\lambda}_P$$

Since the expression in parentheses is negative, $\dot{E}$ and $\dot{\lambda}_P$ have the same sign. So we can conclude that

$$\dot{E} > 0$$

The increasing use of energy over time will, in the present case, lead to the exhaustion of the energy resource. As (8.70) shows, λ_S is a constant—a positive constant because λ_S denotes the shadow price of a valuable resource. The positivity of the λ_S constant means that, in order to satisfy the transversality-condition stipulation

$$S(T)\lambda_S(T) = 0 \qquad [\text{cf. } (8.72)]$$

$S(T)$ must be zero, signifying the exhaustion of the energy resource stock at the terminal time T.

Turning to the other policy, we have pollution abatement at the maximal feasible rate, $A^* = \hat{A}$. Despite the maximal effort, however, we do not expect the antipollution activities to run the pollution stock P all the way down to zero. Upon the assumptions that $U_P < 0$ and $U_{PP} < 0$, we know that, as P steadily diminishes as a result of antipollution activities, U_P will steadily rise from a negative level toward zero. When the P stock becomes sufficiently small, U_P will reach a tolerable level where we can no longer justify the resource cost of further pollution-abatement effort. Therefore, we may expect the terminal pollution stock $P(T)$ to be positive. If so, then $\lambda_P(T) = 0$ by the complementary-slackness condition, and the rest of the story is *qualitatively* the same as the case of $A^* = 0$.

EXERCISE 8.5

1 What will happen to the optimal solution if the control region in the Forster model for control variable A in (8.65) is changed to $A \geq 0$?

2 Let a discount factor $e^{-\rho t}$, $(\rho > 0)$, be incorporated into problem (8.65).

(*a*) Write the current-value Hamiltonian H_c for the new problem.

(*b*) What are the conditions for maximizing H_c?

(*c*) Rewrite these conditions in terms of λ_P and λ_S (instead of m_P and m_S), and compare them with (8.67) and (8.68) to check whether the new conditions are equivalent to the old.

3 In the new problem with the discount factor in Prob. 2, check whether it is possible to have an interior solution for the control variable A. In answering this question, use the regular (present-value) Hamiltonian H, and assume that there exists a tolerably low level of P below which the marginal resource cost of further pollution abatement effort does not compensate for the marginal environmental betterment.

CHAPTER 9

INFINITE-HORIZON PROBLEMS

In our discussion of problems of the calculus of variations, it was pointed out that the extension of the planning horizon to infinity entails certain mathematical complications. The same can be said about problems of optimal control. One main issue is the convergence of the objective functional, which, in the infinite-horizon context, is an improper integral. Since we have discussed this problem in Sec. 5.1, there is no need to elaborate on the topic here. But the other major issue—whether finite-horizon transversality conditions can be generalized to the infinite-horizon context—deserves some further discussion, because serious doubts have been cast upon their applicability by some alleged counterexamples in optimal control theory. We shall examine such counterexamples and then argue that the counterexamples may be specious.

9.1 TRANSVERSALITY CONDITIONS

In their celebrated book, Pontryagin et al. concern themselves almost exclusively with problems with a finite planning horizon. The only digression into the case of an improper-integral functional is a three-page discussion of an autonomous problem in which the "boundary condition at infinity" is assumed to take the specific form[1]

$$\lim_{t \to \infty} y(t) = y_\infty \qquad (y_\infty \text{ given}) \tag{9.1}$$

In other words, the infinite horizon is specified to be accompanied by a fixed terminal state. For that case, it is shown that the maximum-principle condition—including the transversality condition for what we have been referring to as the horizontal-terminal-line problem—apply as they do to a finite-horizon problem. For the latter, the transversality condition is

$$[H]_{t=T} = 0$$

When the problem is autonomous, moreover, the maximized value of the Hamiltonian is known to be constant over time, so the condition $H = 0$ can be checked not only at $t = T$, but at any point of time at all. Accordingly, the transversality condition pertaining to an infinite-horizon autonomous problem with the boundary condition (9.1) can be expressed not only as

$$\lim_{t \to \infty} H = 0 \tag{9.2}$$

but more broadly as

$$H = 0 \qquad \text{for all } t \in [0, \infty) \tag{9.3}$$

The fact that the transversality condition for the horizontal-terminal-line problem can be adapted from the finite-horizon to the infinite-horizon context makes it plausible to expect that a similar adaptation will work for other problems. If the terminal state is free as $t \to \infty$, for instance, we may expect the transversality condition to take the form

(9.4)
$$\lim_{t \to \infty} \lambda(t) = 0 \qquad [\text{transversality condition for free terminal state}]$$

Similarly, in case the variable terminal state is subject to a prespecified minimum level y_{min} as $t \to \infty$, we may expect the infinite-horizon transversality condition to be

$$\lim_{t \to \infty} \lambda(t) \geq 0 \qquad \text{and} \qquad \lim_{t \to \infty} \lambda(t)[y(t) - y_{min}] = 0 \tag{9.5}$$

The Variational View

That (9.4) is a reasonable transversality condition can be gathered from an adaptation of the procedure used in Sec. 7.3 to derive the finite-horizon

[1] L. S. Pontryagin et al., *The Mathematical Theory of Optimal Processes*, Interscience, New York, 1962, pp. 189–191.

transversality conditions for problem (7.20). There, we first combined—via a Lagrange multiplier—the original objective functional V with the equation of motion for y into a new functional

$$V = \int_0^T H(t, y, u, \lambda)\, dt - \int_0^T \lambda(t)\dot{y}\, dt \qquad \text{[from (7.22')]}$$

Then, after integrating the second integral by parts, we rewrote V as

$$V = \int_0^T \left[H(t, y, u, \lambda) + y(t)\dot{\lambda} \right] dt - \lambda(T) y_T + \lambda(0) y_0 \qquad \text{[from (7.22'')]}$$

To generate neighboring paths for comparison with the optimal control path and the optimal state path, we adopted perturbing curves $p(t)$ for u and $q(t)$ for y, to write

$$u(t) = u^*(t) + \epsilon p(t) \qquad \text{and} \qquad y(t) = y^*(t) + \epsilon q(t)$$

[from (7.24) and (7.25)]

Similarly, for variable T and y_T, we wrote

$$T = T^* + \epsilon\, \Delta T \qquad \text{and} \qquad y_T = y_T{}^* + \epsilon\, \Delta y_T \quad \text{[from (7.26)]}$$

As a result, the functional V was transformed into a *function* of ϵ. The first-order condition for maximizing V then emerged as

$$\frac{dV}{d\epsilon} = \int_0^T \left[\left(\frac{\partial H}{\partial y} + \dot{\lambda} \right) q(t) + \frac{\partial H}{\partial u} p(t) \right] dt + [H]_{t=T} \Delta T - \lambda(T)\, \Delta y_T = 0$$

[from (7.30)]

When adapted to the infinite-horizon framework, this equation becomes

(9.6)

$$\frac{dV}{d\epsilon} = \underbrace{\int_0^\infty \left[\left(\frac{\partial H}{\partial y} + \dot{\lambda} \right) q(t) + \frac{\partial H}{\partial u} p(t) \right] dt}_{\Omega_1} + \underbrace{\lim_{t \to \infty} H\, \Delta T}_{\Omega_2} - \underbrace{\lim_{t \to \infty} \lambda(t)\, \Delta y_T}_{\Omega_3}$$

$$= 0$$

In order to satisfy this condition, each of the three component terms $(\Omega_1, \Omega_2, \Omega_3)$ must vanish individually. It is the vanishing of the last two terms, Ω_2 and Ω_3, that gives rise to transversality conditions.

The Variability of T and y_T

For an infinite-horizon problem, the terminal time T is not fixed, so that ΔT is nonzero. Thus, to make the Ω_2 term in (9.6) vanish, we must impose the condition

$$(9.7) \qquad \lim_{t \to \infty} H = 0 \qquad \text{[infinite-horizon transversality condition]}$$

This condition appears to be identical with (9.2). Unlike the latter, however, (9.7) plays the role of a general transversality condition for infinite-horizon problems, regardless of whether the terminal state is fixed. It therefore carries greater significance.

We have based the justification of this condition on the simple fact that the terminal time is perforce not fixed in infinite-horizon problems. But a full-fledged proof for this condition has been provided by Michel.[2] The proof is quite long, and it will not be reproduced here. Instead, we shall add a few words of economic interpretation to clarify its intuitive appeal. If, following Dorfman (see Sec. 8.1), we let the state variable represent the capital stock of a firm, the control variable the business policy decision, and the F function the profit function of the firm, then the Hamiltonian function sums up the overall (current plus future) profit prospect associated with each admissible business policy decision. So long as H remains positive, there is yet profit to be made by the appropriate choice of the control. To require $H \to 0$ as $t \to \infty$ means that the firm should see to it that all the profit opportunities will have been taken advantage of as $t \to \infty$. Intuitively, such a requirement is appropriate whether or not the firm's terminal state (capital stock) is fixed.

With regard to the Ω_3 term in (9.6), on the other hand, it does matter whether y_T is fixed or free. Suppose, first, that the terminal state is fixed. Then we have $\lim_{t \to \infty} y_T = y_\infty$, same as in (9.1). Since this implies $\Delta y_T = 0$ at time infinite, the Ω_3 term will vanish without requiring any restriction on the terminal value of λ. In contrast, with a free terminal state, we no longer have $\Delta y_T = 0$ at time infinite. Consequently, to make Ω_3 vanish, it is necessary to impose the condition $\lim_{t \to \infty} \lambda(t) = 0$, and this provides the rationale for the transversality condition (9.4).

Despite the preceding considerations, many writers consider the question of infinite-horizon transversality conditions to be in an unsettled state. The transversality condition (9.7) is not in dispute. However, the transversality condition (9.4)—$\lim_{t \to \infty} \lambda(t) = 0$—has been thrown into doubt by several writers who claim to have found counterexamples in which that condition is violated. If true, such counterexamples would, of course, disqualify that condition as a necessary condition. We shall argue, however, that those are not genuine counterexamples, because the problems involved do not have a truly free terminal state, so that (9.4) ought not to have been applied in the first place.

[2]Phillipe Michel, "On the Transversality Condition in Infinite Horizon Optimal Problems," *Econometrica*, July 1982, pp. 975–985.

9.2 SOME COUNTEREXAMPLES REEXAMINED

The Halkin Counterexample

Perhaps the most often-cited counterexample is the following autonomous control problem constructed by Halkin:[3]

$$
\begin{aligned}
&\text{Maximize} && \int_0^\infty (1-y)u\,dt \\
(9.8)\qquad &\text{subject to} && \dot{y} = (1-y)u \\
& && y(0) = 0 \\
&\text{and} && u(t) \in [0,1]
\end{aligned}
$$

Inasmuch as the integrand function F is identical with the f function in the equation of motion, the objective function can be rewritten as

$$
(9.9)\qquad \int_0^\infty \dot{y}\,dt = [y(t)]_0^\infty = \lim_{t\to\infty} y(t) - y(0) = \lim_{t\to\infty} y(t)
$$

Viewed in this light, the objective functional is seen to depend exclusively on the terminal state at the infinite horizon. For such a problem, known as a *terminal control problem*, the initial and intermediate state values attained on an optimal path do not matter at all, except for their role as stepping stones leading to the final state value. Hence, to maximize (9.9) is tantamount to requiring the state variable to take one specific value—the upper bound of y—as its terminal value.

To find the upper bound of y, let us first find the $y(t)$ path. The equation of motion for y, which can be written as

$$
\dot{y} + u(t)y = u(t)
$$

is a first-order linear differential equation with a variable coefficient and a

---

[3]Hubert Halkin, "Necessary Conditions for Optimal Control Problems with Infinite Horizons," *Econometrica*, March 1974, pp. 267–272, especially p. 271. A slightly different version of the Halkin counterexample is cited in Kenneth J. Arrow and Mordecai Kurz, *Public Investment, the Rate of Return, and Optimal Fiscal Policy*, Johns Hopkins Press, Baltimore, MD, 1970, p. 46 (footnote). The Halkin counterexample has also been publicized in Akira Takayama, *Mathematical Economics*, 2d ed., Cambridge University Press, Cambridge, 1985, p. 625; and Ngo Van Long and Neil Vousden, "Optimal Control Theorems," Essay 1 in John D. Pitchford and Stephen J. Turnovsky, eds., *Applications of Control Theory to Economic Analysis*, North-Holland, Amsterdam, 1977, p. 16.

variable term. Its general solution is[4]

$$(9.10)\qquad y(t) = ce^{-\int u\,dt} + 1 \qquad (c \text{ arbitrary})$$
$$= ke^{-\int_0^t u\,dt} + 1 \qquad (k \text{ arbitrary})$$

Next, by setting $t = 0$ in the general solution and making use of the initial condition $y(0) = 0$, we find

$$0 = y(0) = ke^0 + 1 = k + 1$$

so that $k = -1$. Therefore, the definite solution is

$$(9.10')\qquad y(t) = 1 - e^{-\int_0^t u\,dt} \equiv 1 - e^{-Z(t)}$$

where $Z(t)$ is a shorthand symbol for the definite integral of u.

Since $u(t) \in [0, 1]$, $Z(t)$ is nonnegative. Hence, the value of $e^{-Z(t)}$ must lie in the interval $(0, 1]$, and the value of $y(t)$ must lie in the interval $[0, 1)$. It follows that *any* $u(t)$ path that makes $Z(t) \to \infty$ as $t \to \infty$—thereby making $e^{-Z(t)} \to 0$ and $y(t) \to 1$—will maximize the objective functional in (9.10′) and be optimal. In other words, there is no unique optimal control. Halkin himself suggests the following as an optimal control:

$$u^* = 0 \qquad \text{for } t \in [0, 1]$$
$$u^* = 1 \qquad \text{for } t \in (1, \infty)$$

This would work because $\int_1^\infty 1\,dt$ is divergent, resulting in an infinite $Z(t)$. But, as Arrow and Kurz point out (*op. cit.*, p. 46), any constant time path for u,

$$u^*(t) = u_0 \qquad \text{for all } t \qquad (0 < u_0 < 1)$$

would serve just as well. The fact that Arrow and Kurz chose $u(t) = u_0$, where u_0 is an *interior* point in the control region [0, 1], makes it possible to use the derivative condition $\partial H/\partial u = 0$. Since the Hamiltonian for this problem is

$$(9.11)\qquad H = (1 - y)u + \lambda(1 - y)u = (1 + \lambda)(1 - y)u$$

the condition $\partial H/\partial u = 0$ means $(1 + \lambda)(1 - y) = 0$, or more simply $(1 + \lambda) = 0$, inasmuch as y is always less than one. Hence, we find $\lambda^*(t) = -1$,

[4]See Alpha C. Chiang, *Fundamental Methods of Mathematical Economics*, 3d ed., McGraw-Hill, New York, 1984, Sec. 14.3, for the standard formula for solving such a differential equation. The rationale behind the conversion of the indefinite integral $\int u\,dt$ to the definite integral $\int_0^t u\,dt$ in (9.10) is discussed in *op. cit.*, Sec. 13.3 (last subsection). The constant k differs from the constant c, because it has absorbed another constant that arises in connection with the lower limit of integration of the definite integral.

which contradicts the transversality condition (9.4), and seemingly makes this problem a counterexample.

As pointed out earlier, however, the structure of the Halkin problem is such that it requires the choice of the upper bound of y as the terminal state, regardless of the $y^*(t)$ path chosen. Put differently, there is an implicit fixed terminal state, even though this is not openly declared in the problem statement. In view of this, the correct transversality condition is not $\lim_{t\to\infty} \lambda(t) = 0$ as in (9.4), but rather $\lim_{t\to\infty} H = 0$ as given in (9.2). More specifically, we should use the condition $H = 0$ for all t as in (9.3), because the problem is autonomous.

Using the transversality condition $H = 0$ for all t, we can approach the Halkin problem as follows. First, noting that the Hamiltonian is linear in u, we only need to consider the boundary values of the control region as candidates for $u^*(t)$, namely, either $u = 0$ or $u = 1$. The consequence of choosing $u = 0$ at any point of time, however, is to make $\dot{y} = 0$ (by the equation of motion) at that point of time, thereby preventing y from growing. For this reason, we must avoid choosing $u(t) = 0$ for all t.[5] Suppose, instead, we choose the control

$$u^*(t) = 1 \qquad \text{for all } t$$

Then we have the equations of motion

$$\dot{\lambda} = -\frac{\partial H}{\partial y} = (1+\lambda)u = 1 + \lambda$$

$$\dot{y} = \frac{\partial H}{\partial \lambda} = (1-y)u = 1 - y$$

whose general solutions are, respectively,

$$\lambda(t) = Ae^{t} - 1 \qquad (A \text{ arbitrary})$$

$$y(t) = Be^{-t} + 1 \qquad (B \text{ arbitrary})$$

By the initial condition $y(0) = 0$, we can determine that $B = -1$, so the definite solution of the state variable is

$$y(t) = 1 - e^{-t} \qquad \text{with } \lim_{t\to\infty} y(t) = 1$$

This latter fact shows that the control we have chosen, $u^*(t) = 1$, can maximize the objective functional just as those of Halkin and Arrow and Kurz. What about the arbitrary constant A? To definitize A, we call upon

[5]The choice of $u = 0$ for some limited initial period of time does no harm, since only the terminal state counts in the present problem. The solution suggested by Halkin does contain the feature that $u = 0$ initially, for $t \in [0, 1]$.

the transversality condition $H = 0$ for all t. From (9.11), we see that, with $u = 1$ and $y < 1$ for all t, the only way to satisfy the condition $H = 0$ is to have $\lambda = -1$ for all t. Hence, $A = 0$.

Even though our optimal control differs from that of Halkin and that of Arrow and Kurz, the conclusion that $\lambda^*(t) = -1$ is the same. Yet, we do not view this as a violation of an expected transversality condition. Rather, it represents a perfectly acceptable conclusion drawn from the correct transversality condition. In this light, Halkin's problem does not constitute a valid counterexample.

Some may be inclined to question this conclusion on grounds that the requirement on the terminal state, being based on the maximization of (9.9), is in the nature of an outcome of optimization rather than an implicit restriction on the terminal state. A moment's reflection will reveal, however, that such an objection is valid only for a static optimization problem, where the objective is merely to choose one particular value of y from its permissible range. But in the present context of dynamic optimization, the optimization process calls for the choice of, not a single value of a variable, but an entire path. In such a context, any restriction on the state variable at a single point of time (here, the terminal time) must, as a general rule, be viewed as a restriction on the problem itself. The fact that the problem is a special case, with the objective functional depending exclusively on the terminal state, in no way disqualifies it as a dynamic optimization problem; nor does it alter the general rule that any requirement placed on the terminal state should be viewed as a restriction on the problem.

Other Counterexamples

In another alleged counterexample, Karl Shell[6] shows that the transversality condition (9.4) does not apply when the problem is to maximize an integral of per-capita consumption $c(t)$ over an infinite period of time: $\int_0^\infty c(t)\,dt$. But we can again show that the problem has an implicit fixed terminal state, so that (9.4) should not have been expected to apply in the first place.

The Shell problem is based on the neoclassical production function $Y = Y(K, L)$, which, on the assumption of linear homogeneity, can be rewritten in per-capita (per-worker) terms as

$$y = \phi(k) \qquad \text{with } \phi'(k) > 0,\ \phi''(k) < 0$$

where $y \equiv Y/L$ and $k \equiv K/L$. Let n denote the rate of growth of labor, and

[6]Karl Shell, "Applications of Pontryagin's Maximum Principle to Economics," in H. W. Kuhn and G. P. Szegö, eds., *Mathematical Systems Theory and Economics*, Vol. 1, Springer-Verlag, New York, 1969, pp. 241–292, especially pp. 273–275.

let δ denote the rate of depreciation of capital. Then the rate of growth of k, the capital-labor ratio, is

$$\dot{k} = \phi(k) - c - (n + \delta)k \tag{9.12}$$

which is similar to the fundamental differential equation of the Solow growth model.[7] In a steady state, we should have $\dot{k} = 0$. By setting $\dot{k} = 0$ and transposing, we obtain the following expression for the steady-state per-capita consumption:

$$c = \phi(k) - (n + \delta)k \qquad [\text{steady-state value of } c] \tag{9.13}$$

Note that the steady-state value of c is a function of the steady-state value of k. The highest possible steady-state c is attained when $dc/dk = \phi'(k) - (n + \delta) = 0$, that is, when

$$\phi'(k) = n + \delta \tag{9.14}$$

The condition given in (9.14) is known as the *golden rule of capital accumulation*.[8] Since $\phi(k)$ is monotonic, there is only one value of k that will satisfy (9.14). Let that particular value of k—the *golden-rule* value of k—be denoted by $\hat{k}$, and let the corresponding *golden-rule* value of c be denoted by $\hat{c}$. Then

$$\hat{c} = \phi(\hat{k}) - (n + \delta)\hat{k} \qquad [\text{by (9.13)}] \tag{9.15}$$

This value of c represents the maximum sustainable per-capita consumption stream, since it is derived from the maximization condition $dc/dk = 0$.

Now consider the problem of maximizing $\int_0^\infty c(t)\,dt$. Since this integral obviously does not converge for $c(t) \geq 0$, Shell adopts the Ramsey device of rewriting the integrand as the deviation from "Bliss"—with the "Bliss" identified in this context with $\hat{c}$. In other words, the problem is to

$$\begin{aligned} &\text{Maximize} && \int_0^\infty (c - \hat{c})\,dt \\ &\text{subject to} && \dot{k} = \phi(k) - c - (n + \delta)k \\ &&& k(0) = k_0 \\ &\text{and} && 0 \leq c(t) \leq \phi[k(t)] \end{aligned} \tag{9.16}$$

---

[7] For details of the derivation, see the ensuing discussion leading to (9.25).

[8] See E. S. Phelps, "The Golden Rule of Capital Accumulation: A Fable for Growthmen," *American Economic Review*, September 1961, pp. 638–643. This rule is christened the "golden rule" because it is predicated upon the premise that each generation is willing to abide by the golden rule of conduct ("Do unto others as you would have others do unto you.") and adopt a uniform savings ratio to be adhered to by all generations.

Although the state variable, k, has a given initial value k_0, no terminal value is explicitly mentioned. The control variable, c, is confined at every point of time to the control region $[0, \phi(k)]$, which is just another way of saying that the marginal propensity to consume is restricted to the interval $[0, 1]$. For this autonomous problem, the Hamiltonian is

$$H = c - \hat{c} + \lambda[\phi(k) - c - (n + \delta)k] \tag{9.17}$$

Shell shows that the optimal path of the costate variable has the limit value 1 as $t \to \infty$, and the optimal state path is characterized by $\lim_{t \to \infty} k(t) = \hat{k}$. The fact that $\lambda(t)$ optimally tends to a unit value instead of zero leads Shell to consider this a counterexample against the transversality condition (9.4).

However, the Shell problem in fact also contains an implicit fixed terminal state, so that the appropriate transversality condition is $H = 0$ for all t, rather than $\lim_{t \to \infty} \lambda(t) = 0$. To see this, recall that the very reason for rewriting the objective functional as in (9.16) is to ensure convergence. Since the integrand expression $(c - \hat{c})$, a continuous function of time, maintains the same sign for all t and offers no possibility of cancellation between positive and negative terms associated with different points of time, the only way for the integral to converge is to have $(c - \hat{c}) \to 0$ as $t \to \infty$.[9] Hence, $c(t)$ must tend to $\hat{c}$ and, in line with (9.15), $k(t)$ must tend to $\hat{k}$, as $t \to \infty$. What this does is to fix the terminal state, if only implicitly, in the form of $\lim_{t \to \infty} k(t) = \hat{k}$, thereby disqualifying (9.4) as a transversality condition. The important point is that this terminal state does not come about as the result of optimization; rather, it is a condition dictated by the convergence requirement of the objective functional.

We can easily check the correct transversality condition $H = 0$ for $t \to \infty$. As t becomes infinite, c and k will take the values $\hat{c}$ and $\hat{k}$, respectively; moreover, $\hat{c}$ and $\hat{k}$ will be related to each other by (9.15). Hence, from (9.17), it is clear that H will tend to zero. Since, for this autonomous problem, H should be zero for all t, we can actually solve for λ by setting (9.17) equal to zero, to get

$$\lambda(t) = \frac{\hat{c} - c(t)}{\phi[k(t)] - c(t) - (n + \delta)k(t)} \tag{9.18}$$

By taking the limit of this expression, we can verify Shell's result that $\lambda(t) \to 1$ as $t \to \infty$.[10] From our perspective, however, this result by no

---

[9]See the discussion of Condition II in Sec. 5.1.

[10]To do so, we can resort to L'Hôpital's rule, since both the numerator and the denominator in (9.18) approach zero as t becomes infinite. Recalling that (9.14) is satisfied as $t \to \infty$, we have

$$\lim_{t \to \infty} \lambda(t) = \lim_{t \to \infty} \frac{-\dot{c}}{f'(k)\dot{k} - \dot{c} - (n + \delta)\dot{k}} = \lim_{t \to \infty} \frac{-\dot{c}}{-\dot{c}} = 1$$

means contradicts any applicable transversality condition. Rather, it is perfectly consistent with, and indeed follows from, the correct transversality condition, $H = 0$ for all t.

Another intended counterexample is found in the regional investment model of John Pitchford.[11] Suppressing population growth, he considers the problem of choosing I_1 and I_2, the rate of investment in each of two regions, so as to

$$
\begin{aligned}
&\text{Maximize} && \int_0^\infty [c(t) - \hat{c}]\, dt \\
&\text{subject to} && \dot{K}_1 = I_1 - \delta K_1 \\
&&& \dot{K}_2 = I_2 - \delta K_2 \\
&&& c = \phi_1(K_1) + \phi_2(K_2) - I_1 - I_2 \geq 0 \\
&&& K_1(0) > 0 \qquad K_2(0) > 0 \\
&\text{and} && I_1 \geq 0 \qquad I_2 \geq 0
\end{aligned} \tag{9.19}
$$

The symbol $\hat{c}$ again denotes the maximum sustainable level of consumption. The model is essentially similar to the Shell counterexample, except that there are here two control variables, I_1 and I_2, and two state variables, K_1 and K_2, as well as certain inequality constraints (a subject to be discussed in the next chapter). Pitchford also assumes implicitly that the terminal states are constrained to be nonnegative.

The solution to this problem involves the terminal states $\hat{K}_1 > 0$ and $\hat{K}_2 > 0$, defined by the relations $\phi_1'(\hat{K}_1) = \phi_2'(\hat{K}_2) = \delta$ [cf. (9.14); here, $n = 0$]. Since the terminal value of each of the costate variables is found to be one, the transversality condition implied by (9.5) for nonnegative terminal capital values, namely, $\lim_{t \to \infty} \lambda_i(t) K_i(t) = 0$, is not satisfied (Pitchford, *op. cit.*, p. 143). This, of course, would not have come as a surprise, if the objective functional had been required to converge. For then $c(t)$ must tend to $\hat{c}$, and K_i must tend to $\hat{K}_i$, thereby making the problem one with fixed terminal states, to which a different type of transversality condition should apply. However, not wanting to rule out the possible equilibria $(0, 0)$, $(0, \hat{K}_2)$, and $(\hat{K}_1, 0)$ a priori, Pitchford decides to proceed on the assumption that the integral is not required to converge—which, however, makes one wonder why he would then still employ the Ramsey device of expressing the integrand as $c(t) - \hat{c}$, a device explicitly purported to ensure convergence. Because of this assumption, it takes a rather lengthy excursion before he is

[11] John D. Pitchford, "Two State Variable Problems," Essay 6 in John D. Pitchford and Stephen J. Turnovsky, eds., *Applications of Control Theory to Economic Analysis*, North-Holland, Amsterdam, 1977, pp. 127–154, especially pp. 134–143.

led back to the conclusion that the objective functional has to converge after all. The upshot is then the same: The terminal states are implicitly fixed. And the model is not a genuine counterexample.

The Role of Time Discounting

In a noteworthy observation, Pitchford points out (*op. cit.*, p. 128, footnote) that all the known counterexamples apparently share one common feature: No time discounting is present in the objective functional. While Pitchford offers no reason for this phenomenon, he reports a finding of Weitzman[12] that, for *discrete*-time problems, the transversality condition involving the costate variables indeed becomes necessary when there is time discounting and the objective functional converges. Moreover, Pitchford conjectures that a similar result would hold for the *continuous*-time case. The relevant work on the continuous-time case comes later in a paper by Benveniste and Scheinkman.[13]

What our foregoing discussion can offer is a simple intuitive explanation of the link between the absence (presence) of time discounting on the one hand and the failure (applicability) of the transversality condition (9.4) or (9.5) on the other. Consider the case of maximizing $\int_0^\infty [c(t) - \hat{c}]\,dt$. If this functional is required to converge, and given that no possibility exists for cancellation between positive and negative terms, the integrand $[c(t) - \hat{c}]$ must tend to zero as $t \to \infty$. Since the only way for this to occur is for $c(t)$ to approach $\hat{c}$, the terminal state for the problem is in effect fixed, whether or not explicitly acknowledged in the problem statement. But if time discounting is admitted, so that the integral is, say, $\int_0^\infty [c(t) - \hat{c}]e^{-\rho t}\,dt$, then convergence will merely call for the vanishing of $[c(t) - \hat{c}]e^{-\rho t}$ as $t \to \infty$, which can be achieved as long as $[c(t) - \hat{c}]$ tends to some finite number, not necessarily zero. Inasmuch as $c(t)$ is no longer obliged to approach a unique value $\hat{c}$ for convergence, and nor is $K(t)$ obliged to approach a unique value K, the problem with time discounting would indeed be one with a free terminal state, with (9.4) or (9.5) as its expected transversality condition. The Shell and Pitchford models omit time discounting, and thus do not fall into this category. The case where the objective functional does contain a discount factor is illustrated by the neoclassical optimal growth model which we shall discuss in Sec. 9.3.

---

[12]M. L. Weitzman, "Duality Theory for Infinite Horizon Convex Models," *Management Science*, 1973, pp. 783–789.

[13]L. M. Benveniste and J. A. Scheinkman, "Duality Theory for Dynamic Optimization Models of Economics: The Continuous Time Case," *Journal of Economic Theory*, 1982, pp. 1–19.

The Transversality Condition as Part of a Sufficient Condition

Although the transversality condition $\lim_{t \to \infty} \lambda(t) = 0$ is not universally accepted as a necessary condition for infinite-horizon problems, a limit condition involving λ does enter as an integral part of a concavity sufficient condition for such problems. In an amalgamated form—combining the concavity provisions of Mangasarian as well as those of Arrow—the sufficiency theorem states that the conditions in the maximum principle are sufficient for the global maximization of V in the infinite-horizon problem

$$\begin{aligned} &\text{Maximize} && V = \int_0^\infty F(t, y, u)\, dt \\ &\text{subject to} && \dot{y} = f(t, y, u) \\ &\text{and} && y(0) = y_0 \qquad (y_0 \text{ given}) \end{aligned}$$

provided

(9.20) either $H \equiv F(t, y, u) + \lambda f(t, y, u)$ is concave in (y, u) for all $t \in [0, T]$, or $H^0 \equiv F(t, y, u^*) + \lambda f(t, y, u^*)$ is concave in y for all t, for given λ

and

$$\lim_{t \to \infty} \lambda(t)[y(t) - y^*(t)] \geq 0 \tag{9.21}$$

where $y^*(t)$ denotes the optimal state path and $y(t)$ is any other admissible state path. The terminal state can be either fixed or free.

Note that in (9.20) we have merged Mangasarian's concavity conditions on the F and f functions and nonnegativity condition on $\lambda(t)$ into a single concavity condition on the Hamiltonian H. The concavity of H is in (y, u), jointly. In contrast, Arrow's condition is that H^0 be concave in the y variable alone.

The limit expression in (9.21) is the infinite-horizon counterpart of the expression $\lambda^*(T)(y_T - y_T^*)$ which we have encountered earlier in the proof of the Mangasarian theorem [see footnote related to (8.31) and (8.31')]. In that proof, the expression just cited is shown to vanish, so that the result $V \leq V^*$ emerges, establishing V^* to be a global maximum. But, to establish the maximality of V^*, it is in fact also acceptable to have $\lambda^*(T)(y_T - y_T^*)$ positive. It is along the same line of reasoning that condition (9.21) restricts the limit of $\lambda(t)[y(t) - y^*(t)]$ to be either positive or zero as $t \to \infty$. The reader may also find it of interest to compare (9.21) with (5.44), the corresponding condition in the calculus-of-variations context. At first glance, condition (5.44) may strike one as entirely different from (9.21) because it shows the "$\leq$" rather than the "$\geq$" inequality. However, from (7.57) we recall that $\lambda = -F_{y'}$. Since multiplying an inequality by a negative number reverses the sense of inequality, it becomes clear that (9.21) and (5.44) are indeed exactly the same.

9.3 THE NEOCLASSICAL THEORY OF OPTIMAL GROWTH

The Ramsey model of saving behavior (Sec. 5.3), which deals with the important issue of intertemporal resource allocation, has exerted a strong influence on economic thinking, although this influence did not come into being until after World War II, when that model was "rediscovered" by growth theorists long after its publication. In a more recent development, the same basic issue is formulated as a problem of optimal control rather than the calculus of variations. Moreover, the new treatment—labeled as "the neoclassical theory of optimal growth"—extends the Ramsey model in two major respects: (1) The labor force (identified with the population) is assumed to be growing at an exogenous constant rate $n > 0$ (the Ramsey model has $n = 0$), and (2) the social utility is assumed to be subject to time discounting at a constant rate $\rho > 0$ (the Ramsey model has $\rho = 0$). Our discussion of this subject will be based primarily on a classic paper by David Cass.[14]

The Model

This theory is labeled a "neoclassical" theory, because its analytical framework revolves around the neoclassical production function $Y = Y(K, L)$, assumed to be characterized by constant returns to scale, positive marginal products, and diminishing returns to each input.[15] Such a production function, being linearly homogeneous, can be rewritten in per-worker terms —or per-capita terms, as we shall draw no distinction between the population and the labor force. Letting the lowercases of Y and K be defined, respectively, as

$$y \equiv \frac{Y}{L} \qquad \text{(average product of labor)}$$

$$k \equiv \frac{K}{L} \qquad \text{(capital-labor ratio)}$$

[14]David Cass, "Optimum Growth in an Aggregate Model of Capital Accumulation," *Review of Economic Studies*, July 1965, pp. 233–240.

[15]In discussing the Ramsey model, we denoted output by Q. Here we use instead the notation Y, which is another commonly used symbol for income and output, especially at the macro level.

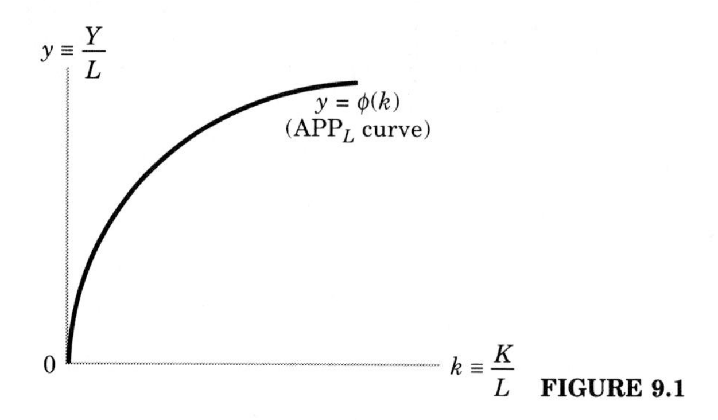

FIGURE 9.1

we can express the production function by

$$y = \phi(k) \qquad \text{with } \phi'(k) > 0 \text{ and } \phi''(k) < 0 \qquad \text{for all } k > 0 \tag{9.22}$$

Additionally, it is stipulated that

$$\lim_{k \to 0} \phi'(k) = \infty \qquad \text{and} \qquad \lim_{k \to \infty} \phi'(k) = 0$$

The graph of $\phi(k)$—the APP_L curve plotted against the capital-labor ratio —has the general shape illustrated in Fig. 9.1.

The total output Y is allocated either to consumption C or gross investment I_g. Therefore, net investment, $I = \dot{K}$, can be expressed as

$$\dot{K} = I_g - \delta K = Y - C - \delta K \qquad [\delta = \text{depreciation rate}]$$

Dividing through by L, and using the symbol $c \equiv C/L$ for per-capita consumption, we have

$$\frac{1}{L}\dot{K} = y - c - \delta k = \phi(k) - c - \delta k \tag{9.23}$$

The right-hand side now contains per-capita variables only, but the left-hand side does not. To unify the two sides, we make use of the relation

$$\begin{aligned} \dot{K} \equiv \frac{dK}{dt} &= \frac{d}{dt}(kL) = k\frac{dL}{dt} + L\frac{dk}{dt} \qquad &[\text{product rule}] \\ &= knL + L\dot{k} &\left[n \equiv \frac{dL/dt}{L}\right] \\ &= L(kn + \dot{k}) \end{aligned}$$

By substituting this last result into (9.23) and rearranging, we finally obtain an equation involving per-capita variables only:

$$\dot{k} = \phi(k) - c - (n + \delta)k \tag{9.24}$$

This equation, describing how the capital-labor ratio k varies over time, is the fundamental differential equation of neoclassical growth theory which we have already encountered in (9.12).

The level of per-capita consumption, c, is what determines the utility or welfare of society at any time. The social utility index function, $U(c)$, is assumed to possess the following properties:

$$U'(c) > 0 \qquad U''(c) < 0 \qquad \text{for all } c > 0$$
$$\lim_{c \to 0} U'(c) = \infty \qquad \text{and} \qquad \lim_{c \to \infty} U'(c) = 0 \tag{9.25}$$

The $U(c)$ function is, of course, to be summed over time in the dynamic optimization problem. But, since the population (labor force) grows at the rate n, Cass decides that the social utility attained at any point of time should be weighted by the population size at that time before summing. Hence, with a discount rate ρ, the objective function takes the form

$$\int_0^\infty U(c)L(t)e^{-\rho t}\,dt = \int_0^\infty U(c)L_0 e^{nt}e^{-\rho t}\,dt \tag{9.26}$$
$$= L_0\int_0^\infty U(c)e^{-(\rho-n)t}\,dt$$

To ensure convergence, Cass stipulates that $\rho - n > 0$. It might be pointed out, however, that this is equivalent to stipulating a single positive discount rate r, where $r \equiv \rho - n$. If, further, we let $L_0 = 1$ by choice of unit, then the functional will reduce to the simple form

$$\int_0^\infty U(c)e^{-rt}\,dt \qquad (r \equiv \rho - n > 0) \tag{9.26'}$$

In other words, weighting social utility by population size and simultaneously requiring the discount rate ρ to exceed the rate of population growth n, is mathematically no different from the alternative of *not* using population weights at all but adopting a new, positive discount rate r. For this reason, we shall proceed on the basis of the simpler alternative (9.26′).

The problem of optimal growth is thus simply to

$$\begin{aligned} &\text{Maximize} && \int_0^\infty U(c)e^{-rt}\,dt \\ &\text{subject to} && \dot{k} = \phi(k) - c - (n+\delta)k \\ &&& k(0) = k_0 \\ &\text{and} && 0 \le c(t) \le \phi[k(t)] \end{aligned} \tag{9.27}$$

This resembles the Shell problem (9.16), but here the integrand function is not $(c - \hat{c})$ (the deviation of c from the golden-rule level), but a utility index

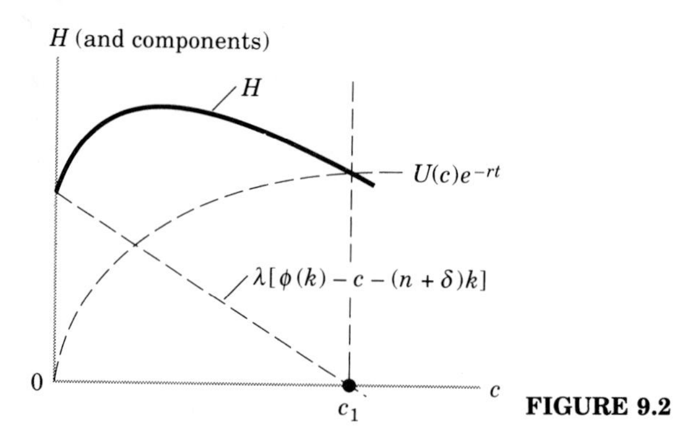

FIGURE 9.2

function of per-capita consumption, discounted at the rate r. There is only one state variable, k, and only one control variable, c.

The Maximum Principle

The Hamiltonian for this problem,

$$H = U(c)e^{-rt} + \lambda[\phi(k) - c - (n + \delta)k] \tag{9.28}$$

is nonlinear in c. More specifically, at any given point of time, the first additive component of H would plot against c as the illustrative broken curve in Fig. 9.2, and the second component would appear as the broken straight line.[16] Their sum, H, is seen to contain a hump, with its peak occurring at a c value between $c = 0$ and $c = c_1$. Since c_1 is the solution of the equation $[\phi(k) - c - (n + \delta)k] = 0$, it follows that $c_1 = \phi(k) - (n + \delta)k$, so that $c_1 < \phi(k)$. Hence, the maximum of H corresponds to a value of c in the interior of the control region $[0, \phi(k)]$. We can accordingly find the maximum of H by setting

$$\frac{\partial H}{\partial c} = U'(c)e^{-rt} - \lambda = 0$$

From this, we obtain the condition

$$U'(c) = \lambda e^{rt} \tag{9.29}$$

which states that, optimally, the marginal utility of per-capita consumption should be equal to the shadow price of capital amplified by the exponential

[16]The shape of the broken curve is based on the specification for $U(c)$ in (9.25). The broken straight line has vertical intercept $\lambda[\phi(k) - (n + \delta)k]$ and slope $-\lambda$. We take λ to be positive at any point of time, since it represents the shadow price (measured in utility) of capital at that time; hence, the negative slope of the broken straight line.

term e^{rt}. Since $\partial^2 H/\partial c^2 = U''(c)e^{-rt}$ is negative by (9.25), H is indeed maximized.

The maximum principle calls for two equations of motion. One of these, $\dot{\lambda} = -\partial H/\partial k$, entails for the present model the differential equation

$$\dot{\lambda} = -\lambda[\phi'(k) - (n + \delta)] \tag{9.30}$$

And the other, $\dot{k} = \partial H/\partial \lambda$, simply restates the constraint

$$\dot{k} = \phi(k) - c - (n + \delta)k \tag{9.31}$$

The three equations (9.29) through (9.31) should in principle enable us to solve for the three variables c, λ, and k. Without knowledge of the specific forms of the $U(c)$ and $\phi(k)$ functions, however, we can only undertake a qualitative analysis of the model.

It is, of course, also possible to work with the current-value Hamiltonian

$$H_c = U(c) + m[\phi(k) - c - (n + \delta)k] \qquad \text{[by (8.10)]} \tag{9.32}$$

In this case, the maximum principle requires that $\partial H_c/\partial c = U'(c) - m = 0$, or

$$m = U'(c) \tag{9.33}$$

This condition does maximize H_c, because $\partial^2 H_c/\partial c^2 = U''(c) < 0$ by (9.25).

The equation of motion for the state variable k can be read directly from the second line of (9.27), but it can also be derived as

$$\dot{k} = \frac{\partial H_c}{\partial m} = \phi(k) - c - (n + \delta)k \tag{9.34}$$

And the equation of motion for the current-value multiplier m is

$$\begin{aligned} \dot{m} &= -\frac{\partial H_c}{\partial k} + rm = -m[\phi'(k) - (n + \delta)] + rm \qquad \text{[by (8.13)]} \\ &= -m[\phi'(k) - (n + \delta + r)] \end{aligned} \tag{9.35}$$

The ensuing discussion will be based on the current-value maximum-principle conditions (9.33) through (9.35). Since there is no explicit t argument in these, we now have an autonomous system. This makes possible a qualitative analysis by a phase diagram, as we did in Sec. 5.4.

Constructing the Phase Diagram

Since the two differential equations (9.34) and (9.35) involve the variables k and m, the normal phase diagram would be in the km space. To use such a diagram, it would be necessary first to eliminate the other variable, c. But since the other equation, (9.33), contains a *function* of c—to wit, $U'(c)$—instead of the plain c itself, the task of eliminating c is more complicated than that of eliminating m. We shall, therefore, depart from Cass's procedure and seek instead to eliminate the m variable. In so doing, we shall create a differential equation in the variable c as a by-product. The analysis can then be carried out with a phase diagram in the kc space.

We begin by differentiating (9.33) with respect to t, to obtain an expression for $\dot{m}$:

$$\dot{m} = U''(c)\dot{c}$$

This, together with the equation $m = U'(c)$, enables us to rid (9.35) of the $\dot{m}$ and m expressions. The result, after rearrangement, is

$$\dot{c} = -\frac{U'(c)}{U''(c)}[\phi'(k) - (n + \delta + r)]$$

which is a differential equation in the variable c. Consequently, we now can work with the differential equation system

$$(9.36) \qquad \begin{aligned} \dot{k} &= \phi(k) - c - (n + \delta)k \\ \dot{c} &= -\frac{U'(c)}{U''(c)}[\phi'(k) - (n + \delta + r)] \end{aligned}$$

To construct the phase diagram, we first draw the $\dot{k} = 0$ curve and the $\dot{c} = 0$ curve. These are defined by the two equations

$$(9.37) \qquad c = \phi(k) - (n + \delta)k \qquad [\text{equation for } \dot{k} = 0 \text{ curve}]$$

$$(9.38) \qquad \phi'(k) = n + \delta + r \qquad [\text{equation for } \dot{c} = 0 \text{ curve}]$$ [17]

The $\dot{k} = 0$ curve shows up in the kc space as a concave curve, as illustrated in Fig. 9.3*b*. As (9.37) indicates, this curve expresses c as the difference between two functions of k: $\phi(k)$ and $(n + \delta)k$. The graph of $\phi(k)$, already encountered earlier in Fig. 9.1, is reproduced in Fig. 9.3*a*. And the $(n + \delta)k$ term simply gives us an upward-sloping straight line. Plotting the difference between these two curves then yields the desired $\dot{k} = 0$ curve in Fig. 9.3*b*. As to the $\dot{c} = 0$ curve, (9.38) requires that the slope of the $\phi(k)$ curve assume the specific value $n + \delta + r$. The $\phi(k)$ curve being monotonic, this requirement can be satisfied only at a single point on that curve, point B,

[17] The $-U'(c)/U''(c)$ term is always positive, by (9.25), and can never be zero. Therefore, the only way for $\dot{c}$ to vanish is to let the bracketed expression in (9.36) be zero.

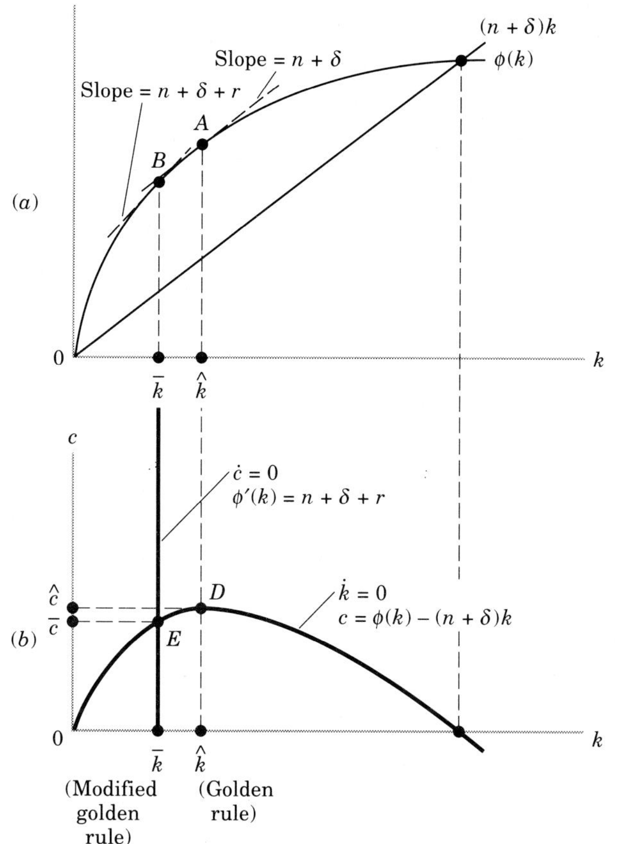

FIGURE 9.3

which corresponds to a unique k value, $\bar{k}$. Thus the $\dot{c} = 0$ curve must plot as a vertical straight line in Fig. 9.3*b*, with horizontal intercept $\bar{k}$.

The intersection of the two curves in Fig. 9.3*b*, at point E, determines the steady-state values of k and c. Denoted by $\bar{k}$ and $\bar{c}$, respectively, these values are referred to in the literature as the *modified-golden-rule* values of capital-labor ratio and per-capita consumption, as distinct from the golden-rule values $\hat{k}$ and $\hat{c}$ discussed earlier in the Shell counterexample. As indicated in (9.14), $\hat{k}$ is defined by $\phi'(\hat{k}) = n + \delta$. Thus it corresponds to point A on the $\phi(k)$ curve in Fig. 9.3*a*, where the tangent to that curve is parallel to the $(n + \delta)k$ line. In contrast, $\bar{k}$ is defined by $\phi'(\bar{k}) = n + \delta + r$, by (9.38), which involves a larger slope of $\phi(k)$. This is why point B (for $\bar{k}$) must be located to the left of point A (for $\hat{k}$). In other words, the modified-golden-rule value of k (based on time discounting at rate r) must be less than the golden-rule value of k (with no time discounting). By the same token, the modified-golden-rule value of c—the height of point E in Fig. 9.3*b*—must be less than the golden-rule value—the height of point D.

Analyzing the Phase Diagram

To prepare for the analysis of the phase diagram, we have, in Fig. 9.4, added vertical sketching bars to the $\dot{k} = 0$ curve, and horizontal ones to the $\dot{c} = 0$ curve. These sketching bars are useful in guiding the drawing of the streamlines—to remind us that the streamlines must cross the $\dot{k} = 0$ curve with infinite slope, and cross the $\dot{c} = 0$ curve with zero slope.

For clues about the general directions the streamlines should take, we partially differentiate the two differential equations in (9.36), to find that

$$\frac{\partial \dot{k}}{\partial c} = -1 < 0 \tag{9.39}$$

$$\frac{\partial \dot{c}}{\partial k} = -\frac{U'(c)}{U''(c)}\phi''(k) < 0 \qquad \text{[by (9.25) and (9.22)]} \tag{9.40}$$

According to (9.39), as c increases (going northward), $\dot{k}$ should follow the $(+,0,-)$ sign sequence. So, the k-arrowheads must point eastward below the $\dot{k} = 0$ curve, and westward above it. Similarly, (9.40) indicates that $\dot{c}$ should follow the $(+,0,-)$ sign sequence as k increases (going eastward). Hence, the c-arrowheads should point upward to the left of the $\dot{c} = 0$ curve, and downward to the right of it.

The streamlines drawn in accordance with such arrowheads yield a saddle-point equilibrium at point E, $(\bar{k}, \bar{c})$, where $\bar{k}$ and $\bar{c}$ denote the intertemporal equilibrium values of k and c, respectively. The constancy of $\bar{k}$ implies that $\bar{y} = \phi(\bar{k})$ is constant, too. Since $k \equiv K/L$ and $y \equiv Y/L$, the simultaneous constancy of $\bar{k}$ and $\bar{y}$ means that, at E, the variables Y, K, and L all grow at the same rate. The fact that Y and K share a common growth rate is especially important as a *sine qua non* of a steady state or

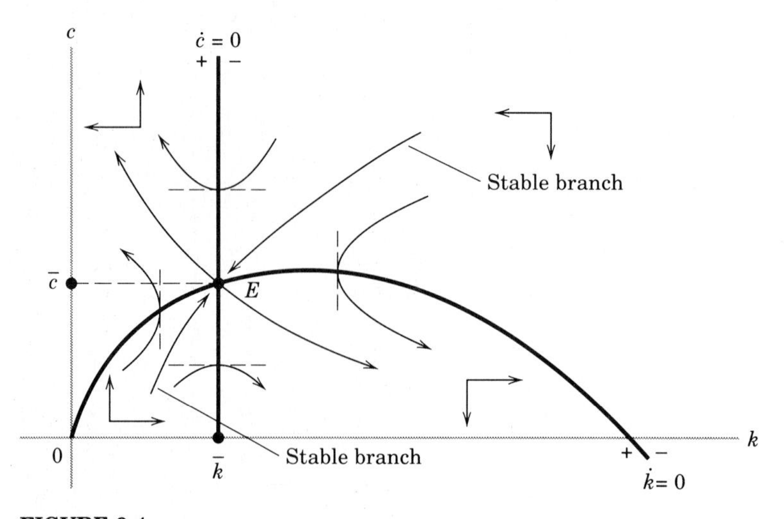

FIGURE 9.4

growth equilibrium. Given the configurations of the $\dot{k} = 0$ and $\dot{c} = 0$ curves in this problem, the steady state is unique.

Note that the only way the economy can ever move toward the steady state is to get onto one of the stable branches—the "yellow brick road"—leading to point E. This means that, given an initial capital-labor ratio k_0, it must choose an initial per-capita consumption level c_0 such that the ordered pair (k_0, c_0)— not shown in the graph—lies on a stable branch. Otherwise, the dynamic forces of the model will lead us into a situation of either (1) ever-increasing k accompanied by ever-decreasing c (along the streamlines that point toward the southeast), or (2) ever-increasing c accompanied by ever-decreasing k (along the streamlines that point toward the northwest). Situation (1) implies progressively severe belt-tightening, culminating in eventual starvation, whereas situation (2) implies growing overindulgence, leading to eventual capital exhaustion. Since neither of these is a viable long-run alternative, the steady state at E, with a sustainable constant level of per-capita consumption, is the only meaningful long-run target for the economy depicted in the present model.

It may be noted that even at E, the per-capita consumption becomes constant, and its level cannot be raised further over time. This is because a static production function, $Y = Y(K, L)$, is assumed in the model. To make possible a rising per-capita consumption, technological progress must be introduced. This will be discussed in the next section.

Transversality Conditions

To select a stable branch from the family of streamlines is tantamount to choosing a particular solution from a family of general solutions by definitizing an arbitrary constant. This is to be done with the help of some boundary conditions. The requirement that we choose a specific initial c_0 such that the ordered pair (k_0, c_0) sits on the stable branch is one way of doing it. An alternative is to look at an appropriate transversality condition. Since we are dealing with general functions and not working with quantitative solutions of differential equations, however, it is difficult to illustrate the use of a transversality condition to definitize an arbitrary constant. Nevertheless, we can, in light of the phase-diagram analysis, verify that the steady-state solution indeed satisfies the expected transversality conditions.

One transversality condition we may expect the steady-state solution to satisfy is that

$$\lambda \to 0 \qquad \text{as } t \to \infty \qquad [\text{by } (9.4)] \tag{9.41}$$

This is because the objective functional does contain a discount factor and the terminal state is free. Since the solution path for λ in the Cass model is

$$\lambda^* = U'(c^*)e^{-rt} \qquad [\text{by } (9.29)]$$

and since the limit of $U'(c^*)$ is finite as $t \to \infty$,[18] this expression does satisfy the transversality condition (9.41).

Another condition we may expect is that, in the solution,

$$H \to 0 \qquad \text{as } t \to \infty \qquad [\text{by (9.7)}] \tag{9.42}$$

For the present problem, the solution path for H takes the form

$$H^* = U(c^*)e^{-rt} + \lambda^*[\phi(k^*) - c^* - (n+\delta)k^*] \qquad [\text{by (9.28)}]$$

Since $U(c^*)$ is finite as $t \to \infty$, the exponential term $U(c^*)e^{-rt}$ tends to zero as t becomes infinite. In the remaining term, we already know from (9.41) that λ^* tends to zero; moreover, the bracketed expression, representing $\dot{k}$ by (9.36), is equal to zero by definition of steady state. Therefore, the transversality condition (9.42) is also satisfied.

Checking the Saddle Point by Characteristic Roots

In Fig. 9.4, the pattern of the streamlines drawn leads us to conclude that the equilibrium at E is a saddle point. Since the sketching of the streamlines is done on a qualitative basis with considerable latitude in the positioning of the curves, it may be desirable to check the validity of the conclusion by another means. We can do this by examining the characteristic roots of the linearization of the (nonlinear) differential-equation system of the model.[19]

On the basis of the two-equation system (9.36):

$$\dot{k} = \phi(k) - c - (n+\delta)k$$
$$\dot{c} = -\frac{U'(c)}{U''(c)}[\phi'(k) - (n+\delta+r)]$$

we first form the Jacobian matrix and evaluate it at the steady-state point E, or $(\bar{k}, \bar{c})$,

$$J_E = \begin{bmatrix} \dfrac{\partial \dot{k}}{\partial k} & \dfrac{\partial \dot{k}}{\partial c} \\ \dfrac{\partial \dot{c}}{\partial k} & \dfrac{\partial \dot{c}}{\partial c} \end{bmatrix}_{(\bar{k},\bar{c})} \tag{9.43}$$

[18] From (9.25), we see that to have $U'(c) \to \infty$, we must have $c \to 0$. In the present problem, c^* does not tend to zero as $t \to \infty$; thus the limit of $U'(c^*)$ is finite as $t \to \infty$.

[19] For more details of the procedure of linearization of a nonlinear system, see Alpha C. Chiang, *Fundamental Methods of Mathematical Economics*, 3d ed., McGraw-Hill, New York, 1984, Sec. 18.6.

The four partial derivatives, when evaluated at E, where $\phi'(\bar{k}) = n + \delta + r$, turn out to be

$$\left.\frac{\partial \dot{k}}{\partial k}\right|_E = \phi'(\bar{k}) - (n + \delta) = r > 0$$

$$\left.\frac{\partial \dot{k}}{\partial c}\right|_E = -1 < 0$$

$$\left.\frac{\partial \dot{c}}{\partial k}\right|_E = -\frac{U'(\bar{c})}{U''(\bar{c})}\phi''(\bar{k}) < 0$$

$$\left.\frac{\partial \dot{c}}{\partial c}\right|_E = \frac{-[U''(\bar{c})]^2 + U'''(\bar{c})U'(\bar{c})}{[U''(\bar{c})]^2}\left[\phi'(\bar{k}) - (n + \delta + r)\right] = 0$$

It follows that the Jacobian matrix takes the form

$$J_E = \begin{bmatrix} r & -1 \\ \dfrac{-U'(\bar{c})}{U''(\bar{c})}\phi''(\bar{k}) & 0 \end{bmatrix} \tag{9.43'}$$

The qualitative information we need about the characteristic roots r_1 and r_2 to confirm the saddle point is conveyed by that result that

$$r_1 r_2 = |J_E| = -\frac{U'(\bar{c})}{U''(\bar{c})}\phi''(\bar{k}) < 0 \tag{9.44}$$

This implies that the two roots have opposite signs, which establishes the steady state to be locally a saddle point.

EXERCISE 9.3

1 Let the production function and the utility function be

$$Y = AK^{\alpha}L^{1-\alpha} \qquad (0 < \alpha < 1)$$

$$U = \hat{U} - \frac{1}{b}c^{-b} \qquad (b > 0)$$

(*a*) Find the $y = \phi(k)$ function and the $U'(c)$ function. Check whether these conform to the specifications in (9.22) and (9.25).
(*b*) Write the specific optimal control problem.
(*c*) Apply the maximum principle, using the current-value Hamiltonian.
(*d*) Derive the differential-equation system in the variables k and c. Solve for the steady-state values $(\bar{k}, \bar{c})$.

2 In the phase-diagram analysis of the Cass model, directional arrowheads in Fig. 9.4 are determined from the signs of the partial derivatives $\partial \dot{k}/\partial c$ and $\partial \dot{c}/\partial k$ in (9.39) and (9.40).

(*a*) Is it possible and/or desirable to accomplish the same purpose by using $\partial \dot{k}/\partial k$ in lieu of $\partial \dot{k}/\partial c$? Why?

(*b*) Is it possible and/or desirable to consider the sign of $\partial \dot{c}/\partial c$ in place of $\partial \dot{c}/\partial k$? Why?

3 In Fig. 9.4, suppose that the economy is currently located on the lower stable branch and is moving toward point E. Let there be a parameter change such that the $\dot{c} = 0$ curve is shifted to the right, but the $\dot{k} = 0$ curve is unaffected.

(*a*) What specific parameter changes could cause such a shift?

(*b*) What would happen to the position of the steady state?

(*c*) Can we consider the preshift position of the economy—call it (k_1, c_1)—as the appropriate initial point on the new stable branch in the postshift phase diagram? Why?

4 Check whether the Cass model (9.27) satisfies the Mangasarian sufficient conditions.

9.4 EXOGENOUS AND ENDOGENOUS TECHNOLOGICAL PROGRESS

The neoclassical optimal growth model discussed in the preceding section provides a steady state in which the per-capita consumption, c, stays constant at $\bar{c}$, leaving no hope of any further improvement in the average standard of living. The culprit responsible for the rigid capping of per-capita consumption is the static nature of the production function, $Y = Y(K, L)$. Since the same production technology holds for all t, no upward shift over time can occur. Once technological progress is allowed, however, we can easily remove the cap on per-capita consumption.

Dynamic Production Functions

For a general representation of a dynamic production function, we can simply write

$$Y = Y(K, L, t) \qquad (Y_t > 0) \tag{9.45}$$

The positive sign of Y_t shows that technological progress does take place, but it offers no explanation of how the progress comes into being. This representation can thus be used only for exogenous technological progress. An alternative way of writing a dynamic production function is to introduce a technology variable explicitly into the function. Let $A = A(t)$ represent the state of the art, with $dA/dt > 0$, then we can write

$$Y = Y(K, L, A) \qquad (Y_A > 0) \tag{9.46}$$

The advantage of (9.46) over (9.45) is that, with an explicit variable A, we now can either leave A as an exogenous variable or make it endogenous by postulating how A is determined in the model.

Neutral Exogenous Technological Progress

Technological progress is said to be "neutral" when it leaves a certain economic variable unaffected under a stipulated circumstance. Specifically, technological progress is *Hicks-neutral* if it leaves the marginal rate of technical substitution ($\text{MRTS} \equiv \text{MPP}_L/\text{MPP}_K$) unchanged at the same K/L ratio. That is, if we hold K/L constant and examine the MRTS's before and after progress, the two MRTS's will be identical. To capture this feature, we can use the following special version of (9.46):

$$Y = A(t)Y(K, L) \qquad \text{[Hicks-neutral]} \tag{9.47}$$

where $Y(K, L)$ is taken to be linearly homogeneous. The reason why the variable A will not disturb the ratio of the marginal products of K and L is that it lies outside of the $Y(K, L)$ expression.

In contrast, technological progress is *Harrod-neutral* if it leaves the output-capital ratio (Y/K) unchanged at the same MPP_K. The special version of (9.46) that displays this feature is

$$Y = Y[K, A(t)L] \qquad \text{[Harrod-neutral]} \tag{9.48}$$

where Y is taken to be linearly homogeneous in K and $A(t)L$. Since $A(t)$ is attached exclusively to L, this type of technological progress is said to be purely labor-augmenting. Considering the way $A(t)$ and L are combined, we may view technology and labor as perfect substitutes in the production process. It is the fact that A is totally detached from the K variable that explains how the Y/K ratio can remain undisturbed at the same MPP_K as technological progress occurs.

A third type of neutrality, *Solow neutrality*, is the mirror image of Harrod neutrality, with the roles of K and L interchanged. The function can then be written as

$$Y = Y[A(t)K, L] \qquad \text{[Solow-neutral]} \tag{9.49}$$

where Y is taken to be linearly homogeneous in AK and L.

Neoclassical Optimal Growth with Harrod-Neutral Progress

In growth models with exogenous technological progress, Harrod neutrality is frequently assumed. The main reason for the popularity of Harrod neutrality is that it is perfectly consistent with the notion of a steady state; it can yield a dynamic equilibrium in which Y and K may grow apace with each other. As a bonus, it turns out that Harrod-neutral technological progress entails hardly any additional analytical complexity compared with the case where a static production function is used.

Let us define *efficiency labor*, η, by

$$\eta \equiv AL \tag{9.50}$$

Then the production function (9.48) becomes

$$Y = Y(K, \eta) \tag{9.51}$$

If we now consider efficiency labor η (rather than natural labor L) as the relevant labor-input variable, then (9.51) can be treated mathematically just as a static production function, despite that its forebear—(9.48)—is really dynamic in nature. By the assumption of linear homogeneity, we can rewrite (9.51) as

$$y_\eta = \phi(k_\eta) \qquad \text{where } y_\eta \equiv \frac{Y}{\eta} \text{ and } k_\eta \equiv \frac{K}{\eta} \tag{9.51'}$$

This is identical in nature with (9.22), and we shall indeed retain all the assumed features of ϕ enumerated in connection with (9.22). The equation of motion for k_η, the new state variable, can be found by the same procedure used to derive (9.24). The result is

$$\dot{k}_\eta = \phi(k_\eta) - c_\eta - (a + n + \delta)k_\eta \tag{9.52}$$

$$\text{where } c_\eta \equiv \frac{C}{\eta} \qquad a \equiv \frac{\dot{A}}{A} \qquad \text{and } n \equiv \frac{\dot{L}}{L}$$

The optimal control problem is therefore

$$\begin{aligned} &\text{Maximize} && \int_0^\infty U(c_\eta)e^{-rt}\,dt \\ &\text{subject to} && \dot{k}_\eta = \phi(k_\eta) - c_\eta - (a + n + \delta)k_\eta \\ & && k_\eta(0) = k_{\eta 0} \\ &\text{and} && 0 \le c_\eta \le \phi(k_\eta) \end{aligned} \tag{9.53}$$

The similarity between this new formulation and the old problem (9.27) should be patently clear.

Since the current-value Hamiltonian in the new problem is

$$H_c' = U(c_\eta) + m'\left[\phi(k_\eta) - c_\eta - (a + n + \delta)k_\eta\right] \tag{9.54}$$

where m' is the new current-value costate variable, the maximum principle calls for the following conditions:

$$m' = U'(c_\eta) \tag{9.55} \quad \text{[cf. (9.33)]}$$

$$\dot{k}_\eta = \phi(k_\eta) - c_\eta - (a + n + \delta)k_\eta \tag{9.56} \quad \text{[cf. (9.34)]}$$

$$\dot{m}' = -m'\left[\phi'(k_\eta) - (a + n + \delta + r)\right] \tag{9.57} \quad \text{[cf. (9.35)]}$$

By eliminating the m' variable, the last three equations can be condensed into the differential-equation system

$$\begin{aligned} \dot{k}_\eta &= \phi(k_\eta) - c_\eta - (a + n + \delta)k_\eta \\ \dot{c}_\eta &= -\frac{U'(c_\eta)}{U''(c_\eta)}\left[\phi'(k_\eta) - (a + n + \delta + r)\right] \end{aligned} \tag{9.58} \quad \text{[cf. (9.36)]}$$

These then give rise to the following pair of equations for the new phase diagram:

$$c_\eta = \phi(k_\eta) - (a + n + \delta)k_\eta \tag{9.59} \quad \left[\text{equation for } \dot{k}_\eta = 0 \text{ curve}\right]$$

$$\phi'(k_\eta) = a + n + \delta + r \tag{9.60} \quad \left[\text{equation for } \dot{c}_\eta = 0 \text{ curve}\right]$$

Since the phase-diagram analysis is qualitatively the same as that of Fig. 9.4, there is no need to repeat it. Suffice it to say that the intersection of the $\dot{k}_\eta = 0$ curve and the $\dot{c}_\eta = 0$ curve delineates a steady state in which Y, K, and η all grow at the same rate. But one major difference should be noted about the new steady state. Previously, we had $c \equiv C/L$ constant in the steady state. Now we have instead

$$c_\eta \equiv \frac{C}{\eta} \equiv \frac{C}{AL} = \text{constant} \tag{9.61}$$

which implies that

$$\frac{C}{L} = c_\eta A \tag{9.61'}$$

Thus, as long as A increases as a result of technological progress, per-capita consumption C/L will rise over time *pari passu*. The cap on the average standard of living has been removed.

Endogenous Technological Progress

While exogenous technological progress has the great merit of simplicity, it shirks the task of explaining the origin of progress. To rectify this shortcoming, it is necessary to endogenize technological progress. As early as the 1960s, economists were already exploring the economics of knowledge creation. The well-known "learning-by-doing" model of Kenneth Arrow,[20] for example, regards gross investment as the measure of the amount of "doing" that results in "learning" (technological progress). As another example, a model of Karl Shell,[21] which we shall now briefly discuss, makes the accumulation of knowledge explicitly dependent on the amount of resources devoted to inventive activity.

The production function in the Shell model,

$$Y(t) = Y[K(t), L(t), A(t)]$$

is the same as (9.46). But now the variable $A(t)$, denoting the stock of knowledge, is given the specific pattern of change

$$\dot{A}(t) = \sigma\alpha(t)Y(t) - \beta A(t) \qquad (0 < \sigma \le 1, 0 \le \alpha \le 1, \beta \ge 0) \tag{9.62}$$

where σ is the research success coefficient, $\alpha(t)$ denotes the fraction of output channeled toward inventive activity at time t, and β is the rate of decay of technical knowledge. Out of the remaining resources, a part will be saved (and invested). The $K(t)$ variable thus changes over time according to

$$\dot{K}(t) = s(t)[1 - \alpha(t)]Y(t) - \delta K(t) \tag{9.63}$$

where s denotes the propensity to save and δ denotes the rate of depreciation. In order to focus attention on the accumulation of knowledge and capital, we assume that L is constant, and (by choice of unit) set it equal to one.

In a decentralized economy, the dynamics of accumulation can be traced from the equation system (9.62) and (9.63) in the two variables A and K. If a government authority seeks to maximize social utility, on the other hand, there then arises the optimal control problem

$$\begin{aligned} &\text{Maximize} && \int_0^\infty U[(1 - s)(1 - \alpha)Y]e^{-\rho t}\,dt \\ &\text{subject to} && \dot{A} = \sigma\alpha Y(K, A) - \beta A \\ & && \dot{K} = s(1 - \alpha)Y(K, A) - \delta K \\ &\text{and} && A(0) = A_0 \qquad K(0) = K_0 \end{aligned} \tag{9.64}$$

The integrand in the objective functional may look unfamiliar, but it is merely another way of expressing $U(C)e^{-\rho t}$, because C is the residual of Y

[20] Kenneth J. Arrow, "The Economic Implications of Learning by Doing," *Review of Economic Studies*, June 1962, pp. 155–173.

[21] Karl Shell, "Towards a Theory of Inventive Activity and Capital Accumulation," *American Economic Review*, May 1966, pp. 62–68.

after deducting what goes to inventive activity and capital accumulation:

$$C = Y - \alpha Y - s(1-\alpha)Y$$
$$= (1-s)(1-\alpha)Y$$

This problem contains two state variables (A and K) and two control variables (α and s).

The maximum principle is, of course, directly applicable to this problem. But, with two state variables, the solution process is not simple. Shell presents no explicit solution of the dynamics of the model, but shows that A and K will approach specific constant limiting values $\bar{A}$ and $\bar{K}$. There is no need to enumerate these limiting values here, for their significance lies not in their exact magnitudes, but in the fact that they are constant. Since $A \to \bar{A}$ and $K \to \bar{K}$ as $t \to \infty$, the specter of a capped standard of living is here to visit us again.

One may wonder whether allowing a growing population and making the social utility a function of per-capita consumption will change the outcome. The answer is no. The difficulty, as Kazuo Sato (a discussant of the Shell paper) points out,[22] may lie in the specification of $\dot{A}$ in (9.62), which ties $\dot{A}$ to Y. If (9.62) is changed to

$$\dot{A} = \sigma\alpha A - \beta A \qquad \text{or} \qquad \frac{\dot{A}}{A} = \sigma\alpha - \beta \tag{9.65}$$

then no cap on the growth of knowledge will exist so long as $\sigma\alpha - \beta > 0$. Interestingly, a similar idea is featured in a recent model of endogenous technological progress by Paul Romer to ensure unbounded progress.

Endogenous Technology à la Romer

Knowledge, according to the Romer model,[23] can be classified into two major components. The first component, which may be broadly labeled as *human capital*, is person-specific. It constitutes a "rival good" in the sense that its use by one firm precludes its use by another. The other component, to be referred to as *technology*, is generally available to the public. It is a "nonrival good" in the sense that its use by one firm does not limit its use by others. Because of the rivalous nature of human capital, the person who invests in the accumulation of human capital receives the rewards arising therefrom. In contrast, the nonrivalry feature of technology implies knowledge spillovers, such that the discoverer of new technology will not be the

[22] *American Economic Review*, May 1966, p. 79.

[23] Paul M. Romer, "Endogenous Technical Change," *Journal of Political Economy*, Vol. 98, No. 5, Part 2, October 1990, pp. S71–S102.

sole beneficiary of that discovery. The inability of the discoverer to reap all the benefits creates an economic externality that causes private efforts at technological improvement to fall short of what is socially optimal.

Both human capital and technology are created by conscious action. To reduce the number of state variables, however, human capital is simply assumed to be fixed and inelastically supplied, although its allocation to different uses is still to be endogenously determined. Romer denotes human capital by H, but to avoid confusion with the symbol for the Hamiltonian, we shall use S (skill or skilled labor) instead, with S_0 as its fixed total. Since S can be used for the production of the final good, Y, or for the improvement of technology, A (state of the art), we have

$$S_Y + S_A = S_0 \tag{9.66}$$

Technology A, on the other hand, is not fixed. It can be created by engaging human capital S_A in research and applying the existing technology A as follows:

$$\dot{A} = \sigma S_A A \qquad \Rightarrow \qquad \frac{\dot{A}}{A} = \sigma S_A \tag{9.67}$$

where σ is the research success parameter. This equation is closely similar to (9.65), except that, here, human capital S_A has replaced α (the fraction of output channeled to research), and $\beta = 0$ (technology does not depreciate). Note that $\dot{A}/A = \sigma S_A > 0$ as long as both σ and S_A are positive. Thus technology can grow without bound. Note also that research activity is assumed to be human-capital-intensive and technology-intensive, with no capital (K) and ordinary unskilled labor (L) engaged in that activity.

In the production of the final good Y, however, K and L do enter as inputs along with human capital S_Y and technology A. The production function for Y is developed in several steps. First, think of technology as made up of an infinite set of "designs" for capital,

$$\{x_1, x_2, \ldots\}$$

including those that have not yet been invented at the present time. If we let $x_i = 0$ for $i \geq A$, then A can serve as the index of the current technology. The production function for the final good is assumed to be of the Cobb-Douglas type:

$$Y = S_Y^{\alpha} L_0^{\beta} \left(x_1^{1-\alpha-\beta} + x_2^{1-\alpha-\beta} + \cdots \right)$$

where L_0 indicates a fixed and inelastically supplied amount of ordinary labor. For simplicity, Romer allows all the designs x_i to enter in an additively separable manner. As the second step, let the design index i be turned into a continuous variable. The production function then becomes

$$Y = S_Y^{\alpha} L_0^{\beta} \int_0^A x(i)^{1-\alpha-\beta}\, di \tag{9.68}$$

Since all $x(i)$ enter symmetrically in the integrand, we may deduce that there is a common level of use, $\bar{x}$, of all $x(i)$. It follows that

$$\begin{aligned}\int_0^A x(i)^{1-\alpha-\beta}\, di &= \int_0^A \bar{x}^{1-\alpha-\beta}\, di \\ &= \bar{x}^{1-\alpha-\beta} \int_0^A di \\ &= A\bar{x}^{1-\alpha-\beta}\end{aligned}$$

Therefore, (9.68) can be simplified to

$$Y = S_Y^{\alpha} L_0^{\beta} A \bar{x}^{1-\alpha-\beta} \tag{9.68'}$$

Next, assume that (1) capital goods are nothing but foregone consumption goods, both types of goods being subject to the same final-good production function, and (2) it takes γ units of capital goods to produce one unit of any type of design. Then the amount of capital actually used will be

$$K = \gamma A \bar{x} \qquad \Rightarrow \qquad \bar{x} = \frac{K}{\gamma A}$$

Substituting this into (9.68′) results in

$$\begin{aligned}Y &= S_Y^{\alpha} L_0^{\beta} A \left(\frac{K}{\gamma A} \right)^{1-\alpha-\beta} \\ &= S_Y^{\alpha} L_0^{\beta} A^{\alpha+\beta} K^{1-\alpha-\beta} \gamma^{\alpha+\beta-1} \\ &= (S_Y A)^{\alpha} (L_0 A)^{\beta} K^{1-\alpha-\beta} \gamma^{\alpha+\beta-1}\end{aligned} \tag{9.68''}$$

In this last function, all four types of inputs are present: human capital S_Y, technology A, labor L_0, and capital K. More importantly, we observe from the $(S_Y A)$ and $(L_0 A)$ expressions that technology can be viewed as human-capital-augmenting as well as labor-augmenting, but it is detached from K. In other words, it is characterized by Harrod neutrality—here endogenously introduced rather than exogenously imposed. Our experience with Harrod neutrality suggests that this production function is consistent with a steady state in which technology—and, along with it, output, capital, and consumption—can all grow without bound.

In this model, consumption C does not need to be expressed in per-capita terms because L_0 (identified with population and measured by head count) is constant. By the same token, K rather than K/L can appropriately serve as the variable for analysis. Net investment, as usual, is just output not consumed. Thus, recalling that $S_Y = S_0 - S_A$, we have

$$\dot{K} = Y - C = \gamma^{\alpha+\beta-1}A^{\alpha+\beta}(S_0 - S_A)^{\alpha}L_0^{\beta}K^{1-\alpha-\beta} - C \tag{9.69}$$

The Optimal Control Problem

Facing this background, society may consider an optimal control problem with two state variables, A and K—with (9.67) and (9.69) as their equations of motion—and two control variables, C and S_A. Adopting the specific constant-elasticity utility function

$$U(C) = \frac{C^{1-\theta}}{1-\theta} \qquad (0 < \theta < 1)$$

the control problem takes the form

$$\text{Maximize} \quad \int_0^{\infty} \frac{C^{1-\theta}}{1-\theta} e^{-\rho t}\, dt$$

$$\text{subject to} \quad \dot{A} = \sigma S_A A \tag{9.70}$$

$$\dot{K} = \gamma^{\alpha+\beta-1}A^{\alpha+\beta}(S_0 - S_A)^{\alpha}L_0^{\beta}K^{1-\alpha-\beta} - C$$

$$\text{and} \quad A(0) = A_0 \qquad K(0) = K_0$$

For convenience, define the shorthand symbol

$$\Delta \equiv \gamma^{\alpha+\beta-1}A^{\alpha+\beta}(S_0 - S_A)^{\alpha}L_0^{\beta}K^{1-\alpha-\beta} \tag{9.71}$$

Then we have the current-value Hamiltonian

$$H_c = \frac{C^{1-\theta}}{1-\theta} + \lambda_A(\sigma S_A A) + \lambda_K(\Delta - C)$$

where λ_A and λ_K represent the shadow prices of A and K, respectively. From this, we get the conditions

$$\frac{\partial H_c}{\partial C} = C^{-\theta} - \lambda_K = 0 \qquad \Rightarrow \qquad \lambda_K = C^{-\theta} \tag{9.72}$$

$$\frac{\partial H_c}{\partial S_A} = \lambda_A \sigma A - \lambda_K \alpha (S_0 - S_A)^{-1}\Delta = 0 \tag{9.73}$$

$$\Rightarrow \qquad \Delta = \frac{\lambda_A \sigma A}{\lambda_K \alpha}(S_0 - S_A)$$

In addition to the $\dot{A}$ and $\dot{K}$ equations given in the problem statement, the maximum principle requires the following equations of motion for the costate variables:

$$\begin{aligned} \dot{\lambda}_A &= -\frac{\partial H_c}{\partial A} + \rho\lambda_A = -\lambda_A \sigma S_A - \lambda_K(\alpha+\beta)A^{-1}\Delta + \rho\lambda_A \\ \dot{\lambda}_K &= -\frac{\partial H_c}{\partial K} + \rho\lambda_K = -\lambda_K(1-\alpha-\beta)K^{-1}\Delta + \rho\lambda_K \end{aligned} \tag{9.74}$$

The Steady State

Since there are four differential equations, the system cannot be analyzed with a phase diagram. And solving the system explicitly for its dynamics is not simple. Romer elects to focus the discussion on the properties of the balanced growth equilibrium inherent in the model: a steady state with Harrod-neutral technological progress. Questions of interest include: What is the rate of growth in that steady state? How is the growth rate affected by the various parameters? What economic policies can be pursued to promote growth?

The basic feature of such a steady state is that the variables Y, K, A, and C all grow at the same rate. We thus have in the steady state

$$\frac{\dot{Y}}{Y} = \frac{\dot{K}}{K} = \frac{\dot{C}}{C} = \frac{\dot{A}}{A} = \sigma S_A \qquad \text{[by (9.67)]} \tag{9.75}$$

But we want to express the growth rate in terms of the parameters only, with S_A substituted out. Since $\lambda_K = C^{-\theta}$ [by (9.72)], we can calculate

$$\frac{\dot{\lambda}_K}{\lambda_K} = \frac{-\theta C^{-\theta-1}\dot{C}}{C^{-\theta}} = -\theta\frac{\dot{C}}{C} = -\theta\sigma S_A \tag{9.76}$$

If we get another $\dot{\lambda}_K/\lambda_K$ expression from the second equation in (9.74), then we can equate it to (9.76) and solve for S_A. But that equation is not convenient to work with. Instead, Romer first calculates $\dot{\lambda}_A/\lambda_A$ from the first equation in (9.74), then takes advantage of the relation $\dot{\lambda}_K/\lambda_K = \dot{\lambda}_A/\lambda_A$ that characterizes the steady state. By dividing the $\dot{\lambda}_A$ expression in (9.74) through by λ_A and simplifying, it is found that

$$\frac{\dot{\lambda}_A}{\lambda_A} = \rho - \sigma\left(\frac{\alpha+\beta}{\alpha}S_0 - \frac{\beta}{\alpha}S_A\right) \tag{9.77}$$

Equating (9.76) and (9.77) and solving for S_A, we then ascertain that S_A

attains in steady state the constant value

$$S_A = \frac{\sigma(\alpha + \beta)S_0 - \alpha\rho}{\sigma(\alpha\theta + \beta)} \tag{9.78}$$

It follows that the parametrically expressed steady-state growth rate is

$$\frac{\dot{Y}}{Y} = \frac{\dot{K}}{K} = \frac{\dot{C}}{C} = \frac{\dot{A}}{A} = \frac{\sigma(\alpha + \beta)S_0 - \alpha\rho}{\alpha\theta + \beta} \tag{9.79}$$

The result in (9.79) shows not only the growth rate, but also how the various parameters affect that rate. Visual inspection is sufficient to establish that human capital S_0 has a positive effect on the growth rate, as does the research success parameter σ, but a negative effect is exerted by the discount rate ρ. More formally, the effects of the various parameters can be found by taking the partial derivatives of (9.79).

To conclude this discussion, we should point out a complication regarding the rate-of-growth expression in (9.79): A rate of growth should be a pure number, and economic commonsense would lead us to expect it to be a positive fraction. Yet the expression in (9.79) comprises not only the pure-number parameters σ, α, β, ρ, and θ, but also the parameter S_0 which has a physical unit. The magnitude of the expression thus becomes problematic. The problem may have started at an early stage with the specification of (9.67), where S_A, a variable with a physical unit, enters into the rate of growth $\dot{A}/A$. If so, the remedy would be to replace S_A with a suitable ratio figure in (9.67).

EXERCISE 9.4

1. Show that, in the Cobb-Douglas production function $Y = A(t)K^{\alpha}L^{\beta}$, the technological progress can be considered to be either (*a*) Hicks-neutral, (*b*) Harrod-neutral, or (*c*) Solow-neutral.
2. Derive the equation of motion (9.52) by following the same procedure used to derive (9.24).
3. Verify the validity of the maximum-principle conditions in (9.55), (9.56), and (9.57).
4. Verify that the differential-equation system (9.58) is equivalent to the set of equations (9.55), (9.56), and (9.57).
5. What would happen if (9.67) were changed to $\dot{A} = \delta S_A \psi(A)$, where ψ is a concave function?
6. In the Romer model, how is the steady-state growth rate specifically affected by the following parameters?
   (*a*) α (*b*) β (*c*) θ

CHAPTER 10

OPTIMAL CONTROL WITH CONSTRAINTS

Constraints have earlier been encountered in Sec. 7.4, where we discussed various types of terminal lines, including truncated ones. Those constraints are only concerned with what happens at the endpoint of the path, and the conditions that are developed to deal with them are in the nature of transversality conditions. In the present chapter, we turn to constraints that apply throughout the planning period $[0, T]$.

As in the calculus of variations, the treatment of constraints in optimal control theory relies heavily on the Lagrange-multiplier technique. But since optimal control problems contain not only state variables, but also control variables, it is necessary to distinguish between two major categories of constraints. In the first category, control variables are present in the constraints, either with or without the state variables alongside. In the second category, control variables are absent, so that the constraints only affect the state variables. As we shall see, the methods of treatment for the two categories are different.

10.1 CONSTRAINTS INVOLVING CONTROL VARIABLES

Within the first category of constraints—those with control variables present—four basic types can be considered: equality constraints, inequality constraints, equality integral constraints, and inequality integral constraints. We shall in general include the state variables alongside the control variables in the constraints, but the methods of treatment discussed in this section are applicable even when the state variables do not appear.

Equality Constraints

Let there be two control variables in a problem, u_1 and u_2, that are required to satisfy the condition

$$g(t, y, u_1, u_2) = c$$

We shall refer to the g function as the constraint function, and the constant c as the constraint constant. The control problem may then be stated as

$$\begin{aligned} &\text{Maximize} && \int_0^T F(t, y, u_1, u_2)\, dt \\ (10.1)\qquad &\text{subject to} && \dot{y} = f(t, y, u_1, u_2) \\ & && g(t, y, u_1, u_2) = c \\ &\text{and} && \text{boundary conditions} \end{aligned}$$

This is a simple version of the problem with m control variables and q equality constraints, where it is required that $q < m$.

The maximum principle calls for the maximization of the Hamiltonian

$$(10.2)\qquad H = F(t, y, u_1, u_2) + \lambda(t) f(t, y, u_1, u_2)$$

for every $t \in [0, T]$. But this time the maximization of H is subject to the constraint $g(t, y, u_1, u_2) = c$. Accordingly, we form the Lagrangian expression

$$\begin{aligned} (10.3)\qquad \mathscr{L} &= H + \theta(t)[c - g(t, y, u_1, u_2)] \\ &= F(t, y, u_1, u_2) + \lambda(t) f(t, y, u_1, u_2) \\ &\quad + \theta(t)[c - g(t, y, u_1, u_2)] \end{aligned}$$

where the Lagrange multiplier θ is made dynamic, as a function of t. This is necessitated by the fact that the g constraint must be satisfied at every t in the planning period. Assuming an interior solution for each u_j, we require that

$$(10.4)\qquad \frac{\partial \mathscr{L}}{\partial u_j} = \frac{\partial F}{\partial u_j} + \lambda \frac{\partial f}{\partial u_j} - \theta \frac{\partial g}{\partial u_j} = 0 \quad \text{for all } t \in [0, T] \quad (j = 1, 2)$$

Simultaneously, we must also set

$$(10.5)\qquad \frac{\partial \mathscr{L}}{\partial \theta} = c - g(t, y, u_1, u_2) = 0 \qquad \text{for all } t \in [0, T]$$

to ensure that the constraint will always be in force. Together, (10.4) and (10.5) constitute the first-order condition for the constrained maximization of H. This must be supported, of course, by a proper second-order condition or a suitable concavity condition.

The rest of the maximum-principle conditions includes:

$$(10.6)\qquad \dot{y} = \frac{\partial \mathscr{L}}{\partial \lambda}\left(= \frac{\partial H}{\partial \lambda}\right) \qquad \text{[equation of motion for } y\text{]}$$

and

$$(10.7)\qquad \dot{\lambda} = -\frac{\partial \mathscr{L}}{\partial y}\left(= -\frac{\partial H}{\partial y} + \theta\frac{\partial g}{\partial y}\right) \qquad \text{[equation of motion for } \lambda\text{]}$$

plus an appropriate transversality condition. Note that the equation of motion for y, (10.6), would turn out the same whether we differentiate the Lagrangian function (the augmented Hamiltonian) or the original Hamiltonian function with respect to λ. On the other hand, it *would* make a difference in the equation of motion for λ, (10.7), whether we differentiate $\mathscr{L}$ or H with respect to y. The correct choice is the Lagrangian expression $\mathscr{L}$. This is because, as the constraint in problem (10.1) specifically prescribes, the y variable impinges upon the range of choice of the control variables, and such effects must be taken into account in determining the path for the costate variable λ.

While it is feasible to solve a problem with equality constraints in the manner outlined above, it is usually simpler in actual practice to use substitution to reduce the number of variables we have to deal with. Substitution is therefore recommended whenever it is feasible.

Inequality Constraints

The substitution method is not as easily applicable to problems with inequality constraints, so it is desirable to have an alternative procedure for such a situation.

We first remark that when the g constraints are in the *inequality* form, there is no need to insist that the number of control variables exceed the number of constraints. This is because inequality constraints allow us much more latitude in our choices than constraints in the form of rigid equalities. For simplicity, we shall illustrate this type of problem with two control variables and two inequality constraints:

$$\begin{aligned}
&\text{Maximize} && \int_0^T F(t, y, u_1, u_2)\, dt \\
&\text{subject to} && \dot{y} = f(t, y, u_1, u_2) \\
(10.8)\qquad & && g^1(t, y, u_1, u_2) \le c_1 \\
& && g^2(t, y, u_1, u_2) \le c_2 \\
&\text{and} && \text{boundary conditions}
\end{aligned}$$

The Hamiltonian defined in (10.2) is still valid for the present problem. But since the Hamiltonian is now to be maximized with respect to u_1 and u_2 subject to the two *inequality* constraints, we need to invoke the Kuhn-Tucker conditions. Besides, for these conditions to be necessary, a constraint qualification must be satisfied. According to a theorem of Arrow, Hurwicz, and Uzawa, any of the following conditions will satisfy the constraint qualification[1]:

(1) All the constraint functions g^i are *concave* in the control variables u_j [here, concave in (u_1, u_2)].

(2) All the constraint functions g^i are *linear* in the control variables u_j [here, linear in (u_1, u_2)]—a special case of (1).

(3) All the constraint functions g^i are *convex* in the control variables u_j. And, in addition, there exists a point in the control region $u_0 \in U$ [here, u_0 is a point (u_{10}, u_{20})] such that, when evaluated at u_0, all constraints g^i are strictly $< c_i$. (That is, the constraint set has a nonempty interior.)

(4) The g^i functions satisfy the *rank condition*: Taking only those constraints that turn out to be effective or binding (satisfied as strict equalities), form the matrix of partial derivative $[\partial g^i/\partial u_j]_e$ (where e indicates "effective constraints only"), and evaluate the partial derivatives at the optimal values of the y and u variables. The rank condition is that the rank of this matrix be equal to the number of effective constraints.

[1]K. J. Arrow, L. Hurwicz, and H. Uzawa, "Constraint Qualifications in Nonlinear Programming," *Naval Research Logistics Quarterly*, January 1961. The summary version here is adapted from Akira Takayama, *Mathematical Economics*, 2d ed., Cambridge University Press, Cambridge, 1985, p. 648. Note that our constraints are written as $g \le c$ (rather than $g \ge c$ as in Takayama).

We now augment the Hamiltonian into a Lagrangian function:

$$(10.9)\quad \mathscr{L} = F(t, y, u_1, u_2) + \lambda(t) f(t, y, u_1, u_2) + \theta_1(t)\left[c_1 - g^1(t, y, u_1, u_2)\right] + \theta_2(t)\left[c_2 - g^2(t, y, u_1, u_2)\right]$$

The essence of $\mathscr{L}$ may become more transparent if we suppress all the arguments and simply write

$$(10.9')\quad \mathscr{L} = F + \lambda f + \theta_1(c_1 - g^1) + \theta_2(c_2 - g^2)$$

The first-order condition for maximizing $\mathscr{L}$ calls for, assuming interior solutions,

$$(10.10)\quad \frac{\partial \mathscr{L}}{\partial u_j} = 0$$

as well as

$$(10.11)\quad \frac{\partial \mathscr{L}}{\partial \theta_i} = c_i - g^i \geq 0 \qquad \theta_i \geq 0 \qquad \theta_i \frac{\partial \mathscr{L}}{\partial \theta_i} = 0$$

$$(i = 1, 2 \text{ and } j = 1, 2) \qquad \text{for all } t \in [0, T]$$

Condition (10.11) differs from (10.5) because the constraints in the present problem are inequalities. The $\partial \mathscr{L}/\partial \theta_i \geq 0$ condition merely restates the ith constraint, and the complementary-slackness condition $\theta_i(\partial \mathscr{L}/\partial \theta_i) = 0$ ensures that those terms in (10.9) involving θ_i will vanish in the solution, so that the value of $\mathscr{L}$ will be identical with that of $H = F + \lambda f$ after maximization.

Note that, unlike in nonlinear programming, we have in (10.10) the first-order conditions $\partial \mathscr{L}/\partial u_j = 0$, *not* $\partial \mathscr{L}/\partial u_j \leq 0$. This is because the u_j variables are not restricted to nonnegative values in problem (10.8). If the latter problem contains additional nonnegativity restrictions

$$u_j(t) \geq 0$$

then, by the Kuhn-Tucker conditions, we should replace the $\partial \mathscr{L}/\partial u_j = 0$ conditions in (10.10) with

$$(10.12)\quad \frac{\partial \mathscr{L}}{\partial u_j} \leq 0 \qquad u_j \geq 0 \qquad u_j \frac{\partial \mathscr{L}}{\partial u_j} = 0$$

It should be pointed out that the symbol $\mathscr{L}$ in (10.12) denotes the same Lagrangian as defined in (10.9), without separate $\theta(t)$ type of multiplier terms appended on account of the additional constraints $u_j(t) \geq 0$. This procedure is directly comparable to that used in nonlinear programming. The alternative approach of adding a new multiplier for each nonnegativity restriction $u_j(t) \geq 0$ is explored in a problem in Exercise 10.1.

Other maximum-principle conditions include the equations of motion for y and λ. These are the same as in (10.6) and (10.7):

$$\dot{y} = \frac{\partial \mathscr{L}}{\partial \lambda} \quad \text{and} \quad \dot{\lambda} = -\frac{\partial \mathscr{L}}{\partial y} \tag{10.13}$$

Wherever appropriate, of course, transversality conditions must be added, too.

Isoperimetric Problem

When an equality integral constraint is present, the control problem is known as an *isoperimetric problem*. Two features of such a problem are worth noting. First, the costate variable associated with the integral constraint is, as in the calculus of variations, constant over time. Second, although the constraint is in the nature of a strict *equality*, the integral aspect of it obviates the need to restrict the number of constraints relative to the number of control variables. We shall illustrate the solution method with a problem that contains one state variable, one control variable, and one integral constraint:

$$\begin{aligned} &\text{Maximize} && \int_0^T F(t, y, u)\, dt \\ &\text{subject to} && \dot{y} = f(t, y, u) \\ &&& \int_0^T G(t, y, u)\, dt = k \qquad (k \text{ given}) \\ &\text{and} && y(0) = y_0 \qquad y(T) \text{ free} \qquad (y_0, T \text{ given}) \end{aligned} \tag{10.14}$$

The approach to be used here is to introduce a new state variable $\Gamma(t)$ into the problem such that the integral constraint can be replaced by a condition in terms of $\Gamma(t)$. To this end, let us define

$$\Gamma(t) = -\int_0^t G(t, y, u)\, dt \tag{10.15}$$

where, the reader will note, the upper limit of integration is the variable t, not the terminal time T. The derivative of this variable is

$$\dot{\Gamma} = -G(t, y, u) \qquad [\text{equation of motion for } \Gamma] \tag{10.16}$$

and the initial and terminal values of $\Gamma(t)$ in the planning period are

$$\Gamma(0) = -\int_0^0 G(t, y, u)\, dt = 0 \tag{10.17}$$

and

$$\Gamma(T) = -\int_0^T G(t, y, u)\, dt = -k \qquad \text{[by (10.14)]} \tag{10.18}$$

From (10.18), it is clear that we can replace the given integral constraint by a terminal condition on the Γ variable.

By incorporating Γ into the problem as a new state variable, we can restate (10.14) as

$$\begin{aligned} &\text{Maximize} && \int_0^T F(t, y, u)\, dt \\ &\text{subject to} && \dot{y} = f(t, y, u) \\ &&& \dot{\Gamma} = -G(t, y, u) \\ &&& y(0) = y_0 \qquad y(T) \text{ free} \qquad (y_0, T \text{ given}) \\ &\text{and} && \Gamma(0) = 0 \qquad \Gamma(T) = -k \qquad (k \text{ given}) \end{aligned} \tag{10.19}$$

This new problem is an unconstrained problem with two state variables, y and Γ. While the y variable has a vertical terminal line, the new Γ variable has a fixed terminal point. Inasmuch as this problem is now an unconstrained problem, we can work with the Hamiltonian without first expanding it into a Lagrangian function.

Note that this procedure of substituting out the constraint can be repeatedly applied to additional integral constraints, with each application resulting in a new state variable for the problem. This is why there is no need to limit the number of integral constraints.

Defining the Hamiltonian as

$$H = F(t, y, u) + \lambda f(t, y, u) - \mu G(t, y, u) \tag{10.20}$$

we have the following conditions from the maximum principle:

$$\begin{aligned} &\max_u H && \text{for all } t \in [0, T] \\ &\dot{y} = \frac{\partial H}{\partial \lambda} && \text{[equation of motion for } y\text{]} \\ &\dot{\lambda} = -\frac{\partial H}{\partial y} && \text{[equation of motion for } \lambda\text{]} \\ &\dot{\Gamma} = \frac{\partial H}{\partial \mu} && \text{[equation of motion for } \Gamma\text{]} \\ &\dot{\mu} = -\frac{\partial H}{\partial \Gamma} && \text{[equation of motion for } \mu\text{]} \\ &\lambda(T) = 0 && \text{[transversality condition]} \end{aligned} \tag{10.21}$$

What distinguishes (10.21) from the conditions for the usual unconstrained problem is the presence of the pair of equations of motion for Γ and μ. Since the Γ variable is an artifact whose mission is only to guide us to add the $-\mu G(t, y, u)$ term to the Hamiltonian—a mission that has already been accomplished—and whose time path is of no direct interest, we can safely omit its equation of motion from (10.21) at no loss. On the other hand, the equation of motion for μ does impart a significant piece of information. Since the Γ variable does not appear in the Hamiltonian, it follows that

$$\dot{\mu} = -\frac{\partial H}{\partial \Gamma} = 0 \qquad \Rightarrow \qquad \mu(t) = \text{constant} \tag{10.22}$$

This validates our earlier assertion that the costate variable associated with the integral constraint is constant over time. But as long as we remember that the μ multiplier is a constant, we may omit its equation of motion from (10.21) as well.

Inequality Integral Constraint

Finally, we consider the case where the integral constraint enters the problem as an inequality, say,

$$\begin{aligned} &\text{Maximize} && \int_0^T F(t, y, u)\, dt \\ &\text{subject to} && \dot{y} = f(t, y, u) \\ &&& \int_0^T G(t, y, u)\, dt \le k \qquad (k \text{ given}) \\ &\text{and} && y(0) = y_0 \qquad y(T)\ \textit{free} \qquad (y_0, T \text{ given}) \end{aligned} \tag{10.23}$$

Taking a cue from the isoperimetric problem, we can again dispose of the inequality integral constraint by making a substitution.

Define a new state variable Γ the same as in (10.15):

$$\Gamma(t) = -\int_0^t G(t, y, u)\, dt$$

where the upper limit of integration is t (not T). The derivative of Γ is simply

$$\dot{\Gamma} = -G(t, y, u) \qquad [\text{equation of motion for } \Gamma] \tag{10.24}$$

and its initial and terminal values are

$$\begin{aligned} \Gamma(0) &= -\int_0^0 G(t, y, u)\, dt = 0 \\ \Gamma(T) &= -\int_0^T G(t, y, u)\, dt \ge -k \qquad [\text{by } (10.23)] \end{aligned} \tag{10.25}$$

Using (10.24) and (10.25), we can restate problem (10.23) as

$$\begin{aligned}
&\text{Maximize} && \int_0^T F(t, y, u)\, dt \\
&\text{subject to} && \dot{y} = f(t, y, u) \\
&(10.26) && \dot{\Gamma} = -G(t, y, u) \\
& && y(0) = y_0 \qquad y(T) \text{ free} \qquad (y_0, T \text{ given}) \\
&\text{and} && \Gamma(0) = 0 \qquad \Gamma(T) \geq -k \qquad (k \text{ given})
\end{aligned}$$

Like the problem in (10.19), this is an unconstrained problem with two state variables. But, unlike (10.19), the new variable Γ in (10.26) has a truncated vertical terminal line.

The Hamiltonian of problem (10.26) is simply

$$H = F(t, y, u) + \lambda f(t, y, u) - \mu G(t, y, u) \tag{10.27}$$

If the constraint qualification is satisfied, then the maximum principle requires that

$$\begin{aligned}
&\max_u H && \text{for all } t \in [0, T] \\
&\dot{y} = \frac{\partial H}{\partial \lambda} && [\text{equation of motion for } y] \\
&\dot{\lambda} = -\frac{\partial H}{\partial y} && [\text{equation of motion for } \lambda] \\
(10.28)\quad &\dot{\Gamma} = \frac{\partial H}{\partial \mu} && [\text{equation of motion for } \Gamma] \\
&\dot{\mu} = -\frac{\partial H}{\partial \Gamma} && [\text{equation of motion for } \mu] \\
&\lambda(T) = 0 && [\text{transversality condition for } y] \\
&\mu(T) \geq 0 \qquad \Gamma(T) + k \geq 0 && \mu(T)[\Gamma(T) + k] = 0 \\
& && [\text{transversality condition for } \Gamma]
\end{aligned}$$

Note, again, that because the Hamiltonian is independent of Γ, we have

$$\dot{\mu} = -\frac{\partial H}{\partial \Gamma} = 0 \qquad \Rightarrow \qquad \mu(t) = \text{constant}$$

It becomes clear, therefore, that the multiplier associated with any integral constraint, whether equality or inequality, is constant over time. In the present problem, moreover, we can deduce from the transversality condition that the constant value of μ is nonnegative. But, as in (10.21), we can in fact omit from (10.28) the equations of motion for Γ and μ, as long as we bear in mind that μ is a nonnegative constant. In contrast, the transversality condition for Γ—involving complementary slackness—should be retained to reflect the inequality nature of the constraint. In sum, the conditions in (10.28) can be restated without reference to Γ as follows:

$$
\begin{aligned}
&\max_u H \qquad \text{for all } t \in [0, T] \\
&\dot{y} = \frac{\partial H}{\partial \lambda} \\
&\dot{\lambda} = -\frac{\partial H}{\partial y} \\
&\lambda(T) = 0 \\
&\mu = \text{constant} \geq 0 \qquad k - \int_0^T G(t, y, u)\, dt \geq 0 \\
&\text{and} \qquad \mu\left[k - \int_0^T G(t, y, u)\, dt\right] = 0 \qquad [\text{by } (10.25)]
\end{aligned}
\tag{10.28'}
$$

In the preceding discussion, the four types of constraints have been explained separately. But in case they appear simultaneously in a problem, we can still accommodate them by combining the procedures appropriate to each type of constraint present. For every constraint in the form of $g(t, y, u) = 0$ or $g(t, y, u) \leq c$, we append a new θ-multiplier term to the Hamiltonian, which is thereby expanded into a Lagrangian. And for every integral constraint, whether equality or inequality, we introduce a new (suppressible) state variable Γ, whose costate variable μ—a constant—is reflected in a $-\mu G(t, y, u)$ term in the Hamiltonian. If a constraint is an inequality, moreover, there will be a complementary-slackness condition on the multiplier θ or μ, as the case may be.

EXAMPLE 1 The political-business-cycle model (7.61):

$$
\begin{aligned}
&\text{Maximize} \qquad \int_0^T v(U, p) e^{rt}\, dt \\
&\text{subject to} \qquad p = \phi(U) + a\pi \\
&\qquad\qquad\qquad \dot{\pi} = b(p - \pi) \\
&\text{and} \qquad\qquad \pi(0) = \pi_0 \qquad \pi(T) \text{ free} \qquad (\pi_0, T \text{ given})
\end{aligned}
\tag{10.29}
$$

contains an equality constraint $p = \phi(U) + a\pi$. But we solved the problem as an unconstrained problem by first substituting out the constraint equation. Now we show how to deal with it directly as a constrained problem.

In this problem, π is the state variable; it has an equation of motion. Earlier, since p was substituted out, U was the only control variable. Now that p is retained in the model, however, it ought to be taken as another control variable. Thus the constraint equation

$$p - \phi(U) - a\pi = 0$$

is in line with the general format of $g(t, y, u_1, u_2) = c$ in (10.1), although there is no explicit t argument in it.

By (10.3), we can write the Lagrangian

$$\mathscr{L} = v(U, p)e^{rt} + \lambda b(p - \pi) + \theta[\phi(U) + a\pi - p] \tag{10.30}$$

If the following specific functions are adopted:

$$v(U, p) = -U^2 - hp \qquad (h > 0) \qquad \text{[from (7.62)]}$$
$$\phi(U) = j - kU \qquad (j, k > 0) \qquad \text{[from (7.63)]}$$

then the Lagrangian becomes

$$\mathscr{L} = (-U^2 - hp)e^{rt} + \lambda b(p - \pi) + \theta[j - kU + a\pi - p] \tag{10.30'}$$

Accordingly, the maximum principle calls for the conditions

$$\frac{\partial \mathscr{L}}{\partial U} = -2Ue^{rt} - \theta k = 0 \qquad \text{[by (10.4)]} \tag{10.31}$$

$$\frac{\partial \mathscr{L}}{\partial p} = -he^{rt} + \lambda b - \theta = 0 \qquad \text{[by (10.4)]} \tag{10.32}$$

$$\frac{\partial \mathscr{L}}{\partial \theta} = j - kU + a\pi - p = 0 \qquad \text{[by (10.5)]} \tag{10.33}$$

$$\dot{\pi} = \frac{\partial \mathscr{L}}{\partial \lambda} = b(p - \pi) \qquad \text{[by (10.6)]} \tag{10.34}$$

$$\dot{\lambda} = -\frac{\partial \mathscr{L}}{\partial \pi} = \lambda b - \theta a \qquad \text{[by (10.7)]} \tag{10.35}$$

plus a transversality condition. These should, of course, be equivalent to the conditions derived earlier in Sec. 7.6 by a different approach.

To verify the equivalence of the two approaches, let us reproduce here for comparison the major conditions from Sec. 7.6:

$$\frac{\partial H}{\partial U} = (-2U + hk)e^{rt} - \lambda bk = 0 \qquad \text{[maximizing the Hamiltonian]} \tag{10.36}$$

$$\dot{\pi} = b[j - kU - (1 - a)\pi] \qquad \text{[state equation of motion]} \tag{10.37}$$

$$\dot{\lambda} = hae^{rt} + \lambda b(1 - a) \qquad \text{[costate equation of motion]} \tag{10.38}$$

We shall first show that (10.31) and (10.32)—conditions on the two control variables U and p—together are equivalent to (10.36). Solving (10.32) for θ, we find

$$\theta = -he^{rt} + \lambda b \tag{10.39}$$

Substituting this into (10.31) results in

$$\frac{\partial \mathscr{L}}{\partial U} = -2Ue^{rt} + hke^{rt} - \lambda bk = 0$$

which is identical with (10.36). Next, solving (10.33) for p, we get

$$p = j - kU + a\pi$$

which, of course, merely restates the constraint. This enables us to rewrite (10.34) as

$$\dot{\pi} = b(j - kU + a\pi - \pi) = b[j - kU - (1 - a)\pi]$$

which is the same as (10.37). Finally, using the θ expression in (10.39), we may rewrite (10.35) as

$$\dot{\lambda} = \lambda b + hae^{rt} - \lambda ba = \lambda b(1 - a) + hae^{rt}$$

which is identical with (10.38). Hence, the two approaches are equivalent.

EXAMPLE 2 In Sec. 7.4, Example 3, we encountered a time-optimal problem

$$\begin{aligned} &\text{Maximize} && \int_0^T -1\,dt \\ &\text{subject to} && \dot{y} = y + u \\ &&& y(0) = 5 \qquad y(T) = 11 \qquad T \text{ free} \\ &\text{and} && u(t) \in [-1, 1] \end{aligned} \tag{10.40}$$

with a constrained control variable. We solved it by resorting to the signum function (7.48) in choosing the optimal control. But it is also possible to view the control set as made up of two inequality constraints

$$-1 \le u(t) \qquad \text{and} \qquad u(t) \le 1 \tag{10.41}$$

and solve the problem accordingly.

Note that the constraint qualification is satisfied here since each constraint function in (10.41) is linear in u.

First, we augment the Hamiltonian of this problem,

$$H = -1 + \lambda(y + u)$$

into a Lagrangian by taking into account the two constraints in (10.41). The result is

$$\mathscr{L} = -1 + \lambda(y+u) + \theta_1(u+1) + \theta_2(1-u) \qquad \text{[by (10.9)]} \tag{10.42}$$

To satisfy the maximum principle, we must have:

$$\frac{\partial \mathscr{L}}{\partial u} = \lambda + \theta_1 - \theta_2 = 0 \qquad \text{for all } t \in [0, T] \tag{10.43}$$

$$\frac{\partial \mathscr{L}}{\partial \theta_1} = u + 1 \geq 0 \qquad \theta_1 \geq 0 \qquad \theta_1(u+1) = 0 \tag{10.44}$$

$$\frac{\partial \mathscr{L}}{\partial \theta_2} = 1 - u \geq 0 \qquad \theta_2 \geq 0 \qquad \theta_2(1-u) = 0 \tag{10.45}$$

$$\dot{y} = \frac{\partial \mathscr{L}}{\partial \lambda} = y + u \tag{10.46}$$

$$\dot{\lambda} = -\frac{\partial \mathscr{L}}{\partial y} = -\lambda \tag{10.47}$$

plus a transversality condition. Even though the two equations of motion (10.46) and (10.47) here are derived from $\mathscr{L}$ instead of H, they are the same as those derived from H in Sec. 7.4. So nothing more needs to be said here about these. What needs to be verified is that conditions (10.43) through (10.45) would lead us to the same choice of control as the signum function:

$$u^* = 1 \text{ if } \lambda > 0 \qquad u^* = -1 \text{ if } \lambda < 0 \tag{10.48}$$

We shall demonstrate here that $\lambda > 0$ indeed implies $u^* = 1$. Let $\lambda > 0$. Then, to satisfy (10.43), we must have $\theta_1 - \theta_2 < 0$ or $\theta_1 < \theta_2$. Since both θ_1 and θ_2 are nonnegative, the condition $\theta_1 < \theta_2$ implies that $\theta_2 > 0$. Thus, by the complementary-slackness condition in (10.45), it follows that $1 - u = 0$, so the optimal control is $u^* = 1$. This completes the intended demonstration. The other case ($\lambda < 0$) is similar, and will be left to the reader as an exercise.

Current-Value Hamiltonian and Lagrangian

When the constrained problem involves a discount factor, it is possible to use the current-value Hamiltonian H_c in lieu of H. In that case, the Lagrangian $\mathscr{L}$ should be replaced by the current-value Lagrangian $\mathscr{L}_c$.

Consider the inequality-constraint problem

$$\text{(10.49)}\qquad \begin{aligned} &\text{Maximize} && \int_0^T \Phi(t, y, u) e^{-\rho t}\, dt \\ &\text{subject to} && \dot{y} = f(t, y, u) \\ & && g(t, y, u) \le c \\ &\text{and} && \text{boundary conditions} \end{aligned}$$

The regular Hamiltonian and Lagrangian are

$$\text{(10.50)}\qquad \begin{aligned} H &= \Phi(t, y, u) e^{-\rho t} + \lambda f(t, y, u) \\ \mathscr{L} &= \Phi(t, y, u) e^{-\rho t} + \lambda f(t, y, u) + \theta[c - g(t, y, u)] \end{aligned}$$

And the maximum principle calls for (assuming interior solution)[2]:

$$\text{(10.51)}\qquad \frac{\partial \mathscr{L}}{\partial u} = 0 \qquad \text{for all } t \in [0, T]$$

$$\text{(10.52)}\qquad \frac{\partial \mathscr{L}}{\partial \theta} = c - g(t, y, u) \ge 0 \qquad \theta \ge 0 \qquad \theta \frac{\partial \mathscr{L}}{\partial \theta} = 0$$

$$\text{(10.53)}\qquad \dot{y} = \frac{\partial \mathscr{L}}{\partial \lambda}$$

$$\text{(10.54)}\qquad \dot{\lambda} = -\frac{\partial \mathscr{L}}{\partial y}$$

plus an appropriate transversality condition.

By introducing new multipliers

$$\text{(10.55)}\qquad \begin{aligned} m &= \lambda e^{\rho t} && (\text{implying } \lambda = m e^{-\rho t}) \\ n &= \theta e^{\rho t} && (\text{implying } \theta = n e^{-\rho t}) \end{aligned}$$

we can introduce the current-value versions of H and $\mathscr{L}$ as follows:

$$\text{(10.56)}\qquad \begin{aligned} H_c &\equiv H e^{\rho t} = \Phi(t, y, u) + m f(t, y, u) \\ \mathscr{L}_c &\equiv \mathscr{L} e^{\rho t} = \Phi(t, y, u) + m f(t, y, u) + n[c - g(t, y, u)] \end{aligned}$$

[2] If the control variable is constrained by a nonnegativity requirement, (10.51) must be changed to

$$\frac{\partial \mathscr{L}}{\partial u} \le 0 \qquad u \ge 0 \qquad u \frac{\partial \mathscr{L}}{\partial u} = 0 \qquad [\text{cf. (10.12)}]$$

It can readily be verified that

$$\frac{\partial \mathcal{L}_c}{\partial u} = \frac{\partial \mathcal{L}}{\partial u} e^{\rho t} \qquad \frac{\partial \mathcal{L}_c}{\partial n} = \frac{\partial \mathcal{L}}{\partial \theta} \qquad \text{and} \qquad \frac{\partial \mathcal{L}_c}{\partial m} = \frac{\partial \mathcal{L}}{\partial \lambda}$$

Therefore, conditions (10.51), (10.52), and (10.53) can be equivalently expressed with $\mathcal{L}_c$ and the new multipliers m and n as follows:

$$\frac{\partial \mathcal{L}_c}{\partial u} = 0 \qquad \text{for all } t \in [0, T] \tag{10.57}$$

$$\frac{\partial \mathcal{L}_c}{\partial n} \geq 0 \qquad n \geq 0 \qquad n \frac{\partial \mathcal{L}_c}{\partial n} = 0 \tag{10.58}$$

$$\dot{y} = \frac{\partial \mathcal{L}_c}{\partial m} \tag{10.59}$$

The only major modification required when we use $\mathcal{L}_c$ is found in the equation of motion for the costate variable, (10.54). The equivalent new statement, using the new variable m, is

$$\dot{m} = -\frac{\partial \mathcal{L}_c}{\partial y} + \rho m \tag{10.60}$$

To verify this, first differentiate the λ expression in (10.55) with respect to t, to obtain

$$\dot{\lambda} = \dot{m} e^{-\rho t} - \rho m e^{-\rho t}$$

Then differentiate $\mathcal{L}$ with respect to y to get

$$-\frac{\partial \mathcal{L}}{\partial y} = -\Phi_y e^{-\rho t} - \lambda f_y + \theta g_y$$

Equating these two expressions in line with (10.54), and multiplying through by $e^{\rho t}$, we find that

$$\dot{m} - \rho m = -\Phi_y - m f_y + n g_y$$

Since the right-hand-side expression is equal to $-\partial \mathcal{L}_c / \partial y$ [from (10.56)], the result in (10.60) immediately follows.

For problems with integral constraints only, no Lagrangian function is needed, or, to put it differently, the Lagrangian reduces to the Hamiltonian. This is because we absorb each integral constraint into the problem via a new state variable—such as Γ in (10.19) and (10.26). The maximum-principle conditions can, accordingly, be stated in terms of the Hamiltonian. If we decide to use H_c in place of H, the procedure outlined in Sec. 8.2 can be applied in a straightforward manner. The major modification is, again, to be found in the costate equation of motion, namely, replacing the condition

$\dot{\lambda} = -\partial H/\partial y$ by the condition $\dot{m} = -\partial H_c/\partial y + \rho m$. With a newly introduced state variable Γ, the problem will have a new costate variable μ. Since its equation of motion is $\dot{\mu} = -\partial H/\partial\Gamma = 0$, μ is a constant.

Sufficient Conditions

The Mangasarian and Arrow sufficient conditions, previously discussed in the context of unconstrained problems, turn out to be valid also for constrained problems when the terminal time T is fixed. This would include cases with a fixed terminal point, a vertical terminal line, or a truncated vertical terminal line.

Let us use the symbol u to represent the vector of control variables. For problems (10.1) and (10.8), for example, we have $u = (u_1, u_2)$. As before, let H^0 denote the maximized Hamiltonian—the Hamiltonian evaluated along the $u^*(t)$ path. But, in the present context, the Hamiltonian is understood to be maximized subject to all the constraints of the $g(t, y, u) = c$ form or the $g(t, y, u) \le c$ form present in the problem. Besides, since every integral constraint present in the problem is reflected in H via the new costate variable μ, it must also be similarly reflected in H^0.

For simplicity, we can consolidate the Mangasarian and Arrow sufficient conditions into a single statement[3]: The maximum-principle conditions are sufficient for the global maximization of the objective functional if

(10.61) either $\mathscr{L}$ is concave in (y, u) for all $t \in [0, T]$; or H^0 is concave in y for all $t \in [0, T]$, for given λ

These conditions are also applicable to infinite-horizon problems, but in the latter case, (10.61) is to be supplemented by a transversality condition

$$\lim_{t\to\infty} \lambda(t)[y(t) - y^*(t)] \ge 0 \qquad \text{[cf. (9.21)]} \tag{10.62}$$

A few comments about (10.61) may be added here. First, as pointed out before, the concavity of $\mathscr{L}$ in (y, u) means concavity in the variables y and u jointly, not separately in y and in u. Second, since H and $\mathscr{L}$ are

[3]For a more comprehensive statement of sufficient conditions, see Ngo Van Long and Neil Vousden, "Optimal Control Theorems," Essay 1 in John D. Pitchford and Stephen J. Turnovsky, eds., *Applications of Control Theory to Economic Analysis*, North-Holland, Amsterdam, 1977, pp. 11–34 (especially pp. 25–28, Theorems 6 and 7).

composed of the F, f, g, and G functions as follows:

$$H = F + \lambda f - \mu G \qquad \text{and} \qquad \mathcal{L} = H + \theta[c - g]$$

it is clear that (10.61) will be satisfied if the following are simultaneously true:

F is concave in (y, u)

λf is concave in (y, u)

μG is convex in (y, u)

and θg is convex in (y, u) for all $t \in [0, T]$

In the case of an inequality integral constraint, however, where μ is a nonnegative constant [by (10.28)], the convexity of μG is ensured by the convexity of G itself. Similarly, in the case of an inequality constraint, where $\theta \geq 0$ [by (10.11)], the convexity of θg is ensured by the convexity of g itself. Finally, if the current-value Hamiltonian and Lagrangian are used, (10.61) can be easily adapted by replacing $\mathcal{L}$ by $\mathcal{L}_c$, and H^0 by H_c^0.

EXERCISE 10.1

1 In Example 1, p is taken to be an additional control variable. Why is it *not* taken to be a new *state* variable?

2 In Example 2, it is demonstrated that if $\lambda > 0$, then the optimal control is $u^* = 1$. By analogous reasoning, demonstrate that, if $\lambda < 0$, then the optimal control is $u^* = -1$.

3 An individual's productive hours per day (24 hours less sleeping and leisure time—normalized to one) can be spent on either *work* or *study*. Work results in immediate earnings; study contributes to human capital $K(t)$ (knowledge) and improves future earnings. Let the proportion of productive hours spent on study be denoted by s. The rate of change of human capital, $\dot{K}$, is assumed to have a constant elasticity α $(0 < \alpha < 1)$ with respect to sK. Current earnings are determined by the level of human capital multiplied by the time spent on work.

(*a*) Formulate the individual's problem of work-study decision for a planning period $[0, T]$ with a discount rate ρ $(0 < \rho < 1)$.
(*b*) Name the state and control variables.
(*c*) What boundary conditions are appropriate for the state variable?
(*d*) Is this problem a constrained control problem? Which type of constraints does it have?
(*e*) How does this problem resemble the Dorfman model of Sec. 8.1? How does it differ from the Dorfman model?

4 Consider the problem

$$\text{Maximize} \quad \int_0^T F(t, y, u)\, dt$$

$$\text{subject to} \quad \dot{y} = f(t, y, u)$$

$$g(t, y, u) \le c$$

$$y(0) = y_0 \qquad y(T) \text{ free} \qquad (y_0, T \text{ given})$$

$$\text{and} \quad 0 \le u(t)$$

We can, as we did in (10.12), apply the Kuhn-Tucker conditions to the nonnegativity restriction on $u(t)$ without introducing a specific multiplier. Alternatively, we may treat the nonnegativity restriction as an additional constraint of the $g(t, y, u) \le c$ type. Write out the maximum-principle conditions under both approaches, and compare the results. [Use θ' do denote the new multiplier for the $0 \le u(t)$ constraint and $\mathscr{L}'$ to denote the corresponding new Lagrangian.]

5 In a maximization problem let there be two state variables (y_1, y_2), two control variables (u_1, u_2), one inequality constraint, and one inequality integral constraint. The initial states are fixed, but the terminal states are free at a fixed T.

(*a*) Write the problem statement.
(*b*) Define the Hamiltonian and the Lagrangian.
(*c*) List the maximum-principle conditions, assuming interior solutions.

10.2 THE DYNAMICS OF A REVENUE-MAXIMIZING FIRM

In the Evans model of the dynamic monopolist (Sec. 2.4), the stated objective is to maximize the total profit. This, of course, is nothing extraordinary because profit maximization has long been the accepted hypothesis among economists. A well-known model by Baumol takes the view, however, that instead of maximizing profits, modern firms may actually try to maximize sales revenue, subject to a minimum requirement on the rate of return.[4] The primary basis for this view is the separation of ownership and management in the typical corporate form of the firm, where the managers may, in their own interest, seek to maximize sales, despite the stockholders'

---

[4] William J. Baumol, "On the Theory of Oligopoly," *Econometrica*, August 1958, pp. 187–198. See also his *Business Behavior, Value and Growth*, revised edition, Harcourt, Brace & World, New York, 1967. This model is discussed as an example of nonlinear programming in Alpha C. Chiang, *Fundamental Methods of Mathematical Economics*, 3d ed., McGraw-Hill, New York, 1984, Sec. 21.6.

objective of maximizing profit instead. To placate the stockholders, however, the managers must attempt to attain at least a minimum acceptable rate of return.

Since the Baumol model is static, its validity has been questioned in the dynamic context. In particular, since profits serve as the vehicle of growth, and growth makes possible greater sales, it would seem that in the dynamic context profit maximization may be a prerequisite for sales maximization. To address the issue of whether the Baumol proposition retains its validity in a dynamic world, Hayne Leland has developed an optimal control model.[5]

The Model

Consider a firm that produces a single good with a neoclassical production function $Q = Q(K, L)$, which is linearly homogeneous and strictly quasiconcave. The usual properties listed below are supposed to hold:

$$\begin{aligned} &\frac{Q}{L} = \phi(k) \qquad \left(k \equiv \frac{K}{L}\right) \qquad \phi(0) = 0 \\ &Q_K = \phi'(k) \qquad Q_K > 0 \qquad Q_{KK} < 0 \\ &Q_L = \phi(k) - k\phi'(k) \qquad Q_L > 0 \qquad Q_{LL} < 0 \end{aligned} \tag{10.63}$$

All prices—including the wage rate W and the price of capital goods—are assumed to be constant, with the price of the firm's product normalized to one. Thus the revenue and (gross) profits of the firm are, respectively,

$$R = Q(K, L) \cdot 1 = Q(K, L)$$

$$\pi = R - WL = Q(K, L) - WL$$

To satisfy the stockholders, the managers must keep in mind a minimum acceptable rate of return on capital, r_0. This means that the managerial behavior is constrained by the inequality

$$\frac{\pi}{K} \geq r_0 \qquad \text{or} \qquad Q(K, L) - WL - r_0 K \geq 0 \tag{10.64}$$

It remains to specify the decision on investment and capital accumulation. For simplicity, it is postulated that the firm always reinvests a fixed proportion of its profits π. Define α as the fraction of profits reinvested,

[5] Hayne E. Leland, "The Dynamics of a Revenue Maximizing Firm," *International Economic Review*, June 1972, pp. 376–385.

divided by the constant price of capital goods. Then we have

$$\dot{K}(=I) = \alpha\pi = \alpha[Q(K,L) - WL] \tag{10.65}$$

with initial capital $K(0) = K_0$.

Drawing together the various considerations, we can state the problem of this revenue-maximizing firm as one with an inequality constraint:

$$\begin{aligned} &\text{Maximize} && \int_0^T Q(K,L)e^{-\rho t}\,dt \\ &\text{subject to} && \dot{K} = \alpha[Q(K,L) - WL] \\ &&& WL + r_0 K - Q(K,L) \le 0 \\ &\text{and} && K(0) = K_0 \qquad K(T) \text{ free} \qquad (K_0, T \text{ given}) \end{aligned} \tag{10.66}$$

There are only two variables in this model, K and L. The presence of an equation of motion for K identifies K as a state variable; this leaves L as the sole control variable.

Note that the constraint function $WL + r_0 K - Q(K, L)$ is convex in the control variable L, and the constraint set does contain a point such that the strict inequality holds (the rate of return exceeds the r_0 level). Thus the constraint qualification is satisfied.

The Maximum Principle

The current-value Hamiltonian of this problem is

$$H_c = Q(K,L) + m\alpha[Q(K,L) - WL]$$

By augmenting the Hamiltonian with the information in the inequality constraint, we can write the current-value Lagrangian [see (10.56)] as

$$\mathcal{L}_c = Q(K,L) + m\alpha[Q(K,L) - WL] + n[Q(K,L) - WL - r_0 K]$$

From (10.57) and (10.58), we then have the following first-order conditions for maximizing $\mathcal{L}_c$:

$$\frac{\partial \mathcal{L}_c}{\partial L} = (1 + m\alpha + n)Q_L - (m\alpha + n)W = 0 \quad \text{for all } t \in [0, T] \tag{10.67}$$

$$\frac{\partial \mathcal{L}_c}{\partial n} = Q(K,L) - WL - r_0 K \ge 0 \quad n \ge 0 \quad \text{and} \quad n[Q(K,L) - WL - r_0 K] = 0 \tag{10.68}$$

To determine the dynamics of the system, we rely on the two equations of motion

$$\dot{K} = \frac{\partial \mathcal{L}_c}{\partial m} = \alpha[Q(K,L) - WL]$$

and

$$\dot{m} = -\frac{\partial \mathscr{L}_c}{\partial K} + \rho m = -(1 + m\alpha + n)Q_K + nr_0 + \rho m \qquad \text{[by (10.60)]}$$

Finally, the transversality condition

$$m(T) = 0$$

should be imposed since $K(T)$ is free.

Qualitative Analysis

Rather than employ a specific production function to obtain a quantitative solution, Leland carries out the analysis of the model qualitatively. Also, instead of working with the original input variables K and L, he uses the properties of the production function in (10.63) to convert the maximum-principle conditions into terms of the capital-labor ratio, k. Accordingly, condition (10.67) is restated as

(10.69)

$$\frac{\partial \mathscr{L}_c}{\partial L} = (1 + m\alpha + n)[\phi(k) - k\phi'(k)] - (m\alpha + n)W = 0$$

By dividing the first inequality in (10.68) by $L \neq 0$ and using (10.63), we can write a new version of the condition:

$$(10.70) \qquad \phi(k) - W - r_0 k \geq 0 \qquad n \geq 0 \qquad n[\phi(k) - W - r_0 k] = 0$$

And the two equations of motion become

$$(10.71) \qquad \dot{K} = \alpha L[\phi(k) - W]$$

$$(10.72) \qquad \dot{m} = -(1 + m\alpha + n)\phi'(k) + nr_0 + \rho m$$

The k variable now occupies the center stage except in (10.71), where the original variables K and L still appear.

Certain specific levels of the capital-labor ratio k have special roles to play in the model. Two of these are:

$\hat{k} \equiv$ the profit-maximizing level of k

$$\left[\hat{k} \text{ satisfies } Q_L = W \text{ or } \phi(k) - k\phi'(k) = W\text{—by (10.63)}\right]$$

$k^0 \equiv$ the k that yields the rate of return r_0

$$\left[k^0 \text{ satisfies } \frac{\pi}{K} = r_0 \text{ or } \phi(k) - W = r_0 k\text{—by (10.64)}\right]$$

In Fig. 10.1, we plot the APP_L (Q/L) and the MPP_L (Q_L) curves against k. The APP_L curve has been encountered previously in Figs. 9.1 and 9.3a; the MPP_L curve should lie below the APP_L curve by the vertical distance

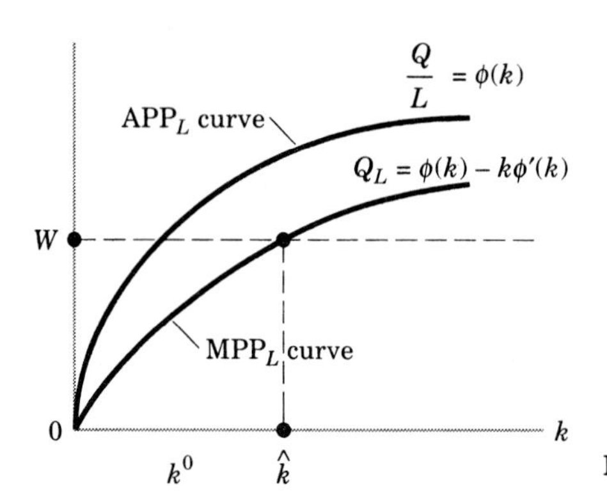

FIGURE 10.1

$k\phi'(k)$. The intersection of the Q_L curve and the W line determines the magnitude of $\hat{k}$. While the position of k^0 can only be arbitrarily indicated, it is reasonable to expect k^0 to be located to the left of $\hat{k}$.

Since it is not permissible to choose a $k < k^0$, Leland concentrates on the situation of $k > k^0$, the case where the rate of return exceeds the minimum acceptable level r_0. The complementary-slackness condition in (10.70) then leads to $n = 0$. This will simplify (10.69) to

$$\text{(10.73)} \qquad \left.\frac{\partial \mathscr{L}_c}{\partial L}\right|_{n=0} = (1 + m\alpha)[\phi(k) - k\phi'(k)] - m\alpha W = 0$$

This equation can be plotted as a curve in the km space if we can find an expression for its slope so that we can ascertain its general configuration. Such a slope expression is available by invoking the implicit-function rule.

Call the expression in (10.73) between the two equals signs $F(m, k)$. Then the slope of the curve is

(10.74)
$$\frac{dm}{dk}\left(\text{for } \frac{\partial \mathscr{L}_c}{\partial L} = 0\right) = -\frac{\partial F/\partial k}{\partial F/\partial m} = \frac{(1 + m\alpha)k\phi''(k)}{\alpha[\phi(k) - k\phi'(k) - W]}$$

The following observations may be gathered from this result:

(1) The numerator is negative, because $\phi''(k) < 0$, as shown by the curvature of the APP_L curve in Fig. 10.1.

(2) For any $k < \hat{k}$, the denominator is also negative. This can also be seen in Fig. 10.1 since the Q_L curve lies below the W line to the left of $\hat{k}$. Thus, for any $k \in (k^0, \hat{k})$, the $\partial \mathscr{L}_c/\partial L = 0$ curve is positively sloped.

(3) As k approaches $\hat{k}$ from the left, the denominator tends to zero, so $dm/dk \to \infty$.

(4) As k approaches zero, the numerator tends to zero, so $dm/dk \to 0$.

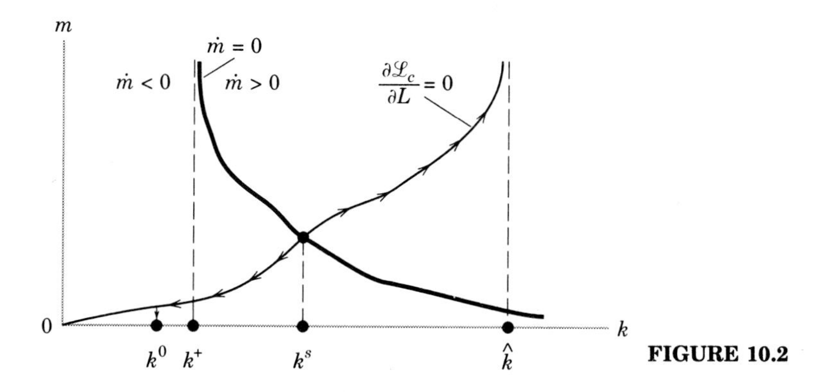

FIGURE 10.2

These observations enable us to sketch the $\partial\mathscr{L}_c/\partial L = 0$ curve the way it is shown in Fig. 10.2.

By concentrating on $k > k^0$, which implies $n = 0$, we also get a simpler version of $\dot{m}$ from (10.72). If we set that $\dot{m}$ equal to zero, then we have the equation

$$\dot{m}|_{n=0} = -(1 + m\alpha)\phi'(k) + \rho m = 0 \tag{10.75}$$

which can also be graphed in the km space. To use the implicit-function rule again, let us call the expression in (10.75) between the two equals signs $G(m, k)$. Then the slope of the $\dot{m} = 0$ curve is

$$\frac{dm}{dk} \text{ (for } \dot{m} = 0) = -\frac{\partial G/\partial k}{\partial G/\partial m} = \frac{(1 + m\alpha)\phi''(k)}{-\alpha\phi'(k) + \rho} \tag{10.76}$$

While the numerator is negative, the denominator can take either sign. Let

$$k^+ \equiv \text{the } k \text{ value that satisfies } \rho = \alpha\phi'(k)$$

Then we see, first, that since $\phi'(k)$ decreases with k (see Fig. 10.1), for any $k > k^+$ the denominator in (10.76) is positive, so that dm/dk is negative and the $\dot{m} = 0$ curve is downward sloping. Second, as $k \to k^+$ from the right, the slope of the $\dot{m} = 0$ curve tends to infinity. These explain the general shape of the $\dot{m} = 0$ curve in Fig. 10.2. The two curves intersect at $k = k^s$. Thus,

$$k^s \equiv \text{the } k \text{ level that satisfies } \partial\mathscr{L}_c/\partial L = \dot{m} = 0, \text{ when } n = 0$$

Such an intersection exists if and only if k^+ occurs to the left of $\hat{k}$.

The $\dot{m} = 0$ curve is where the multiplier m is stationary. What happens to m at points off the $\dot{m} = 0$ curve? The answer is provided by the derivative

$$\frac{\partial \dot{m}}{\partial k} = -(1 + m\alpha)\phi''(k) > 0$$

The message here is that as k increases, $\dot{m}$ will increase, too. Consequently, we have $\dot{m} < 0$ to the left of the $\dot{m} = 0$ curve, and $\dot{m} > 0$ to its right. This is why we have made the arrowheads on the $\partial \mathscr{L}_c / \partial L = 0$ curve point southward to the left, and northward to the right, of the $\dot{m} = 0$ curve.

From Fig. 10.2, it should be clear that the stationary point formed by the intersection of the two curves may not occur at $\hat{k}$, the profit-maximizing level of k. That is, placing the revenue-maximizing firm in the dynamic context does not force it to abandon its revenue orientation in favor of the profit orientation. In fact, Leland shows by further analysis that there is an eventual tendency for the revenue-maximizing firm to gravitate toward k^0, where the rate of return is at the minimum acceptable level r_0. The only circumstance under which the firm would settle at the profit-maximizing level $\hat{k}$ is when k^+ exceeds $\hat{k}$.

The reader may have noticed that condition (10.71) has played no role in the preceding analysis. This is because the analysis has been couched in terms of the capital-labor ratio k, whereas (10.71) involves K and L as well as k. A simple transformation of (10.71) will reveal what function it can perform in the model. Since $k \equiv K/L$, we have $K = kL$, so that

$$\dot{K} = \dot{k}L + k\dot{L}$$

Equating this with the $\dot{K}$ expression in (10.71) and rearranging, we find that

$$\frac{\dot{L}}{L} = \frac{\alpha[\phi(k) - W] - \dot{k}}{k}$$

Once we have found an optimal time path for k, then this equation can yield a corresponding optimal path for the control variable L.

10.3 STATE-SPACE CONSTRAINTS

The second category of constraints consists of those in which no control variables appear. What such constraints do is to place restrictions on the state space, and demarcate the permissible area of movement for the variable y. A simple example of this type of constraints relevant to many economic problems is the nonnegativity restriction

$$y(t) \geq 0 \qquad \text{or} \qquad -y(t) \leq 0 \qquad \text{for all } t \in [0, T]$$

But more generally, the constraint may take the form

$$h(t, y) \leq c \qquad \text{for all } t \in [0, T]$$

In either case, the control variable u is absent in the constraint function.

By coincidence, it may happen that when we ignore the state-space constraint and solve the given problem as an unconstrained one, the optimal path $y^*(t)$ lies entirely in the permissible area. In that event, the constraint

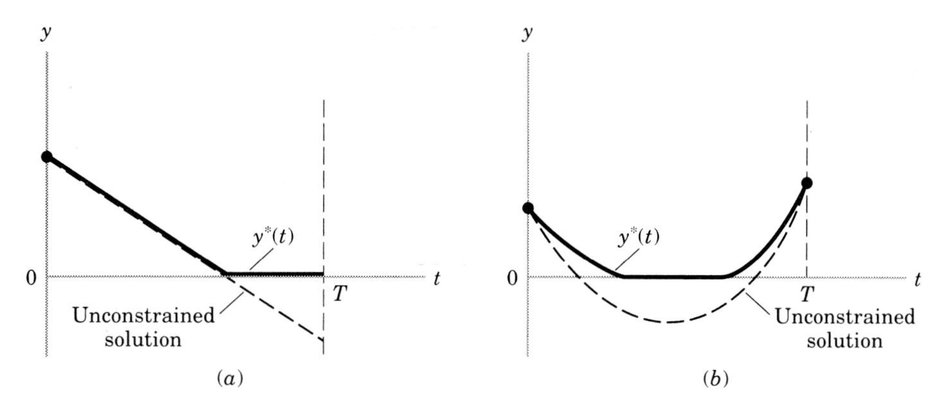

FIGURE 10.3

is trivial. With a meaningful constraint, however, we would expect the unconstrained optimal path to violate the constraint, such as the broken curve in Fig. 10.3*a* which fails the nonnegativity restriction. The true optimal solution, exemplified by the solid curve, can contain one or more segments lying on the constraint boundary itself. That is, there may be one or more "constrained intervals" (or "blocked intervals") within $[0, T]$ when the constraint is binding (effective), with $h(t, y) = c$. Sometimes the true optimal path may have some segment(s) in common with the unconstrained solution path; but the two paths can also be altogether different, such as illustrated in Fig. 10.3*b*.

Although we cannot in general expect the unconstrained solution to work, it is not a bad idea to try it anyway. Should that solution turn out to satisfy the constraint, then the problem would be solved. Even if not, useful clues will usually emerge regarding the nature of the true solution.

Dealing with State-Space Constraints

Let the problem be

$$
\begin{aligned}
&\text{Maximize} && \int_0^T F(t, y, u)\, dt \\
&\text{subject to} && \dot{y} = f(t, y, u) \\
& && h(t, y) \le c \\
&\text{and} && \text{boundary conditions}
\end{aligned}
\tag{10.77}
$$

Instinctively, we may expect the method of Sec. 10.1 to apply to this problem. If so, we have the Lagrangian

$$\mathscr{L} = F(t, y, u) + \lambda f(t, y, u) + \theta[c - h(t, y)] \tag{10.78}$$

with maximum-principle conditions (assuming interior solution)

$$\frac{\partial \mathcal{L}}{\partial u} = F_u + \lambda f_u = 0$$

$$\frac{\partial \mathcal{L}}{\partial \theta} = c - h(t, y) \geq 0 \qquad \theta \geq 0 \qquad \theta \frac{\partial \mathcal{L}}{\partial \theta} = 0$$

$$\dot{y} = \frac{\partial \mathcal{L}}{\partial \lambda} = f(t, y, u) \tag{10.79}$$

$$\dot{\lambda} = -\frac{\partial \mathcal{L}}{\partial y} = -F_y - \lambda f_y + \theta h_y$$

plus a transversality condition if needed

Why cannot these conditions be used as we did before?

For one thing, in previously discussed constrained problems, the solution is predicated upon the continuity of the y and λ variables, so that only the control variables are allowed to jump (e.g., bang-bang). But here, with pure state constraints, the costate variable λ can also experience jumps at the junction points where the constraint $h(t, y) \leq c$ turns from inactive (nonbinding) to active (binding) status, or vice versa. Specifically, if τ is a junction point between an unconstrained interval and a constrained interval (or the other way around), and if we denote by $\lambda^-(\tau)$ and $\lambda^+(\tau)$ the value of λ just before and just after the jump, respectively, then the jump condition is[6]

$$\lambda^+(\tau) = \lambda^-(\tau) + bh_y \qquad (b \geq 0) \tag{10.80}$$

Since the value of b is indeterminate, however, this condition can only help in ascertaining the direction of the jump. Note that it is possible for b to be zero, which means that λ may not be discontinuous at a junction point. This latter situation can occur when the constraint function $h(t, y)$ has a sharp point at τ, that is, when $\dot{h}$ is discontinuous at τ.

At any rate, the solution procedure requires modification. We now need to watch out for junction points, and make sure not to step into the forbidden part of the state space.

An Alternative Approach

While it is possible to proceed on the basis of the conditions in (10.79), it would be easier to take an alternative approach in which the change in the status of the constraint $h(t, y) \leq c$ at junction points is taken into account

[6] For a more detailed discussion of jump conditions, see Atle Seierstad and Knut Sydsæter, *Optimal Control Theory with Economic Applications*, Elsevier, New York, 1987, Chap. 5, especially pp. 317–319.

in a more explicit way.[7] The major consideration of this latter approach is that since $h(t, y)$ is not allowed to exceed c, then whenever $h(t, y) = c$ (the constraint becomes binding) we must forbid $h(t, y)$ to increase. This can be accomplished simply by imposing the condition

$$\frac{dh}{dt} \le 0 \qquad \text{whenever } h(t, y) = c$$

Note that the derivative dh/dt (a total derivative), unlike h itself, is a function of t and y, as well as u, because

$$\frac{d}{dt}h(t, y) = \frac{\partial h}{\partial t} + \frac{\partial h}{\partial y}\frac{dy}{dt} = h_t + h_y f(t, y, u) \equiv \dot{h}(t, y, u)$$

Therefore, the new constraint $dh/dt \le 0$ whenever $h(t, y) = c$, or, more explicitly,

(10.81)
$$\dot{h}(t, y, u) \equiv h_t + h_y f(t, y, u) \le 0 \qquad \text{whenever } h(t, y) = c$$

fits nicely into the $g(t, y, u) \le c$ category discussed in Sec. 10.1. The only difference is that (10.81) is not meant for all $t \in [0, T]$, but needs to be put in force only when $h(t, y) = c$.

With this new constraint, the problem statement now takes the form

$$\begin{aligned} &\text{Maximize} && \int_0^T F(t, y, u)\, dt \\ &\text{subject to} && \dot{y} = f(t, y, u) \\ (10.82) \qquad & && \dot{h}(t, y, u) = h_t + h_y f(t, y, u) \le 0 \\ & && \qquad \text{whenever } h(t, y) = c \\ &\text{and} && \text{boundary conditions} \end{aligned}$$

In order to make possible a comparison of the new maximum-principle conditions with those in (10.79), we shall adopt distinct multiplier symbols Λ and Θ here. Let the Lagrangian be written as

(10.83)
$$\mathscr{L}' = F(t, y, u) + \Lambda f(t, y, u) - \Theta \dot{h}$$

Then, as part of the maximum-principle conditions, we require (assuming that the constraint qualification is satisfied) that

$$\frac{\partial \mathscr{L}'}{\partial u} = F_u + \Lambda f_u - \Theta h_y f_u = 0$$

[7] See Magnus R. Hestenes, *Calculus of Variations and Optimal Control Theory*, Wiley, New York, 1966, Chap. 8, Theorem 2.1.

and

$$\frac{\partial \mathscr{L}'}{\partial \Theta} = -\dot{h} = -h_t - h_y f(t, y, u) \geq 0 \qquad \Theta \geq 0 \qquad \Theta \frac{\partial \mathscr{L}'}{\partial \Theta} = 0$$

While the set of conditions on Θ seems to serve the purpose of restating the constraint, it does not make clear that this set applies only when $h(t, y) = c$. To remedy this, we append the complementary-slackness condition

$$h(t, y) \leq c \qquad \Theta[c - h(t, y)] = 0$$

Then, $h(t, y) < c$ (constraint not binding) would mean $\Theta = 0$, which would cause the last term in $\mathscr{L}'$ to drop out, and thereby nullify the conditions regarding $\partial \mathscr{L}'/\partial \Theta$. Conversely, when $h(t, y) = 0$ (constraint binding), we intend the complementary-slackness condition to imply $\Theta > 0$. Thus this is a stronger form of complementary slackness. [In the normal interpretation of complementary slackness, $h(t, y) = 0$ is consistent with $\Theta = 0$ as well as $\Theta > 0$.]

In addition to the preceding, the maximum principle for the present approach also places a restriction on the way the Θ multiplier changes over time: At points where Θ is differentiable, $\dot{\Theta}$ must be nonpositive whenever $h(t, y) = c$. This restriction will be explained later.

Collecting all the results together, we have (assuming interior solution) the following set of conditions:

$$\frac{\partial \mathscr{L}'}{\partial u} = F_u + \Lambda f_u - \Theta h_y f_u = 0$$

$$\frac{\partial \mathscr{L}'}{\partial \Theta} = -\dot{h} = -\left[h_t + h_y f(t, y, u)\right] \geq 0 \qquad \Theta \geq 0 \qquad \Theta \frac{\partial \mathscr{L}'}{\partial \Theta} = 0$$

$$h(t, y) \leq c \qquad \Theta[c - h(t, y)] = 0$$

$$\dot{\Theta} \leq 0 \qquad [= 0 \text{ when } h(t, y) < c] \tag{10.84}$$

$$\dot{y} = \frac{\partial \mathscr{L}'}{\partial \Lambda} = f(t, y, u)$$

$$\dot{\Lambda} = -\frac{\partial \mathscr{L}'}{\partial y} = -F_y - \Lambda f_y + \Theta\left[h_{yt} + h_y f_y + h_{yy} f\right]$$

plus a transversality condition if needed

Note that, after deriving the $\dot{h}$ expression from the constraint function $h(t, y)$, we can directly apply the conditions in (10.84) without first transforming problem (10.77) into the form of (10.82).

Here, as in (10.79), we assume that the maximization of the Lagrangian with respect to u yields an interior solution. If the control variable is itself subject to a nonnegativity restriction $u(t) \geq 0$, then the $\partial \mathscr{L}'/\partial u = 0$

condition should be replaced by the Kuhn-Tucker conditions

$$\frac{\partial \mathscr{L}'}{\partial u} \leq 0 \qquad u \geq 0 \qquad u\frac{\partial \mathscr{L}'}{\partial u} = 0$$

These conditions allow for the possibility of a boundary solution. Boundary solutions can also occur, of course, if there is a closed control region for u.

A distinguishing feature of this approach is that the $f(t, y, u)$ function enters into the Lagrangian $\mathscr{L}'$ twice, with two different multipliers, Λ (a costate variable) and Θ (a Lagrange multiplier for a constrained problem). Accordingly, the partial derivative f_u also appears twice in the $\partial \mathscr{L}'/\partial u = 0$ condition—the first line in (10.84)—once with Λ and once with Θ. In contrast, the $\partial \mathscr{L}/\partial u = 0$ condition in (10.79) does not contain the θ multiplier. Thus, under the new approach, the behavior of Θ and its effects on the system are more explicitly depicted. When the state-space constraint is nonbinding, with $h(t, y) < c$, Θ takes a zero value, and (10.84) reduces to the regular maximum-principle conditions. But when the constraint changes its status from nonbinding (inactive) to binding (active), the conditions in (10.84) will become fully operative, prescribing how the Θ multiplier affects the system, and how Θ itself must change over time.

While the approach summarized in (10.84), with more explicit information about junction points, is easier to use than the approach in (10.79), the two approaches are in fact equivalent. The reader is asked to demonstrate their equivalence by following the set of steps outlined in Exercise 10.3, Prob. 1. In the process, it will become clear why the $\dot{\Theta} \leq 0$ condition is needed.

In another exercise problem, the reader is asked to write out the maximum-principle conditions for the special case of nonnegativity constraint $y(t) \geq 0$.

An Example

Let us consider the problem[8]

$$
\begin{aligned}
&\text{Maximize} && \int_0^3 (4 - t)u\,dt \\
&\text{subject to} && \dot{y} = u \\
(10.85)\qquad & && y - t \leq 1 \\
& && y(0) = 0 \qquad y(3) = 3 \\
&\text{and} && u \in [0, 2]
\end{aligned}
$$

[8] This problem is given as an exercise in Atle Seierstad and Knut Sydsæter, op. cit. p. 329. We provide and discuss here its complete solution.

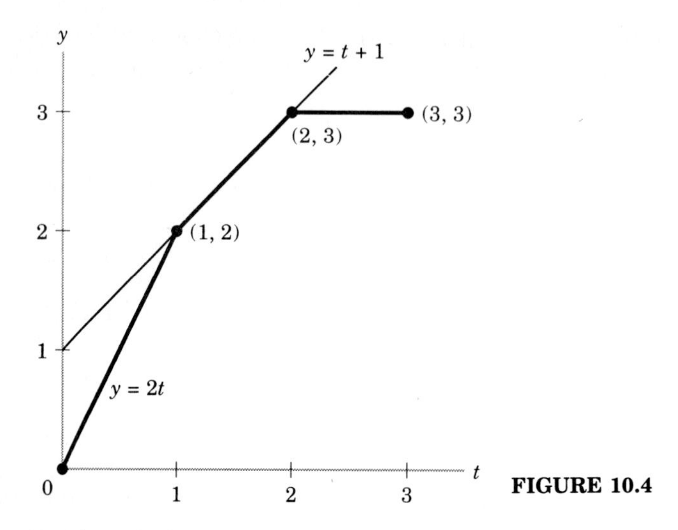

FIGURE 10.4

which contains a state-space constraint, $y - t \leq 1$. If we disregard this constraint and form the Hamiltonian

$$H = (4 - t)u + \lambda u$$

we see that H is linear in u, with slope

$$\frac{\partial H}{\partial u} = 4 - t + \lambda$$

Given the closed control region $[0, 2]$, we may expect a boundary solution to arise. Specifically, from the integrand function we gather that in order to maximize the objective functional, we should let u be as high as possible. Thus a reasonable conjecture about the unconstrained solution is $u(t) = 2$. This would imply that $\dot{y} = u = 2$ and $y(t) = 2t + k$. After using the initial condition $y(0) = 0$, we obtain the simple linear path

$$y(t) = 2t$$

This path is the one that allows the fastest rise in y, because it corresponds to the highest value of u, and hence the highest value of $\dot{y}$. As Fig. 10.4 shows, however, this path, steadily rising from the point of origin, can stay below the constraint boundary $y = t + 1$ only up to the point $(1, 2)$, that is, up to $t = 1$, at which time the constraint becomes binding. Then we have to change course.

Let us now turn to (10.84) for guidance as to what new course to take. First, the state-space constraint $y - t \leq 1$ implies that

$$h(t, y) = y - t \qquad c = 1 \qquad \dot{h} = \dot{y} - 1 = u - 1$$

Since the constraint function is linear in u (it does not contain u), the constraint qualification is satisfied. The Lagrangian is, by (10.83),

$$\mathscr{L}' = (4 - t)u + \Lambda u - \Theta(u - 1)$$

which is linear in u. The maximum principle requires, by (10.84), that

$$\mathscr{L}' \text{ be maximized with respect to } u \qquad \text{[corner solution]}$$

$$\frac{\partial \mathscr{L}'}{\partial \Theta} = 1 - u \geq 0 \qquad \Theta \geq 0 \qquad \Theta(1 - u) = 0$$

$$y - t \leq 1 \qquad \Theta(1 - y + t) = 0$$

$$\dot{\Theta} \leq 0 \qquad (= 0 \text{ when constraint nonbinding})$$

$$\dot{y} = u$$

$$\dot{\Lambda} = 0 \qquad \Rightarrow \qquad \Lambda = \text{constant}$$

No transversality condition is needed because the endpoints are fixed.

Taking the clue from the earlier-discussed unconstrained solution, we initially adopt for the first time interval $[0, 1)$ the control $u = 2$, which implies the state path $y = 2t$. Thus, we have

$$u^*[0, 1) = 2 \qquad y^*[0, 1) = 2t \tag{10.86}$$

As soon as we hit the constraint boundary, Θ becomes positive, and the $\partial \mathscr{L}'/\partial \Theta$ condition tells us to set

$$u = 1 \qquad \text{[by complementary slackness]}$$

This means that the equation of motion becomes $\dot{y} = 1$, and yields the path $y(t) = t + k_1$. To definitize the arbitrary constant k_1, we reason from the continuity of y that this second segment must start from the point $(1, 2)$ where the first segment ends. This fact enables us to definitize the path to

$$y(t) = t + 1 \qquad \text{[same as the constraint boundary]}$$

Thus, as shown in Fig. 10.4, the y path now begins to crawl along the boundary. But this path obviously cannot take us to the destination point $(3, 3)$, so we need at least one more new segment in order to complete the journey. To determine the third segment, we first try to ascertain the proper control, and then use the equation $\dot{y} = u$.

For the first segment, we chose $u^* = 2$ to ensure the fastest rise in y, and adhered to that u value for as long as we could, until the boundary constraint forced us to change course. The constraint boundary later becomes the new best path, and we should again stay on that path for as long as feasible, until we are forced by the consideration of the destination location to veer off toward the point (3, 3). Since $u(=\dot{y})$ cannot be negative, it is not feasible to overshoot the $y = 3$ level and then back down. So the second segment should aim at $y = 3$ as its target, terminating at the point (2, 3). Thus we have

$$u^*[1,2) = 1 \qquad y^*[1,2) = t + 1 \tag{10.87}$$

It then follows that the third segment is a horizontal straight line with slope $u = 0$:

$$u^*[2,3] = 0 \qquad y^*[2,3] = 3 \tag{10.88}$$

By piecing together (10.86), (10.88), and (10.89), we finally obtain the complete picture of the optimal control and state paths.

In the present example, the first segment of the optimal path coincides with that of the unconstrained problem. But this is not always the case. In problems where nonlinearity exists, the true optimal path may have no shared segment with the unconstrained path. This type of outcome is illustrated in Fig. 10.3*b*.[9]

EXERCISE 10.3

1. Establish the equivalence of (10.79) and (10.84) by the following procedure:
   (*a*) Equate $\partial\mathcal{L}/\partial u$ and $\partial\mathcal{L}'/\partial u$ (both $= 0$), and solve for Λ.
   (*b*) Differentiate Λ totally with respect to t, to get an expression for $\dot{\Lambda}$.
   (*c*) Expand the expression obtained in part (*b*) by using the $\dot{\lambda}$ expression in (10.79).
   (*d*) Substitute the result from part (*a*) into the $\dot{\Lambda}$ expression in (10.84).
   (*e*) Equate the two $\dot{\Lambda}$ expressions in parts (*c*) and (*d*), then simplify, and deduce that $\dot{\Theta} \le 0$.
2. (*a*) Write out the maximum-principle conditions under each of the two approaches when the state-space constraint is $y(t) \ge 0$.
   (*b*) In a problem with $y(t) \ge 0$, assume that the terminal time T is fixed. Is there need for a transversality condition? If so, what kind of transversality condition is appropriate?

---

[9] For an example, see Morton I. Kamien and Nancy L. Schwartz, *Dynamic Optimization: The Calculus of Variations and Optimal Control in Economics and Management*, 2d ed., Elsevier, New York, 1991, pp. 231–234.

10.4 ECONOMIC EXAMPLES OF STATE-SPACE CONSTRAINTS

In this section, we present two economic examples in which a nonnegativity state-space constraint appears.

Inventory and Production

The first example is concerned with a firm's decision regarding inventory and production.[10] To begin with, let the demand for the firm's product be given generally as $D(t) > 0$. To meet this demand, the firm can either draw down its inventory, X, or produce the demanded quantity as current output, Q, or use a combination thereof. Production cost per unit is c, and storage cost for inventory is s per unit, both assumed to be constant over time. The firm's objective is to minimize the total cost $\int_0^T (cQ + sX)\,dt$ over a given period of time $[0, T]$.

It is obvious that output cannot be negative, and the same is true of inventory. Thus, $Q(t) \geq 0$ and $X(t) \geq 0$. These two variables are related to each other in that inventory accumulation (decumulation) occurs whenever output exceeds (falls short of) the demanded quantity. Thus,

$$\dot{X}(t) = Q(t) - D(t)$$

This suggests that Q can be taken as a control variable that drives the state variable X.

Assuming a given initial inventory X_0, we can express the problem as

$$\begin{aligned}
&\text{Maximize} && \int_0^T (-cQ - sX)\,dt \\
&\text{subject to} && \dot{X} = Q - D(t) \\
&(10.89) && -X \leq 0 \\
& && X(0) = X_0 \qquad X(T) \geq 0 \text{ free} \qquad (X_0, T \text{ given}) \\
&\text{and} && Q \in [0, \infty)
\end{aligned}$$

Note that minimization of total cost has been translated into maximization

[10] This problem is discussed in Greg Knowles, *An Introduction to Applied Optimal Control*, Academic, New York, 1981, pp. 135–137.

of the negative of total cost. Note also that the nonnegativity state-space constraint automatically necessitates the truncation of the vertical terminal line.

The unconstrained view of this problem shows that the Hamiltonian

$$H = -cQ - sX + \lambda(Q - D)$$

is linear in the control variable Q, with

$$\frac{\partial H}{\partial Q} = -c + \lambda$$

The rule for the choice of Q is therefore

$$\lambda \begin{Bmatrix} < \\ = \\ > \end{Bmatrix} c \qquad \Rightarrow \qquad Q^* \begin{Bmatrix} = 0 \\ \text{is indeterminate} \\ \text{is unbounded} \end{Bmatrix}$$

The third possibility (unbounded Q) is not feasible, whereas the second possibility (indeterminate Q) is not helpful. Observe, however, that even in the case of $\lambda = c$, we can still select $Q^* = 0$, as in the case where $\lambda < c$. In fact, $Q^* = 0$ would make a lot of sense since Q enters into the objective functional negatively, so that choosing the minimum admissible value of Q serves the purpose of the problem best. However, then the equation of motion becomes

$$\dot{X} = -D(t) < 0 \qquad [\text{when } Q = 0]$$

and the firm's inventory is bound to be exhausted sooner or later. When the firm runs into the constraint boundary $X = 0$, it must reorient the control.

Before applying conditions (10.84), we first verify that the constraint qualification is satisfied. This is indeed so, because the constraint function $-X$ is linear in Q (it contains no Q). From the constraint $-X \leq 0$, we readily find that

$$h = -X \qquad c = 0 \qquad \text{and} \qquad \dot{h} = -\dot{X} = -(Q - D)$$

Thus, by (10.83), the Lagrangian is

$$\mathscr{L}' = -cQ - sX + \Lambda(Q - D) + \Theta(Q - D)$$

which is linear in the control variable Q. The conditions in (10.84) require that

$$\mathscr{L}' \text{ be maximized with respect to } Q \qquad \text{[corner solution]}$$

$$\frac{\partial \mathscr{L}'}{\partial \Theta} = Q - D \geq 0 \qquad \Theta \geq 0 \qquad \Theta(Q - D) = 0$$

$$X \geq 0 \qquad \Theta X = 0$$

$$\dot{\Theta} \leq 0 \qquad [= 0 \text{ when } X > 0]$$

$$\dot{X} = \frac{\partial \mathscr{L}'}{\partial \Lambda} = Q - D$$

$$\dot{\Lambda} = -\frac{\partial \mathscr{L}'}{\partial X} = s$$

$$\Lambda(T) \geq 0 \qquad X(T) \geq 0 \qquad \Lambda(T)X(T) = 0$$

Acting on the clue from the unconstrained view of the problem, we first choose

$$Q = 0$$

which simplifies the state equation of motion to

$$\dot{X} = -D$$

This result indicates that the firm should produce nothing and meet the demand exclusively by inventory decumulation. While following this rule, the inventory existing at any time will be

$$X = X_0 - \int_0^t D(t)\, dt$$

But such a policy can continue only up to the time $t = \tau$, when the given inventory X_0 is exhausted. The value of τ, the junction point, can be found from the equation

$$\int_0^\tau D(t)\, dt = X_0 \qquad \text{[exhaustion of inventory]}$$

If we assume that $\tau < T$, then at $t = \tau$ we need to revise the zero-production policy.

The rest of the story is quite simple. Having no inventory left, the firm must start producing. In terms of the maximum principle, the activation of the constraint makes the Θ multiplier positive, so that $\partial \mathscr{L}'/\partial \Theta = 0$, or

$$Q = D \qquad \text{for } t \in [\tau, T]$$

Under this new rule, the firm should meet the entire demand from current production. Inasmuch as $\dot{X} = Q - D$, it follows that

$$\dot{X} = 0 \qquad \text{for } t \in [\tau, T]$$

meaning that no change in inventory should be contemplated. Since $X(\tau) = 0$, inventory should remain at the zero level from the point τ on. The complete optimal control and state paths are therefore

$$\begin{aligned} &Q^*[0,\tau) = 0 \qquad && X^*[0,\tau) = X_0 - \int_0^t D(t)\,dt \\ &Q^*[\tau,T] = D(t) && X^*[\tau,T] = 0 \end{aligned} \tag{10.90}$$

Capital Accumulation under Financial Constraint

William Schworm has analyzed a firm that has no borrowing facilities, and whose investment must be financed wholly by its own retained earnings.[11] This model can serve as another illustration of a state-space constrained problem.

The firm is assumed to have gross profit $\pi(t, K)$ and investment expenditures $I(t)$. It cannot sell used capital. Hence, $I(t) \geq 0$. Besides, it cannot borrow funds. Thus its cash flow $\phi(t)$ is dependent only on the profit proceeds and the investment outlay:

$$\phi(t) = \pi(t, K) - I(t)$$

It is the goal of the firm to maximize its present value

$$\int_0^\infty \phi(t) e^{-\rho t}\,dt$$

where the symbol ρ denotes the rate of return the stockholders can earn on alternative investments as well as the rate of return the firm can earn on its retained earnings.

Starting from a given initial level, the firm's retained earnings $R(t)$ can be augmented only by any returns received (at the rate ρ) on R, and by any positive cash flow. Consequently, we have

$$\begin{aligned} \dot{R}(t) &= \rho R(t) + \phi(t) \\ &= \rho R(t) + \pi(t, K) - I(t) \end{aligned}$$

[11] William E. Schworm, "Financial Constraints and Capital Accumulation," *International Economic Review*, October 1980, pp. 643–660.

Assuming no depreciation, we also have

$$\dot{K}(t) = I(t)$$

These two differential equations suggest that R and K can serve as state variables of the model, whereas I can play the role of the control variable. Both state variables R and K must, of course, be nonnegative. But whereas the assumption of $I(t) \geq 0$ automatically takes care of the nonnegativity of K, the nonnegativity of retained earnings, R, needs to be explicitly prescribed in the problem. That is why the problem features a state-space constraint.

Collecting all the elements mentioned above, we can state the problem as follows:

$$\begin{aligned}
&\text{Maximize} && \int_0^{\infty} [\pi(t, K) - I(t)] e^{-\rho t}\, dt \\
&\text{subject to} && \dot{R}(t) = \rho R(t) + \pi(t, K) - I(t) \\
&&& \dot{K}(t) = I(t) \\
&&& -R(t) \leq 0 \\
&&& R(0) = R_0 \qquad K(0) = K_0 \\
&\text{and} && I(t) \in [0, \infty)
\end{aligned} \tag{10.91}$$

Being given in the general-function form, this model can only be analyzed qualitatively. Actually, Schworm first transforms the model into one where $R(t)$ is a constrained control (rather than state) variable. But we shall treat it as one with a state-space constraint and derive some of Schworm's main results by using the maximum-principle conditions in (10.84).

Again, we can verify that the constraint qualification is satisfied, because the constraint function, $-R(t)$, is linear in the control variable I. The constraint function also supplies the information that

$$h = -R \qquad \Rightarrow \qquad \dot{h} = -\dot{R} = -[\rho R + \pi(t, K) - I]$$

Thus, by (10.83), we have the Lagrangian

$$\begin{aligned}
\mathscr{L}' = {} & [\pi(t, K) - I] e^{-\rho t} + \Lambda_R[\rho R + \pi(t, K) - I] \\
& + \Lambda_K I + \Theta[\rho R + \pi(t, K) - I]
\end{aligned}$$

where Λ_R and Λ_K are the costate variables for R and K, respectively. The conditions in (10.84) stipulate that

$$(10.92)\qquad \frac{\partial \mathcal{L}'}{\partial I} = -e^{-\rho t} - \Lambda_R + \Lambda_K - \Theta \le 0 \qquad I \ge 0 \qquad I\frac{\partial \mathcal{L}'}{\partial I} = 0$$

$$\frac{\partial \mathcal{L}'}{\partial \Theta} = \rho R + \pi(t, K) - I \ge 0 \qquad \Theta \ge 0 \qquad \Theta\frac{\partial \mathcal{L}'}{\partial \Theta} = 0$$

$$R(t) \ge 0 \qquad \Theta R(t) = 0$$

$$(10.93)\qquad \dot{\Theta} \le 0 \qquad [= 0 \text{ when } R > 0]$$

$$\dot{R} = \frac{\partial \mathcal{L}'}{\partial \Lambda_R} = \rho R + \pi(t, K) - I$$

$$\dot{K} = \frac{\partial \mathcal{L}'}{\partial \Lambda_K} = I$$

$$(10.94)\qquad \dot{\Lambda}_R = -\frac{\partial \mathcal{L}'}{\partial R} = -\rho(\Lambda_R + \Theta)$$

$$(10.95)\qquad \dot{\Lambda}_K = -\frac{\partial \mathcal{L}'}{\partial K} = -\pi_K\left(e^{-\rho t} + \Lambda_R + \Theta\right)$$

plus transversality conditions

Note that, by virtue of the nonnegativity restriction on the control variable I, the Kuhn-Tucker conditions are applied in (10.92).

While Schworm discusses both the cases of $I(t) > 0$ and $I(t) = 0$, we shall consider here only the optimization behavior for the case of $I(t) > 0$. With I positive, complementary slackness mandates that $\partial \mathcal{L}'/\partial I = 0$, or

$$\Lambda_K = e^{-\rho t} + \Lambda_R + \Theta \qquad [\text{by } (10.92)]$$

Differentiating this equation with respect to t, we get

$$\begin{aligned}\dot{\Lambda}_K &= -\rho e^{-\rho t} + \dot{\Lambda}_R + \dot{\Theta} \\ &= -\rho e^{-\rho t} - \rho(\Lambda_R + \Theta) + \dot{\Theta} \qquad [\text{by } (10.94)] \\ &= -\rho\left(e^{-\rho t} + \Lambda_R + \Theta\right) + \dot{\Theta}\end{aligned}$$

Moreover, by equating this $\dot{\Lambda}_K$ expression with (10.95), we find that

$$(10.96)\qquad -\pi_K\left(e^{-\rho t} + \Lambda_R + \Theta\right) = -\rho\left(e^{-\rho t} + \Lambda_R + \Theta\right) + \dot{\Theta}$$

This result embodies the optimization rule for the firm when $I > 0$.

It is of interest that, except for the $\dot{\Theta}$ term on the right, the two sides of (10.96) are identical in structure. From (10.93) we know that $\dot{\Theta} = 0$ when $R > 0$. We can thus conclude from (10.96) that whenever $R > 0$, the firm should see to it that

(10.97) $\pi_K = \rho$ [investment rule when $R > 0$]

That is, when retained earnings are positive (financial constraint nonbinding), the firm should simply follow the usual myopic rule that the marginal profitability of capital be equal to the market rate of return.

If, on the other hand, $R = 0$ (financial constraint binding) in some time interval (t_1, t_2), then it follows that $R = 0$ and $\dot{R} = 0$ in that interval. Hence, the $\dot{R}(t)$ equation in (10.91) will reduce to

(10.98) $I(t) = \pi(t, K)$ [investment rule when $R = \dot{R} = 0$]

This means that in a constrained interval, the firm should invest its current profit. We see, therefore, that when the R constraint is binding, investment becomes further constrained by the availability of current profit.

Rules (10.97) and (10.98), though separate, are to be used in combination, with the firm switching from one rule to the other as the status of the R constraint changes. By so doing, the firm will be able to get as close as possible to the unconstrained optimal capital path.

Schworm also derives other analytical conclusions. For those, the reader is referred to the original paper.

EXERCISE 10.4

1 Draw appropriate graphs to depict the Q^* and X^* paths given in (10.90).

2 In problem (10.89), the demand is generally expressed as $D(t)$. Would different specifications of $D(t)$ affect the following?

(*a*) τ

(*b*) $Q^*[0, \tau)$

(*c*) $Q^*[\tau, T]$

(*d*) $X^*[0, \tau)$

(*e*) $X^*[\tau, T]$

10.5 LIMITATIONS OF DYNAMIC OPTIMIZATION

While by no means exhaustive, the present volume has attempted to introduce—in a readable way, we hope—most of the basic topics in the calculus of variations and optimal control theory. The reader has undoubtedly noticed that even in fairly simple problems, the solution and analysis procedure may be quite lengthy and tedious. It is for this reason that simple specific functions are often invoked in economic models to render the solution more tractable, even though such specific functions may not be totally satisfactory from an economic point of view.

It is for the same reason that writers often assume that the parameters in the problem, including the discount rate, remain constant throughout the planning period. Although there do exist in real life some economic parameters that are remarkably stable over time, certainly not all of them are. The constancy assumption becomes especially problematic in infinite-horizon problems, where the parameters are supposed to remain at the same levels from here to eternity. Yet the cost—in terms of analytical complexity—of relaxing this and other simplifying assumptions can be extremely high. Thus we have here a real dilemma. Of course, in practical applications involving specific parameter values, the simple way out of the dilemma is to reformulate the problem whenever there occur significant changes in parameter values. But the constancy assumption may have to be retained in the new formulation.

Owing to the complexity of multiple differential equations, economic models also tend to limit the number of variables to be considered. With the advent of powerful computer programs for solving mathematical problems, though, this limitation may be largely overcome in due time.

After the reader has spent so much time and effort to master the various facets of the dynamic-optimization tool, we really ought not to end on a negative note. So by all means go ahead and have fun playing with Euler equations, Hamiltonians, transversality conditions, and phase diagrams to your heart's content. But do please bear in mind what they can and cannot do for you.

ANSWERS TO SELECTED EXERCISE PROBLEMS

EXERCISE 1.2

3 The terminal curve should be upward sloping.

EXERCISE 1.3

1 $F[t, y(t), y'(t)] = 1$

2 (b) and (d)

EXERCISE 1.4

1 $V^*(D) = 8$; optimal path DZ is $DGIZ$.
$V^*(E) = 8$; optimal path EZ is $EHJZ$.
$V^*(F) = 10$; optimal path FZ is $FHJZ$.

4 $V^*(I) = 3$; IZ $\quad V^*(J) = 1$; JZ $\quad V^*(K) = 2$; KZ
$V^*(A) = 23$; $ACFHKZ$

EXERCISE 2.1

2 $dI/dx = 4x^3(b - a)$

4 $dI/dx = 2e^{2x}$

6 $y^*(t) = \frac{1}{24}t^3 + \frac{23}{24}t + 1$

8 $y^*(t) = e^t + e^{-t} + \frac{1}{2}te^t$

EXERCISE 2.2

1 $y^*(t) = 2t$

3 $y^*(t) = \frac{1}{2}t^2 + \frac{5}{2}t + 2$

5 $y^*(t) = e^{(1+t)/2} + e^{(1-t)/2}$

EXERCISE 2.3

1 $y^*(t) = t$

3 $y^*(t) = z^*(t) = \dfrac{e^t - e^{-t}}{e^{\pi/2} - e^{-\pi/2}}$

EXERCISE 2.4

1 $P_s = \dfrac{a + 2\alpha ab + \beta b}{2b(1 + \alpha b)} > 0 \qquad \overline{P} = P_s =$ static monopoly price

4 The lowest point of the price curve occurs at

$$t_0 = (\ln A_2 - \ln A_1)/2r \gtrless 0 \qquad \text{as } A_2 - A_1 \gtrless 0$$

Only case (c) can possibly (but not necessarily) involve a price reversal in the time interval $[0, T]$. The other cases have rising price paths.

EXERCISE 2.5

2 (a) Not different.

3 (b) $A_1 \gtrless 0$, $A_2 > 0$

EXERCISE 3.2

1 (a) $y^*(t) = 4$

3 (a) $y^*(t) = t + 4$, $T^* = 1$

EXERCISE 3.3

1 $y^*(t) = \frac{1}{2}(t^2 - 3t + 1)$

2 (a) No. $\qquad (b)$ $y^*(t) = \frac{1}{2}t^2 - \frac{7}{4}t + 1$

3 (a) Only one condition is needed, to determine T^*.

EXERCISE 3.4

1 (a) $\pi'(L_T) = 2\rho\sqrt{bk} > 0$. Thus $L_T{}^*$ is located such that the slope of the $\pi(L)$ curve is $2\rho\sqrt{bk}$, on the positively sloped segment.

(*b*) An increase in ρ (or b, or k) pushes the location of L_T^* to the left.

2 (*a*) $L_T^* = \dfrac{m}{n} - \dfrac{\rho}{n}\sqrt{bk}$

(*b*) $\pi'(L) = 0$ when $L = \dfrac{m}{n}$

(*c*) L_T^* is located to the left of m/n, such that the $\pi(L)$ curve has a positive slope.

EXERCISE 4.2

1 (*a*) There is no y term in F, and F is strictly convex in y'.

(*c*) Sufficient for a unique minimum.

3 (*a*) The determinantal test (4.9) fails because $|D_2| = 0$.

(*b*) The determinantal test (4.12) is satisfied for positive semidefiniteness because $|\tilde{D}_1| = 8$ and 2, and $|\tilde{D}_2| = 0$. The characteristic roots are $r_1 = 10$ and $r_2 = 0$.

(*c*) Sufficient for a minimum.

EXERCISE 4.3

1 Problem 6: $F_{y'y'} = 4$ for all t; satisfies the Legendre condition for a minimum.

Problem 9: $F_{y'y'} = 0$ for all t; satisfies the Legendre condition for a maximum as well as for a minimum.

3 $F_{\pi'\pi'} = 2\left(\dfrac{1 + \alpha\beta^2}{\beta^2 j^2}\right)e^{-\rho t} > 0$ for all t; satisfies the Legendre condition for a minimum.

EXERCISE 5.2

1 $K^*(t) = (K_0 - 25 + \rho)\exp\left(\dfrac{\rho - \sqrt{\rho^2 + 4}}{2}t\right) + 25 - \rho$

3 $C_{K'K'}K'' - \rho C_{K'} + \pi_K = 0$

4 $K^*(t) = (K_0 - 25)\exp\left(\dfrac{\rho - \sqrt{\rho^2 + 4}}{2}t\right) + 25$

EXERCISE 5.3

1 (*a*) $C^*(t) = A^{-1/(b+1)}e^{rt/(b+1)} = C^*_0 e^{rt/(b+1)}$

(*b*) $K^{*\prime}(t) = \dfrac{B - \hat{U} + (C^*)^{-b}/b}{(C^*)^{-(b+1)}} = \dfrac{1}{b}C^*$ [since $B = \hat{U}$]

(*c*) $K^*(t) = \dfrac{r}{b+1} K_0 e^{rt/(b+1)}$

(*d*) $K^*(t) = K_0 e^{rt/(b+1)}$

EXERCISE 5.4

2 (*a*) No, (5.40) is valid with or without saturation.
(*b*) Equation (5.41) is unaffected, but (5.42) should be changed to $Q'(K) = 0$ because now μ can never be zero.

3 The new equilibrium is still a saddle point, but it involves a positive (rather than zero) marginal utility in equilibrium.

5 (*a*) Only the streamlines lying above the stable branch remain relevant; the others cannot take us to the new level of K_T.
(*c*) Yes.

EXERCISE 6.1

3 $y^*(t) = -\dfrac{\lambda}{4}t^2 + c_1 t + c_2$

4 $y^*(t) = A_1 e^t + A_2 e^{-t} + c_1 \qquad z^*(t) = A_1 e^t - A_2 e^{-t} + c_1 t + c_2$

5 The general solution is the same as in Example 2.

EXERCISE 6.2

1 $\mathscr{F} = B - U(C) + D(L) + \lambda[-C + Q(K, L) - K']$
$-U'(C) - \lambda = 0 \quad \Rightarrow \quad \lambda = -U'(C) = -\mu$

2 (*a*) Three variables and two constraints.
(*b*) $\mathscr{F} = (\pi - C)e^{-\rho t} + \lambda_1[-\pi + \alpha K - \beta K^2] + \lambda_2[-C + aK'^2 + bK']$
(*c*) For π: $e^{-\rho t} - \lambda_1 = 0 \quad \Rightarrow \quad \lambda_1 = e^{-\rho t}$
For C: $-e^{-\rho t} - \lambda_2 = 0 \quad \Rightarrow \quad \lambda_2 = -e^{-\rho t}$
For K: $\lambda_1(\alpha - 2\beta K) - \dfrac{d}{dt}[\lambda_2(2aK' + b)] = 0$
$\Rightarrow \quad \lambda_1(\alpha - 2\beta K) - \dfrac{d\lambda_2}{dt}(2aK' + b) - \lambda_2(2aK'') = 0$

EXERCISE 6.3

2 Similarly to (6.10), λ may now be ≥ 0. A complementary-slackness condition will come into play.

4 $r_{Qs} = g - E\rho \qquad r_{Qm} = r_{Qs} + \dfrac{r_E}{E-1} > r_{Qs} \quad [E > 1 \text{ for positive MR}]$

6 (*b*) Maximize $\int_0^\infty N(q, q')e^{-\rho t}\, dt$

subject to $\int_0^\infty q'\, dt = S_0$

EXERCISE 7.2

2 $\lambda^*(t) = 3e^{4-t} - 3 \qquad u^*(t) = 2 \qquad y^*(t) = 7e^t - 2$

4 $u^*(t) = \lambda^*(t) = A_1 e^{\sqrt{2}t} + A_2 e^{-\sqrt{2}t}$

$y^*(t) = (\sqrt{2} - 1)A_1 e^{\sqrt{2}t} - (\sqrt{2} + 1)A_2 e^{-\sqrt{2}t}$

where $A_1 = \dfrac{-e^{-2\sqrt{2}}}{(1 - \sqrt{2})e^{-2\sqrt{2}} - (\sqrt{2} + 1)}$

$A_2 = \dfrac{1}{(1 - \sqrt{2})e^{-2\sqrt{2}} - (\sqrt{2} + 1)}$

EXERCISE 7.4

1 $\lambda^*(t) = 2 \qquad u^*(t) = 1 \qquad y^*(t) = t + 4$

3 $u^* = -1 \qquad \lambda^* = -\frac{1}{2} \qquad y^*(t) = -2t + 8 \qquad T^* = 4$

4 $u^* = \sqrt{1/38} \qquad y^*(t) = \sqrt{1/38}\, t \qquad \lambda^* = \sqrt{1/39}$

EXERCISE 7.6

1 (*a*) The cyclical pattern will be eliminated.

3 $\dfrac{\partial}{\partial r}\left(\dfrac{dU^*}{dt}\right) = -(T - t)\dfrac{1}{2}khbae^{B(T-t)} < 0$ for all t (except $t = T$)

EXERCISE 7.7

1 (*a*) Yes.

(*b*) No, E^*_2 is characterized by "marginal utility > marginal disutility."

3 (*a*) $H = U[C(E), P(E)]e^{-\rho t} - \lambda E$

$$\frac{\partial H}{\partial E} = [U_C C'(E) + U_P P'(E)]e^{-\rho t} - \lambda = 0$$

(*b*) The λ path is still a constant (zero) path.

(*c*) $E^*(t) = E^*$ as in (7.81).

(*d*) The condition in part (*a*) becomes

$$(\Psi \equiv) U_C C' + U_P P' - ce^{\rho t} = 0$$

so

$$\frac{dE}{dt} = \frac{-\partial\Psi/\partial t}{\partial\Psi/\partial E} = \frac{\rho c e^{\rho t}}{U_{CC}C'^2 + U_C C'' + U_{PP}P'^2 + U_P P''} < 0$$

EXERCISE 8.2

1 $[H_c - m\phi']_{t=T}e^{-\rho T} = 0$

3 $[H_c]_{t=T} \geq 0 \qquad T^* \leq T_{max} \qquad (T^* - T_{max})[H_c]_{t=T} = 0$

EXERCISE 8.3

1 The Mangasarian conditions are satisfied, based on F being strictly concave in u ($F_{uu} = -2$), f being linear in (y, u), and (8.26) being irrelevant. The Arrow condition is satisfied because $H^0 = \lambda y + \frac{1}{2}\lambda^2$, which is linear in y for given λ.

4 The Mangasarian and the Arrow conditions are both satisfied. (Here, H^0 is independent of y, and hence linear in y for given λ.)

6 Both the Mangasarian and Arrow conditions are satisfied.

EXERCISE 8.4

2 For a positive y_0, the y^* value at first increases at a decreasing rate, to reach a peak when arc CD crosses the y axis. It then decreases at a decreasing rate until τ is reached, and then decreases at an increasing rate, to reach zero at $t = T$.

3 (*a*) The initial point cannot be on arc $A0$ or $B0$; there must be a switch.
(*b*) The initial point lies on arc $A0$; no switch.
(*c*) The initial point lies on arc $B0$; no switch.

5 (*a*) The switching point, F, has to satisfy simultaneously

$$y = -\tfrac{1}{2}z^2 \qquad (z > 0) \qquad [\text{equation for arc } B0]$$

and

$$y = \tfrac{1}{2}z^2 + k \qquad (k < 0) \qquad [\text{equation for parabola containing arc } EF]$$

The solution yields $y = \frac{1}{2}k$ and $z = \sqrt{-k}$ (the negative root is inadmissible).
(*b*) $\tau = \sqrt{-k} - z_0$
(*c*) The optimal total time $= 2\sqrt{-k} - z_0$.

EXERCISE 8.5

1 An unbounded solution can occur when $A^* \neq 0$.

3 It is not possible to have an interior solution for the control variable A. [*Hint:* An interior solution A^* would mean that $\beta\lambda_P + \lambda_S = 0$, which would imply $\lambda_P =$ constant, and $U_P e^{-\rho t} = \delta\lambda_P$ from the costate equation of motion. The existence of a tolerably low level of P would imply $\lambda_P = 0$ for all t, which would lead to a contradiction.]

EXERCISE 9.3

1 (*a*) $\phi(k) = AK^{\alpha} \qquad U'(c) = c^{-(1+b)}$

(*c*) $H_c = \hat{U} - \dfrac{1}{b}c^{-b} + m[Ak^{\alpha} - c - (n+\delta)k]$

(*d*) $\dot{k} = Ak^{\alpha} - c - (n+\delta)k \qquad \dot{c} = \dfrac{c}{1+b}[A\alpha k^{\alpha-1} - (n+\delta+r)]$

$$\bar{k} = \left(\frac{n+\delta+r}{A\alpha}\right)^{1/(\alpha-1)}$$

$$\bar{c} = A\left(\frac{n+\delta+r}{A\alpha}\right)^{\alpha/(\alpha-1)} - (n+\delta)\left(\frac{n+\delta+r}{A\alpha}\right)^{1/(\alpha-1)}$$

3 (*b*) E will give way to a new steady state, E', on the $\dot{k} = 0$ curve to the right of E, with a higher $\bar{k}$.

(*c*) No.

EXERCISE 9.4

1 (*b*) To write it as $Y = K^{\alpha}[A(t)L^{\beta}]$ will conform to the format of (9.48).

5 The marginal productivity of human capital in research activities could not grow in proportion to A.

EXERCISE 10.1

3 (*a*) Maximize $\displaystyle\int_0^T (1-s)Ke^{-\rho t}\,dt$

subject to $\dot{K} = A(sK)^{\alpha} \qquad (A > 0)$

$K(0) = K_0 \qquad K(T)$ free $\qquad (K_0, T$ given$)$

and $0 \le s \le 1$

EXERCISE 10.3

1 (*a*) $\Lambda = \lambda + \Theta h_y$

(*b*) $\dot{\Lambda} = \dot{\lambda} + \dot{\Theta} h_y + \Theta(h_{ty} + h_{yy} f)$

(*c*) $\dot{\Lambda} = -F_y - \lambda f_y + \theta h_y + \dot{\Theta} h_y + \Theta(h_{ty} + h_{yy} f)$

(*e*) $(\theta + \dot{\Theta})h_y = 0 \quad \Rightarrow \quad \dot{\Theta} = -\theta \leq 0$ [since $\theta \geq 0$]

EXERCISE 10.4

2 (*a*) Higher (lower) demand would lower (raise) τ. But it is also conceivable that a mere rearrangement of the time profile of $D(t)$ may leave τ unchanged.

(*c*) Yes, because $Q^*[\tau, T]$ is identical with $D(t)$.

(*e*) $X^*[\tau, T] = 0$ regardless of $D(t)$, but the length of the period $[\tau, T]$ would be affected by any change in τ.

INDEX